D0860442

Convergence, Approximation, and Differential Equations

Convergence, Approximation, and Differential Equations

Eugene A. Herman
Grinnell College

John Wiley & Sons

New York   Chichester   Brisbane
Toronto   Singapore

Copyright © 1986, by John Wiley & Sons, Inc.

All rights reserved. Published simultaneously in Canada.

Reproduction or translation of any part of
this work beyond that permitted by Sections
107 and 108 of the 1976 United States Copyright
Act without the permission of the copyright
owner is unlawful. Requests for permission
or further information should be addressed to
the Permissions Department, John Wiley & Sons.

Library of Congress Cataloging in Publication Data:

Herman, Eugene A.
   Convergence, approximation, and differential
equations.

   Bibliography: p.
   Includes indexes.
   1. Convergence.   2. Approximation theory.
3. Differential equations.   I. Title.
QA295.H39   1986   515'.24   85-22620
ISBN 0-471-81762-7

10 9 8 7 6 5 4 3 2 1

For Augusta Herman
and in memory of Leon Herman

Preface

This book is, by design, unconventional. Its departures from standard practice have two related purposes: to permit significant uses of the computer that are integrated with traditional topics of theoretical and applied mathematics, and to permit efficient reorganizations of the mathematics curriculum that reflect new priorities currently being discussed nationwide. A description of the course for which the book was developed will explain its advantages as well as the situations in which its use is most appropriate.

We use the book at Grinnell College for our standard fourth-semester course, which we teach to a heterogeneous audience of math and science students. These students have taken two semesters of calculus, including multivariable calculus but not infinite series or numerical methods, and a one-semester course in linear algebra and differential equations. Also, all the students have studied some computer programming, although their backgrounds range from as little as a few weeks of BASIC in high school to as much computer science as we offer. Their background in the sciences shows a similar range—from negligible to half a dozen or more substantial courses.

Our goals for the course are as diverse as our audience. Since this course is our students' introduction to concepts of convergence, we want the treatment to be sound but not much more sophisticated than it would be at the first-year level. However, we want to benefit from having waited until the fourth semester by building on the first three semesters to give our students a richer understanding of why convergence is an important and useful subject. We chose, therefore, to build on the students' knowledge of calculus and differential equations by emphasizing the use of series in representing functions—especially their use in representing solutions of differential equations—and by emphasizing the use of numerical methods for approximating integrals, solutions of differential equations, and sums of series.

But we expect additional benefits from these choices of topics. By using differential equations as a natural bridge between significant problems in the sciences and the theory of convergence, we hope that science students will learn more about the utility of mathematics, and mathematics students will learn more about the problems that motivate mathematics. Furthermore, by introducing methods of numerical approximation, we hope not only to show students the concrete and computational side of convergence but also to introduce them to an important component of modern mathematical problem solving in the sciences. In short, we want the course to be a mixture of theory, computation, and applications, with each aspect supporting the others. It is this mixture that I have tried to attain in the book.

Although the impetus for the book came from a particular course, I have tried to make it flexible enough to meet a variety of needs. The primary prerequisite for reading this book is an understanding of the differential and integral calculus for functions of one variable. Even students lacking a background in differential equations can use the book, since I have included a substantial appendix containing the usual beginning material on differential equations with the usual kinds of homework exercises. I have not, however, included an appendix on linear algebra because only the last half of the last chapter depends on it, and that part of the book may easily be omitted.

The book is especially flexible in accommodating differing computing backgrounds and resources. It is not even necessary to require that students write programs in the course, as there are calculator exercises for almost all of the numerical algorithms. However, one of my reasons for writing this book was to help instructors integrate computing into the standard mathematics curriculum. The obstacles to achieving this goal are disappearing. Students, especially those who have had a few semesters of college, are now quite likely to know how to write simple programs in a high-level language. Also, they are quite likely to have access to a computer on which they can run their programs. Any of the popular microcomputers, for example, would be sufficiently powerful for the programming assignments in this book. To address a remaining obstacle, the lack of classroom materials that are flexible enough to use with whatever programming language and computer are available, I have expressed the numerical algorithms in the book in a stylized format that resembles a computer program but is not tied to any particular language. This device not only makes the book's many programming exercises accessible to a wide audience, but also helps students with little programming experience write successful programs.

I have also tried to make the book flexible enough to be used for other types of courses. For example, by covering Appendix A and omitting or going quickly over familiar material in Chapters 1 and 3, an instructor could use it as an introductory one-variable analysis text. For this purpose, the book has the advantage of containing concrete material that can be used to motivate the theory. Or, by covering all of Chapters 2, 4, and 5 and slighting Chapters 1 and 3, an instructor could use it as a one-variable applied analysis text. For this purpose, the book has the advantage of containing (in Appendix A) the basic calculus theory that needs to be referred to so often in such a course.

The book does not, however, pretend to be a text on numerical analysis. It may give students enough of a sense of the subject that they will be able to wait until graduate school to delve more deeply into the subject. Or it may simply give them enough concrete computational experience that they are able to go into a solid undergraduate numerical analysis course with confidence. My approach to the subject is to consider only a few key approximation problems but to take a fairly consistent and serious approach to them, so that some of the basic ideas begin to emerge. For example, in Chapter 2, the chapter devoted to numerical methods, only three approximation problems are considered; but for each problem one or two crude algorithms are developed, one or two more sophisticated algorithms follow these, and one proof of convergence is presented. To help students get a feel for both the power and limitations of these algorithms, I usually assign two or three "computer projects" taken from the programming assignments in Chapter 2. Each project requires that the student convert the assigned algorithm into a program, apply the program to several of the exercises associated with that computer assignment, and write up an analysis of the results. Typically, the students find the programming easy and the analysis hard. To use the output of a numerical computation to reach a justifiable conclusion about an approximation seems to require the same kind of careful, thorough analysis as many of the paper-and-pencil exercises; hence, it effectively complements students' other work.

Another way in which I have tried to help students benefit from this modest introduction to numerical analysis is to stress simple inequalities that are basic to both numerical analysis and the theory of convergence. For example, two themes in the book are the search for inequalities of the form

$$|x_n - x| \le \varepsilon_n$$

for convergence and approximation involving numbers, and

$$|f_n(x) - f(x)| \le \varepsilon_n$$

for uniform convergence and approximation involving functions. An added advantage of such inequalities is that they are more concrete and contain more information than the usual epsilon-delta arguments.

The topics, goals, and approaches I have been describing are not, of course, especially new. My choices have been influenced, for example, by the report, *Recommendations for a General Mathematical Sciences Program*, produced by the Committee on the Undergraduate Program in Mathematics in 1981 under the auspices of the Mathematical Association of America (MAA). In fact, the report's recommendations for second-semester calculus have quite a bit in common with the choices I have made. We have found, however, that they can be implemented more effectively in the second year. I have also been influenced by the MAA's "Panel on Discrete Mathematics in the First Two Years." One of the two approaches the Panel is exploring is an integrated program in discrete and continuous mathematics, which this book may help make possible.

Such intellectual guidance is helpful, but practical assistance is necessary. Financial support was given me by the Alfred P. Sloan Foundation, Grinnell College, and John Wiley & Sons, Inc. The Sloan Foundation, through its program in the *New*

Liberal Arts, provided funds that allowed the College to release me from some of my teaching duties at a critical time; the College and John Wiley generously paid for my student assistants. These assistants, Brenda Johnson, Laura Clymore, Valerie Mauck, and Rachel Barnes, did a superb job of typing, proofreading, correcting, and researching. I also benefitted from the help of an unpaid student assistant, the extraordinary Albert Goodman. Albert, while a student in my course during the last semester in which I used the prepublication manuscript, read the entire book and gave me daily lists of corrections and suggestions of a caliber that not many professional reviewers achieve. The manuscript Albert worked from, however, was already fairly solid, thanks to the criticisms of my colleagues Henry Walker, Charles Jepsen, and especially Arnold Adelberg. If I have managed to be clear and precise without being pedantic, then Arnold's trenchant comments were effective. And if I have written a lucid paragraph here and there, then the guidance of Matilda Liberman, the most clear-thinking writing teacher I have known, must have been at least partially successful.

Finally, I was helped immensely in preparing the final manuscript by the professional review of Kenneth Bube of UCLA—a review that was thorough, knowledgeable, and encouraging. I am grateful to all these for not merely helping but for giving their time and expertise so generously.

E. A. H.

Recommended Sections

The following sections are reasonable to cover in a one-semester course for students who are familiar with the material on differential equations in Appendix D, but are not yet familiar with concepts of convergence.

1.1–1.6 (cover Section 1.6 in less depth than the others)

2.1–2.3, 2.5, 2.7, 2.8

3.1–3.6, 3.8, 3.9 (cover Sections 3.8 and 3.9 in less depth)

4.1–4.3

5.1–5.4

The instructor may deviate considerably from the recommendations for sections from Chapters 2, 4, and 5, but the sections not listed above contain more difficult material.

Contents

CHAPTER 1   **Sequences**   **1**

    1.1   Limits of Numerical Sequences   **2**

    1.2   Properties of Limits   **10**

    1.3   Difference Equations   **15**

    1.4   Taylor Polynomials   **23**

    1.5   Lagrange Polynomials   **32**

    1.6   Uniform Convergence   **39**

CHAPTER 2   **Numerical approximation**   **49**

    2.1   The Bisection Method   **49**

    2.2   The Secant Method and Newton's Method   **55**

    2.3   Sources of Error   **63**

    2.4   A Hybrid Method   **70**

    2.5   The Trapezoid Rule and Simpson's Rule   **73**

    2.6   The Romberg Method   **83**

    2.7   The Euler Method   **88**

    2.8   Runge–Kutta Methods   **97**

CHAPTER 3   **Series**   **105**

3.1   The Sum of a Series   105

3.2   Taylor Series   111

3.3   Two Fundamental Convergence Criteria   117

3.4   Comparison Tests   121

3.5   Absolute and Conditional Convergence   126

3.6   Approximate Sum of a Series   131

3.7   Speeding Up Convergence   137

3.8   Uniform Convergence of Series   144

3.9   Power Series   147

3.10 A Fundamental Theorem on
      Differential Equations   155

CHAPTER 4   **Series and differential equations**   **161**

4.1   Power Series Solutions   164

4.2   Regular Singular Points, Part 1   174

4.3   Bessel Functions: An Application   184

4.4   Regular Singular Points, Part 2   194

4.5   Asymptotic Series   204

CHAPTER 5   **Fourier series and least squares**   **219**

5.1   Definitions and Examples   225

5.2   Convergence Theorems for Fourier Series   236

5.3   Other Types of Fourier Series   245

5.4   Partial Differential Equations: An Application   250

5.5   Inner Products   262

5.6   Orthogonal Sequences   271

5.7   Fourier Series Revisited   278

5.8   Systems of Linear Equations   283

APPENDIX A **Continuous functions and mean value theorems**   **299**

    A.1 Continuous Functions and the Intermediate Value Theorem   **299**

    A.2 Two Fundamental Theorems on Continuous Functions   **302**

    A.3 Derivatives and Mean Value Theorems   **305**

    A.4 Integrals and Mean Value Theorems   **309**

APPENDIX B **The binomial theorem**   **319**

APPENDIX C **Complex numbers**   **321**

APPENDIX D **Differential equations**   **327**

    D.1 First-Order Equations   **327**

    D.2 Methods for Solving First-Order Equations   **330**

    D.3 Second-Order Equations   **335**

    D.4 Methods for Solving Second-Order Equations   **341**

APPENDIX E **The principle of mathematical induction**   **349**

References   **351**

Solutions to Selected Exercises   **353**

Index   **371**

CHAPTER 1

Sequences

This is a book about sequences and their limits. Its two major themes are the practical uses of sequences in computing approximate solutions of problems and the more abstract uses of sequences in constructing representations of functions. Sequences arise in the study of approximations because exact solutions that cannot be computed may be expressible as limits of sequences of successively better approximate solutions that can be computed. For example, if we cannot evaluate a given integral exactly or solve a given equation exactly, we may be able to satisfy the requirements of the problem by computing an approximate solution selected from a sequence of successively better approximate solutions. Sequences arise in the study of functions because functions that are difficult to analyze or manipulate may be expressible as limits of sequences of simpler functions that can be more readily analyzed or manipulated. For example, we will see that even familiar functions as $\sin x$, $\log x$, and e^x can be understood more deeply and manipulated more flexibly when they are expressed as limits of simpler functions such as polynomials.

In Chapter 1, as in the rest of the book, we shift back and forth between our two themes, using each to deepen our understanding of the other. Section 1.1 is devoted to examples of sequences and their limits, and it concludes with a precise definition of the concept of "limit." But, although any precise and general definition of limit must be quite abstract, the idea of the definition is shown to come from the practical realm of approximation. Section 1.2 presents basic theorems on the properties of limits, but one of the key theorems is an inequality that is useful for both the theory of limits and the estimation of errors in an approximation. Sections 1.3, 1.4, and 1.5 cover topics that provide a foundation for our study of approximations in Chapter 2. Section 1.3 is devoted to "recursively" defined sequences, which are sequences that are defined in a manner that is especially useful for describing methods of approximation, while Sections 1.4 and 1.5 describe two specific methods for approximating functions by sequences of polynomials. But these practical topics also have implications for the theory of sequences. For example, the method of approximation in Section 1.4 leads to an important method for representing functions in Chapter 3. Also, the study of sequences of polynomials in Sections 1.4 and 1.5 suggests some general questions about the behavior of arbitrary sequences of functions, a topic taken up in Section 1.6.

Thus, Chapter 1 can be thought of as an introduction to the twin themes of approximation and representation by sequences. In this chapter, basic concepts are defined, typical methods are introduced, a few fundamental results are proved, and

1

interrelationships begin to appear. But there is another important purpose to this introduction. Chapter 1 also serves as a banquet table of individual sequences whose diverse characteristics are to be savored and remembered.

1.1 Limits of Numerical Sequences

A *sequence* of numbers is an infinite list

$$x_1, x_2, \ldots, x_n, \ldots \tag{1.1}$$

in which each x_n is a number. We will usually assume that the number x_n are real, not complex. By the nth *term* of the sequence, we mean the specific number x_n. For example, the fourth term of the sequence

$$\frac{1}{2}, \frac{2}{3}, \ldots, \frac{n}{n+1}, \ldots \tag{1.2}$$

is the number $\frac{4}{5}$.

Rather than writing a sequence as an explicit list, as in (1.1) or (1.2), it is more usual to use the abbreviated notation

$$\{x_n\}_{n+1}^{\infty} \quad \text{or simply} \quad \{x_n\}$$

For example, the sequence (1.2) is usually written as $\{n/(n+1)\}_{n=1}^{\infty}$, or simply $\{n/(n+1)\}$. We also permit sequences whose first term does not correspond to $n = 1$. For example, $\{n/(n+1)\}_{n=4}^{\infty}$ is a legitimate sequence whose first term is $\frac{4}{5}$, not $\frac{1}{2}$. However, as we will see, it usually does not matter whether the sequence starts at $n = 1$ or at some other value of n, and so we will usually find the simpler notation $\{x_n\}$ sufficiently clear.

A picture can help us visualize the behavior of a sequence. In Figure 1.1, for example, we see the first few terms of $\{n/(n+1)\}$ on a number line and marching to the right toward the number 1, as n gets larger. In Figure 1.2, this behavior is illustrated somewhat differently. The points (n, x_n), where $x_n = n/(n+1)$, are plotted in the plane, and we see them approaching the asymptote $y = 1$ as n gets larger.

We could consider Figure 1.2 to be a picture of the graph of the function f defined by $f(n) = n/(n+1)$, for $n = 1, 2, 3, \ldots$. This interpretation suggests a more precise definition of the concept of a sequence: A sequence is a function f from the set of positive integers to the set of real numbers. In this book, however, our more informal description of a sequence as an "infinite list" will be sufficient.

For us, the most important concept associated with a sequence is its limit. Although we make the definition of limit precise at the end of this section, we first compute several examples of limits. For now, we rely on the following informal description of the limit of the sequence.

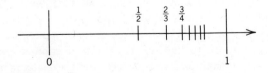

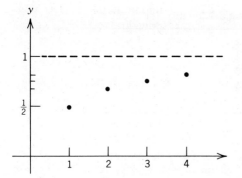

Figure 1.2 Some points $(n,\ x_n)$ where $x_n = n/(n+1)$

The number x is the *limit* of the sequence $\{x_n\}$ if the terms x_n get arbitrarily close to x for n sufficiently large. In symbols we express this by

$$\lim_{n\to\infty} x_n = x \quad \text{or simply} \quad x_n \to x$$

For example, the sequence $\{n/(n+1)\}$ pictured in Figures 1.1 and 1.2 has the limit 1. In symbols,

$$\lim_{n\to\infty} n/(n+1) = 1 \quad \text{or simply} \quad n/(n+1) \to 1$$

Not all sequences, however, have limits. The terms of $\{(-1)^n n\}$, for example, oscillate between the far positive and negative ends of the number line, as n gets larger, and so they get close to no limit.

EXAMPLE 1.1   To find the limit of a sequence whose terms are given by a rational function of n, divide numerator and denominator by an appropriate power of n:

$$\frac{3n^2 - 2n + 5}{4n^2 + n - 6} = \frac{3 - 2/n + 5/n^2}{4 + 1/n - 6/n^2} \to \frac{3 - 0 + 0}{4 + 0 - 0} = \frac{3}{4}$$

In Example 1.8 we verify that $1/n \to 0$, and in Section 1.2 we verify that limits of sums, differences, products, and quotients behave as we just now assumed they do.

EXAMPLE 1.2   The sequence $\{\sqrt{n^2 + n} - n\}$ seems to have the limit $\infty - \infty$; but this is an undefined expression. Instead, we find the limit by rationalizing the terms of the sequence.

$$\sqrt{n^2 + n} - n = \frac{(\sqrt{n^2 + n} - n)(\sqrt{n^2 + n} + n)}{(\sqrt{n^2 + n} + n)}$$

$$= \frac{n^2 + n - n^2}{\sqrt{n^2 + n} + n}$$

$$= \frac{n}{\sqrt{n^2 + n} + n} = \frac{1}{\sqrt{1 + 1/n} + 1} \to \frac{1}{2}$$

Table 1.1 Initial Terms of $(0.5)^n$
0.5
0.25
0.125
0.0625
0.03125

Table 1.2 Initial Terms of 2^n
2
4
8
16
32

EXAMPLE 1.3   We can guess at the limit of the sequence $\{a^n\}$, for various values of a, by using a calculator to compute the first few terms (Tables 1.1 and 1.2). It seems reasonable to guess that $a^n \to 0$ when $0 < a < 1$. We confirm this guess by using the binomial theorem (Appendix B).

$$(p + q)^n = \sum_{k=0}^{n} \binom{n}{k} p^{n-k} q^k$$

$$= p^n + np^{n-1}q + \frac{n(n-1)}{2} p^{n-2}q^2 + \cdots + npq^{n-1} + q^n \qquad (1.3)$$

If p and q are both positive, $(p + q)^n$ is larger than every term in the sum. For example

$$(p + q)^n > np^{n-1}q \qquad (1.4)$$

$$(p + q)^n > \frac{n(n-1)}{2} p^{n-2}q^2 \qquad (1.5)$$

We now use inequality (1.4) with $p = 1$ and $q = (1 - a)/a$. Then $a = 1/(1 + q)$ and so

$$a^n = \frac{1}{(1 + q)^n} = \frac{1}{(p + q)^n} < \frac{1}{np^{n-1}q} = \frac{1}{nq}$$

Since $1/nq \to 0$, it seems reasonable to conclude that, since a^n is smaller than the nth term of a sequence with limit 0, $\{a^n\}$ also should have limit 0. This argument is confirmed in Section 1.2.

EXAMPLE 1.4   As in Example 1.3, we guess that $na^n \to 0$ when $0 < a < 1$ and we confirm the guess by using the binomial theorem. By inequality (1.5) with $p = 1$

and $q = (1 - a)/a$, we have

$$na^n = \frac{n}{(1 + q)^n} < \frac{2n}{n(n - 1)q^2} = \frac{2}{(n - 1)q^2}$$

Since $2/(n - 1)q^2 \to 0$, we conclude that $na^n \to 0$.

EXAMPLE 1.5 The sequence $\{a^n/n!\}$ has the limit 0 for every value of a. To confirm this fact, we first choose a fixed positive integer $N > |a|$. Then, if $n \geq N$

$$\frac{|a^n|}{n!} = \frac{|a|^N|a|^{n-N}}{n!} = \frac{|a|^N}{N!} \overbrace{\left(\frac{|a|}{N + 1}\right)\left(\frac{|a|}{N + 2}\right) \cdots \left(\frac{|a|}{n}\right)}^{n - N \text{ factors}} < \frac{|a|^N}{N!}$$

Next, we choose a real number b so that $|a| < |b| < N$, and we apply the preceding inequality to $|b^n|/n!$. Thus, if $n \geq N$

$$\frac{|a^n|}{n!} = \left|\frac{a}{b}\right|^n \frac{|b|^n}{n!} < \left|\frac{a}{b}\right|^n \frac{|b|^N}{N!}$$

But $|a/b|^n \to 0$ by Example 1.3 since $|a/b| < 1$, and therefore $a^n/n! \to 0$.

We will encounter many sequences that, rather than being defined by explicit formulas as in the preceding examples, are defined recursively. In a "recursively" defined sequence, later terms are expressed as a function of earlier terms. The *Fibonacci* sequence, introduced in the next example, is a recursively defined sequence with many interesting applications, a few of which appear in the exercises for Section 1.3.

EXAMPLE 1.6 Let $F_0 = 0$, $F_1 = 1$, and $F_{n+2} = F_{n+1} + F_n$ for $n \geq 0$. The first few terms of the Fibonacci sequence are thus

$$0, 1, 1, 2, 3, 5, 8, \ldots$$

which get arbitrarily large. The pattern in Table 1.3, however, suggests that the sequence of ratios of successive terms may have a limit. We return to this example in Section 1.3.

Table 1.3 F_{n+1}/F_n for $n \geq 1$

1.0
2.0
1.5
1.6667
1.6
1.625
1.6154
1.6190
1.6176

Table 1.4

n	x_n where $x_{n+2} = (x_{n+1} + x_n)/2$
1	0.0
2	1.0
3	0.5
4	0.75
5	0.625
6	0.6875
7	0.65625
8	0.671875
9	0.6640625

EXAMPLE 1.7   The sequence defined recursively by

$$x_1 = 0, \quad x_2 = 1 \qquad x_{n+2} = \frac{x_{n+1} + x_n}{2} \quad \text{for } n \geq 1$$

seems to be approaching a limit of around 0.66 or 0.67, and its terms seem to be oscillating about the limit (see Table 1.4 and Figure 1.3). In fact, each term is the average of the two preceding terms and can therefore be pictured as the midpoint between them. So it is reasonable to guess that $\{x_n\}$ has a limit x such that

$$x_1 < x_3 < x_5 < \cdots < x < \cdots < x_6 < x_4 < x_2$$

We resolve this limit problem in Section 1.3. But, assuming that the limit x exists, we can at least find approximations to x. For example, let us approximate x with an error of less than, say, 0.1. From Table 1.4, note that $x_6 - x_5 = 0.0625 < 0.1$; so, since we are assuming that $x_5 < x < x_6$, we see that

$$|x_5 - x| < 0.1$$

and moreover

$$|x_n - x| < 0.1 \quad \text{for all } n \geq 5$$

That is, the term 0.625 is within the distance 0.1 of the limit, as are 0.6875 and all the remaining terms of the sequence. Similarly, we can approximate x with an error of less than 0.01 by noting that

$$|x_8 - x_9| < 0.01$$

and so

$$|x_n - x| < 0.01 \quad \text{for all } n \geq 8$$

Example 1.7 hints at the major themes of this book. For one, we see that we may be able to approximate a limit to a high degree of accuracy even if we cannot find the

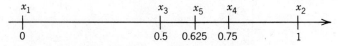

Figure 1.3 $x_{n+2} = (x_{n+1} + x_n)/2$

limit exactly. For another, we see that we need some theory to assure us that the limit exists and to guide us in seeking an approximation. But Example 1.7 also hints at how we might use our experience in making approximations to develop a deeper understanding of the theory of limits. Up to now, our understanding of the phrase "x is the limit of the sequence $\{x_n\}$" has been confined to the imprecise notion that "the terms x_n get arbitrarily close to x for n sufficiently large." Now we see that to make sure x_n is arbitrarily close to x, we can check that $|x_n - x|$, the distance between x_n and x, is arbitrarily small. And we can make sure this distance is arbitrarily small by checking that it is less than *every* measure of error ε, not just 0.1 and 0.01. Of course, we saw that when the error was smaller, the subscript n needed to be larger; so we should expect our decision on how large an n we must choose to depend on how small ε is. The precise meaning of limit is described in Definition 1.1.

Definition 1.1 $x_n \to x$ if for every positive real number ε, there exists a positive integer N such that

$$|x_n - x| < \varepsilon \quad \text{for all } n \geq N$$

Then we say that x is the *limit* of the sequence $\{x_n\}$ or that $\{x_n\}$ *converges* to x or simply that $\{x_n\}$ is *convergent* (to some unspecified limit). If the sequence $\{x_n\}$ is not convergent, we say it is *divergent*.

EXAMPLE 1.8   Prove that $1/n \to 0$.

Definition 1.1 says that for every positive real number ε, we must find an integer N such that

$$\left| \frac{1}{n} - 0 \right| < \varepsilon \quad \text{for all } n \geq N$$

But we can rewrite the displayed line as

$$\frac{1}{n} < \varepsilon \quad \text{for all } n \geq N$$

or equivalently

$$\frac{1}{\varepsilon} < n \quad \text{for all } n \geq N$$

Therefore, if we choose N to be a fixed integer bigger than $1/\varepsilon$, it follows that $n > 1/\varepsilon$ for all $n \geq N$. This completes the proof.

We next list three elementary properties of limits. The proofs of the first two are left to the exercises; the third is proved in Appendix A (Theorem A.1). The abbreviation "iff" in property (1.7) stands for "if and only if."

$$\text{If } x_n = c \text{ for all } n, \quad \text{then } x_n \to c \tag{1.6}$$

$$x_n \to x \quad \text{iff} \quad x_{n+1} \to x \tag{1.7}$$

$$\text{If the function } f \text{ is continuous at } x \text{ and if } x_n \to x, \quad \text{then } f(x_n) \to f(x) \tag{1.8}$$

Property (1.7) has a subtle consequence. It says that if the sequence

$$x_1, x_2, x_3, \ldots$$

converges to x, then the sequence

$$x_2, x_3, x_4, \ldots$$

also converges to x; and, it says that the converse of this statement is true. But if we apply property (1.7) to the sequence which begins with x_2, we see that the sequence

$$x_3, x_4, x_5, \ldots$$

converges to the same limit as the first two sequences, if it has a limit. So the convergence properties of a sequence do not depend on its first term, its first two terms, its first ten terms, or even its first million terms. As a consequence, we may ignore any finite number of initial terms of a sequence when we are interested only in the limit of the sequence.

Property (1.8) is one that we will use extensively in computing limits. To use it effectively, it is not necessary to review all the facts about continuous functions collected in Appendix A. For our present purposes, we may treat property (1.8) as if it is the definition of f being continuous at x; in fact, according to Theorem A.1, it is logically equivalent to the more standard ε–δ definition. The only other fact about continuous functions that we presently need is that the elementary functions of calculus are continuous wherever they are defined. For example, $\sin x$ is continuous at every x, $\arcsin x$ is continuous at every x in the interval $[-1, 1]$, $\log x$ is continuous at every x in $(0, \infty)$, e^x (which is also denoted by $\exp(x)$) is continuous at every x, and x^p (where p is a positive real number) is continuous at every x in $[0, \infty)$. Throughout this book, we assume that the independent variable x in $\sin x$ and in the other trigonometric functions is measured in radians, not degrees, and we use $\log x$ to denote the natural (base e) logarithm of x, not the common (base 10) logarithm.

EXAMPLE 1.9   By property (1.8)

$$\sin\left(\frac{1}{n}\right) \to 0 \quad \text{since } \frac{1}{n} \to 0 \text{ and } \sin 0 = 0$$

$$\exp\left(\frac{1}{n}\right) = e^{1/n} \to 1 \quad \text{since } \frac{1}{n} \to 0 \text{ and } e^0 = 1$$

$$\log\left(\frac{n}{n+1}\right) \to 0 \quad \text{since } \frac{n}{n+1} \to 1 \text{ and } \log 1 = 0$$

$$\arctan\left(\frac{n}{n+1}\right) \to \frac{\pi}{4} \quad \text{since } \frac{n}{n+1} \to 1 \text{ and } \arctan 1 = \frac{\pi}{4}$$

At times, it will be useful to allow ∞ and $-\infty$ as limits.

Definition 1.2 $x_n \to \infty$ if for every positive real number B, there exists a positive integer N such that

$$x_n > B \quad \text{for all } n \geq N$$

The definition of $x_n \to -\infty$ is similar. In these cases, however, we do not say that $\{x_n\}$ converges; we say that $\{x_n\}$ diverges to ∞ or $-\infty$.

EXERCISES

In Exercises 1–12, either find the limit of the sequence $\{x_n\}$ or explain why there is no limit.

1. $\dfrac{3 - 4n}{n + 5}$

2. $\dfrac{3n^3 + 2n - 1}{2n^3 - 4n^2 + 7}$

3. $\dfrac{n^2 + 5}{2n^3 - 1}$

4. $\dfrac{5n^2 + 2n - 3}{7 - 6n}$

5. $\dfrac{\sqrt{n^2 + 3n}}{2n - 5}$

6. $\sqrt{n + 1} - \sqrt{n}$

7. $\sqrt{n^2 + 2n} - \sqrt{n^2 - n}$

8. $1 - (-1)^n$

9. $\cos(\pi n)$

10. $(1/n)\sin(n\pi/2)$

11. $\cos(\log(n^2/(n^2 - 2)))$

12. $n^p a^n$ where $0 < a < 1$ and $p > 0$. *Hint*: Write $n^p a^n = [n(a^{1/p})^n]^p$

In Exercises 13–17, explore the behavior of the sequence $\{x_n\}$ (typically, by computing several terms on a calculator). If the sequence seems to be converging, guess at the answer to one decimal place.

13. $n^{1/n}$

14. $n \sin(1/n)$

15. $(1 + (1/n))^n$

16. $(\log n)/n$

17. $n!/n^n$

In Exercises 18–23, explore the behavior of the recursively defined sequence $\{x_n\}$ (typically, by computing several terms on a calculator). If the sequence seems to be converging, guess at the answer to one decimal place.

18. $x_{n+1} = \sqrt{3 + x_n}$, $\quad x_1 = 0$

19. $x_{n+1} = (x_n + 3/x_n)/2$, $\quad x_1 = 3$

20. $x_{n+1} = (x_n + 3/x_n)/2$, $\quad x_1 = -3$

21. $x_{n+1} = (2x_n + 2/x_n^2)/3$, $\quad x_1 = 1$

22. $x_{n+1} = (x_n + \sqrt{x_n})/2$, $\quad x_1 = 2$

23. $x_{n+1} = x_n + 1/(n + 1)$, $\quad x_1 = 1$

24. (A method for approximating π) Write a computer program to compute the first 20 terms of the sequences $\{x_n\}$ and $\{y_n\}$, where $x_0 = 2\sqrt{3}$, $y_0 = 3$ and

$$x_{n+1} = \frac{2x_n y_n}{x_n + y_n}, \qquad y_{n+1} = \sqrt{x_{n+1} y_n} \quad \text{for } n \geq 0$$

25. Find N such that $(\tfrac{1}{2})^n < 0.01$ for all $n \geq N$

26. Find N such that $|n/(n + 1) - 1| < 0.001$ for all $n \geq N$

In Exercises 27–35, use Definition 1.1 or Definition 1.2 to prove the assertion.

27. $2/n \to 0$

28. $2/n^2 \to 0$

29. $5/\sqrt{n} \to 0$

30. If $x_n = c$ for all n, then $x_n \to c$.

31. $a^n \to 0$ if $0 < a < 1$

32. $2n \to \infty$

33. $2 - 3n^2 \to -\infty$

34. If $x_n \to x$, then $x_{n+1} \to x$

35. If $x_n \to 0$ and $x_n \geq 0$, then $\sqrt{x_n} \to 0$

1.2 Properties of Limits

To work with limits of sequences effectively, we must know their properties. The theorems in this section describe the most basic ones. The proofs all depend on Definition 1.1; however, one of our goals is to prove properties of limits that are easier to work with than the definition.

To help us manipulate the expressions in Definition 1.1, we need the following elementary properties of absolute values.

$$|x| = \begin{cases} x & \text{if } x \geq 0 \\ -x & \text{if } x < 0 \end{cases} \tag{1.9}$$

$$|-x| = |x| \geq 0 \tag{1.10}$$

$$|x|^2 = x^2 \tag{1.11}$$

$$x \leq |x| \quad \text{and} \quad -x \leq |x| \tag{1.12}$$

$$|x| < r \quad \text{iff} \quad -r < x < r \quad \text{(for every } r > 0\text{)} \tag{1.13}$$

$$|xy| = |x||y| \tag{1.14}$$

$$|x + y| \leq |x| + |y| \quad \text{(triangle inequality)} \tag{1.15}$$

Here is a proof of the triangle inequality from properties (1.11), (1.12), and (1.14):

$$|x + y|^2 = (x + y)^2 = x^2 + 2xy + y^2 \leq x^2 + |2xy| + y^2$$
$$= |x|^2 + 2|x||y| + |y|^2 = (|x| + |y|)^2$$

Then take the square root of both sides.

Our first theorem says that a sequence can have at most one limit.

Theorem 1.1 If $x_n \to x$ and $x_n \to y$, then $x = y$.

Proof Suppose that $x \neq y$. We may then choose ε to be the positive number $|x - y|/2$. When we apply Definition 1.1 to this value of ε, we reach a contradiction as follows. Since $x_n \to x$, there exists an integer N_1 such that $|x_n - x| < \varepsilon$ for all $n \geq N_1$. And, since $x_n \to y$, there exists N_2 such that $|x_n - y| < \varepsilon$ for all $n \geq N_2$. Therefore, if n is chosen so that it is larger than both N_1 and N_2, then, by the triangle inequality,

$$|x - y| = |(x - x_n) + (x_n - y)| \leq |x - x_n| + |x_n - y| < \varepsilon + \varepsilon = |x - y|$$

But this computation shows that $|x - y| < |x - y|$, which is impossible. So our initial supposition, that $x \neq y$, must be false. In other words, $x = y$ must be true.

Theorem 1.2   Suppose $x_n \to x$.

(a) If $x < b$, then there exists N such that $x_n < b$ for all $n \geq N$.

(b) If $x > b$, then there exists N such that $x_n > b$ for all $n \geq N$.

Proof

(a) Since $b - x > 0$, we may choose ε to be the number $b - x$. So, by Definition 1.1, there exists N such that $|x_n - x| < b - x$ for all $n \geq N$. But then, for $n \geq N$,

$$x_n = (x_n - x) + x \leq |x_n - x| + x < (b - x) + x = b$$

(b) This theorem is proved similarly (Exercise 3).

Theorem 1.3

(a) $x_n \to x$ iff $|x_n - x| \to 0$.

(b) If $|x_n - x| \leq y_n$ for all n and if $y_n \to 0$, then $x_n \to x$.

Proof

(a) Since $|x_n - x| = ||x_n - x| - 0|$, the definitions of $x_n \to x$ and $|x_n - x| \to 0$ are exactly the same.

(b) For every $\varepsilon > 0$, we must find a positive integer N such that $|x_n - x| < \varepsilon$ for all $n \geq N$. We may thus treat ε as given; but, since $y_n \to 0$, there exists N such that $|y_n - 0| < \varepsilon$ for all $n \geq N$. Hence, for $n \geq N$,

$$|x_n - x| \leq y_n = |y_n - 0| < \varepsilon$$

Therefore $x_n \to x$.

Theorem 1.3b and its later variations are used extensively throughout the book. Its role in the proof of Theorem 1.4 is typical; rather than use Definition 1.1 over and over, we give simpler, clearer arguments by using Theorem 1.3b in its place. We have already used Theorem 1.3b informally in Examples 1.3, 1.4, and 1.5. The next example is another of this type.

EXAMPLE 1.10   In Example 1.27 we show that

$$|\log(1 + x) - x| \leq x^2/2 \quad \text{if } 0 \leq x$$

Use this fact to prove that $\left(1 + \dfrac{1}{n}\right)^n \to e$.

If we replace x by $1/n$ in the given inequality and then multiply by n, we get

$$\left| n \log\left(1 + \frac{1}{n}\right) - 1 \right| \leq \frac{1}{2n} < \frac{1}{n}$$

or

$$\left| \log\left(1 + \frac{1}{n}\right)^n - 1 \right| \leq \frac{1}{n}$$

Since $1/n \to 0$ (Example 1.8), Theorem 1.3b tells us that $\log(1 + 1/n)^n \to 1$. Then, when we apply the continuous function e^x to this sequence, property (1.8) yields

$$\left(1 + \frac{1}{n}\right)^n = \exp\left(\log\left(1 + \frac{1}{n}\right)^n\right) \to \exp(1) = e$$

In the remainder of this section, we verify that convergent sequences can be combined, using the operations of arithmetic, to form new convergent sequences.

Theorem 1.4   Suppose $x_n \to x$ and $y_n \to y$. Then

(a) $|x_n| \to |x|$
(b) $x_n + y_n \to x + y$
(c) $x_n - y_n \to x - y$
(d) $c x_n \to c x$ for every number c
(e) $x_n y_n \to x y$
(f) $x_n/y_n \to x/y$ provided $y \neq 0$.

We reduce the effort needed to prove Theorem 1.4 by first proving the following special case to which we then apply Theorem 1.3b.

Lemma 1.5   If $x_n \to 0$ and $y_n \to 0$ then

(a) $x_n + y_n \to 0$
(b) $c x_n \to 0$ for every number c.

Proof

(a) If ε is an arbitrary, given positive number, we must find N such that $|x_n + y_n| < \varepsilon$ for all $n \geq N$. But we may apply Definition 1.1 to the sequences $\{x_n\}$ and $\{y_n\}$, and we may choose the value of the number ε in the definition to be half of the given number ε. Hence, there exist integers N_1 and N_2 such that

$$|x_n| < \frac{\varepsilon}{2} \quad \text{for all } n \geq N_1 \quad \text{and} \quad |y_n| < \frac{\varepsilon}{2} \quad \text{for all } n \geq N_2$$

We then choose N to be the larger of N_1 and N_2. Therefore, for $n \geq N$,

$$|x_n + y_n| \leq |x_n| + |y_n| < \frac{\varepsilon}{2} + \frac{\varepsilon}{2} = \varepsilon$$

This proves $x_n + y_n \to 0$.

(b) If $c = 0$, the sequence $\{c x_n\}$ is the zero sequence and so has the limit zero, by property (1.6). Thus, we may assume that $c \neq 0$. If ε is an arbitrary, given positive number, we must find N such that $|c x_n| < \varepsilon$ for all $n \geq N$. So we apply Definition 1.1 to the sequence $\{x_n\}$ and we choose the value of ε in the definition to be $\varepsilon/|c|$, where this last ε is the given one. Hence, there exists N such that

$$|x_n| < \frac{\varepsilon}{|c|} \quad \text{for all } n \geq N$$

Therefore, for $n \geq N$,

$$|c x_n| = |c| \, |x_n| < |c| \frac{\varepsilon}{|c|} = \varepsilon$$

This proves $c x_n \to 0$.

Proof of Theorem 1.4

(a) By Exercise 2, $||x_n| - |x|| \le |x_n - x|$. But by Theorem 1.3a, $|x_n - x| \to 0$. Therefore, by Theorem 1.3b, $|x_n| \to |x|$.

(b) By the triangle inequality

$$|(x_n + y_n) - (x + y)| = |(x_n - x) + (y_n - y)| \le |x_n - x| + |y_n - y|$$

But, by Theorem 1.3a and Lemma 1.5a, $|x_n - x| + |y_n - y| \to 0$. Therefore, by Theorem 1.3b, $x_n + y_n \to x + y$.

(c) This theorem is proved similarly.

(d) See Exercise 4.

(e) By the triangle inequality

$$|x_n y_n - xy| = |(x_n y_n - x_n y) + (x_n y - xy)| \le |x_n y_n - x_n y| + |x_n y - xy|$$
$$= |x_n||y_n - y| + |y||x_n - x|$$

Furthermore, by part (a) of this theorem, $|x_n| \to |x|$. So, since $|x| < |x| + 1$, there exists N such that $|x_n| < |x| + 1$ for all $n \ge N$ (Theorem 1.2a). Hence, for $n \ge N$,

$$|x_n y_n - xy| < (|x| + 1)|y_n - y| + |y||x_n - x|$$

But the right side of this inequality has limit 0, by Theorem 1.3a and Lemma 1.5. Therefore, by Theorem 1.3b, $x_n y_n \to xy$.

(f) First we show that $1/y_n \to 1/y$.

$$\left|\frac{1}{y_n} - \frac{1}{y}\right| = \left|\frac{y - y_n}{y y_n}\right| = \frac{|y - y_n|}{|y||y_n|}$$

Furthermore, $|y_n| \to |y| > |y|/2$. So there exists N such that $|y_n| > |y|/2$ for all $n \ge N$ (Theorem 1.2b). Hence, for $n \ge N$,

$$\left|\frac{1}{y_n} - \frac{1}{y}\right| < \frac{2|y - y_n|}{|y|^2}$$

Therefore, by Lemma 1.5b and Theorem 1.3, $1/y_n \to 1/y$. Finally, by part (e) of this theorem,

$$\frac{x_n}{y_n} = x_n\left(\frac{1}{y_n}\right) \to x\left(\frac{1}{y}\right) = \frac{x}{y}$$

EXAMPLE 1.11 We reconsider Example 1.1. We proved that $1/n \to 0$ in Example 1.8 and noted that $c \to c$ in property (1.6). Therefore, we can use the various parts of Theorem 1.4 to conclude that $1/n^2 \to 0$ (by e), $2/n \to 0$ and $5/n^2 \to 0$ (by d or e), $3 - 2/n + 5/n^2 \to 3$ (by b and c), and so on.

EXAMPLE 1.12 A sequence $\{x_n\}$ is defined recursively by $x_{n+1} = (x_n + 2/x_n)/2$, $x_1 = 3$. Assume that $x_n \to x$. Find x.

We use Theorem 1.4 to conclude that

$$\frac{2}{x_n} \to \frac{2}{x} \quad \text{and so} \quad \left(x_n + \frac{2}{x_n}\right)\left(\frac{1}{2}\right) \to \left(x + \frac{2}{x}\right)\left(\frac{1}{2}\right)$$

But, by property (1.7), $x_{n+1} \to x$; so by Theorem 1.1, $x_{n+1} = (x_n + 2/x_n)/2$ implies

$$x = \left(x + \frac{2}{x} \right)\left(\frac{1}{2} \right)$$

This is an equation in x which we can now solve. After multiplying through by $2x$ and combining like terms, we get that $x^2 = 2$; hence, $x = \pm \sqrt{2}$. In fact, $x = \sqrt{2}$ since x_1 is positive and the formula for x_{n+1} produces a positive term whenever the preceding term is positive. Therefore, the sequence $\{ x_n \}$ has the limit $\sqrt{2}$ *if* it has a limit at all. (In Section 3.3 we develop the theory needed to prove that this sequence does have a limit.)

Sequences of complex numbers can be studied in much the same way as sequences of real numbers. Definition 1.1 can be used exactly as stated, but the absolute value signs must be interpreted as indicating the absolute value of a complex number:

$$|x + iy| = \sqrt{x^2 + y^2}$$

Theorems 1.1, 1.3, and 1.4 are valid for sequences of complex numbers exactly as stated.

EXERCISES

1. Prove that $|x - y| \le |x - z| + |z - y|$. *Hint*: Use the triangle inequality.
2. Prove that $|x - y| \ge ||x| - |y||$. *Hint*: Use the triangle inequality to show that $|x - y| + |y| \ge |x|$.
3. Prove Theorem 1.2b.

In Exercises 4–7, prove the assertions by using the theorems proved in Section 1.2, especially Theorem 1.3b; try not to use Definition 1.1 or property (1.8).

4. If $x_n \to x$, then $cx_n \to cx$
5. If $x_n \to x$ and $y_n \to y$, then $x_n - y_n \to x - y$
6. If $x_n \to x$ and $x > 0$, then $\sqrt{x_n} \to \sqrt{x}$
7. $x_n \to x$ and $y_n \to y$ iff $\sqrt{(x_n - x)^2 + (y_n - y)^2} \to 0$

Use Theorem 1.4 and Example 1.8 to prove the assertions in Exercises 8 and 9.

8. $\dfrac{3n - 4}{2n + 1} \to \dfrac{3}{2}$
9. $\dfrac{4n^2 - n + 5}{n^2 + 6} \to 4$

Assume that the recursively defined sequences in Exercises 10–13 converge, and use Theorem 1.4 to find the limit.

10. $x_{n+1} = \sqrt{3 + x_n}$, $x_1 = 0$
11. $x_{n+1} = (x_n + 3/x_n)/2$, $x_1 = 3$
12. $x_n = F_{n+1}/F_n$ where $\{ F_n \}$ is the Fibonacci sequence (Example 1.6)

13. $x_{n+1} = (2x_n + 2/x_n^2)/3, \qquad x_1 = 1$

14. Use the fact that $|\sin x - x| \le |x|^3/6$ to prove $n \sin\left(\dfrac{1}{n}\right) \to 1$.

15. (a) Use the fact that $|n^{1/n} - 1| \le \sqrt{2/(n-1)}$ to prove $n^{1/n} \to 1$.

  (b) Then deduce that $(\log n)/n \to 0$.

16. Use the fact that $|\log(1+x) - x + x^2/2| \le x^3/3$ when $0 \le x$ to prove that $n^2 \log\left(1 + \dfrac{1}{n}\right) - n \to -\frac{1}{2}$.

17. If $x_n \le b$ for all n and $x_n \to x$, prove that $x \le b$.

18. If $x_n \ge b$ for all n and $x_n \to x$, prove that $x \ge b$.

19. (a) If $x_n \to x$, prove that $(x_1 + x_2 + \cdots + x_n)/n \to x$.

  (b) If $x_n \to x$, prove that $(x_1 x_2 \cdots x_n)^{1/n} \to x$.

1.3 Difference Equations

In Examples 1.6 and 1.7 we encountered sequences, such as Fibonacci's, that are defined recursively rather than explicitly. We will encounter many more later in the book. The methods of numerical approximation in Chapter 2 will usually be described by a recursion formula, and the techniques for solving differential equations in Chapter 4 will usually involve finding an explicit formula for a recursively defined sequence of unknown coefficients. However, even though the present section is largely devoted to techniques for finding explicit formulas for a few special types of recursively defined sequences, its principal value will not be in the use of these techniques. In fact, few of the recursively defined sequences in later chapters are among the special types studied here. Of greater value will be the experience that the material of the section provides for working with, and thinking about, a variety of recursively defined sequences.

Another purpose of the section is to lay some groundwork for the study of differential equations. Although we will rarely use the techniques of this section in Chapter 4, we will build on the ideas developed here. As we go on, similarities between the theorems, methods, and applications of differential equations and the theorems, methods, and applications of the "difference equations" in this section will become increasingly evident.

EXAMPLE 1.13 Solve for x_n in the difference equation

$$x_{n+1} = 3x_n, \qquad n \ge 0, \quad x_0 \text{ arbitrary}$$

The first few terms of $\{x_n\}$ are

$$x_1 = 3x_0, \qquad x_2 = 3x_1 = 3^2 x_0, \qquad x_3 = 3x_2 = 3^3 x_0, \ldots$$

from which we infer the formula $x_n = 3^n x_0$ for all $n \ge 0$.

Note: *Strictly speaking, the equation* $x_{n+1} = 3x_n$ *is not in the form of a difference equation. However, it can be written in the form* $\Delta x_n = 2x_n$, *where* Δx_n *denotes the difference* $x_{n+1} - x_n$; *now it may properly be called a difference equation.*

More generally, the unique solution of the difference equation

$$x_{n+1} = a x_n, \qquad n \geq 0, \quad x_0 \text{ arbitrary} \qquad (1.16)$$

is the sequence $\{x_0\, a^n\}$. Equation (1.16) is called a *first-order, linear, constant coefficient, homogeneous* difference equation.

EXAMPLE 1.14   Solve for x_n in the difference equation

$$x_{n+2} = 3x_n, \qquad n \geq 0, \quad x_0 \text{ and } x_1 \text{ arbitrary}$$

The first few terms are

$$x_2 = 3x_0, \qquad x_3 = 3x_1, \qquad x_4 = 3x_2 = 3^2 x_0, \qquad x_5 = 3x_3 = 3^2 x_1, \ldots$$

from which we infer the formulas $x_{2n} = 3^n x_0$, $x_{2n+1} = 3^n x_1$ for all $n \geq 0$.

Note: *The equation* $x_{n+2} = 3x_n$ *is equivalent to the difference equation* $\Delta^2 x_n = -2\,\Delta x_n + 2x_n$, *where* $\Delta^2 x_n$ *denotes* $\Delta x_{n+1} - \Delta x_n$.

Solving the general *second-order, linear, constant coefficient, homogeneous* difference equation

$$x_{n+2} = a x_{n+1} + b x_n, \qquad n \geq 0, \quad x_0 \text{ and } x_1 \text{ arbitrary} \qquad (1.17)$$

will be not as easy as solving (1.16) was. Still, Examples 1.13 and 1.14 suggest that we might look for solutions of the form $x_n = w^n$. If we substitute w^n for x_n, equation (1.17) becomes

$$w^{n+2} = a w^{n+1} + b w^n$$

or, after dividing by w^n,

$$w^2 = a w + b \qquad (1.18)$$

Equation (1.18) is called the *characteristic equation* associated with the difference equation (1.17). Since (1.18) is a quadratic equation, it has either two distinct real roots or a single real root (called a "double" root since it appears in both linear factors of (1.18)) or two nonreal complex roots. If w denotes a root, then, since our derivation of (1.18) is reversible, the sequence $\{w^n\}$ is a solution of the difference equation (1.17). In particular, if (1.18) has two distinct real roots w_1 and w_2, then $\{w_1^n\}$ and $\{w_2^n\}$ are both solutions of (1.17). In Theorem 1.6 we will also see how to find two solutions of (1.17) when (1.18) has a double root or two complex roots; even if the roots are complex, we will find real solutions of (1.17).

But once we have found two solutions $\{y_n\}$ and $\{z_n\}$ of the difference equation (1.17) it is not hard to find many more. Specifically, it can be shown that, for any constants c_1 and c_2, the sequence

$$\{c_1\, y_n + c_2\, z_n\} \qquad (1.19)$$

(which is called a *linear combination* of $\{y_n\}$ and $\{z_n\}$) is also a solution of (1.17). This may be verified (Exercise 22a) by simply substituting the expression (1.19) for x_n in (1.17). Now that we have an infinite collection of solutions with two independent parameters c_1 and c_2, we might suspect that we have found all the solutions of (1.17). This suspicion is based on the observation that the terms x_0 and x_1 of the general solution $\{x_n\}$ can be thought of as two independent parameters which completely determine $\{x_n\}$; for x_0 and x_1 determine x_2 by (1.17) with $n = 0$, and then x_1 and x_2 determine x_3 by (1.17) with $n = 1$, and so on. In fact, it may be verified (Exercise 22b) that if $\{y_n\}$ and $\{z_n\}$ are solutions of (1.17) and if

$$y_0 z_1 \neq y_1 z_0 \tag{1.20}$$

then the solutions $\{c_1 y_n + c_2 z_n\}$ do constitute the entire collection of solutions of (1.17). This collection is referred to as the *general solution* of (1.17). The condition (1.20) guarantees that the solutions $\{y_n\}$ and $\{z_n\}$ are *linearly independent*; that is, it guarantees that $\{y_n\}$ is not of the form $\{cz_n\}$ and $\{z_n\}$ is not of the form $\{cy_n\}$.

Theorem 1.6

(a) If (1.18) has two distinct real roots w_1 and w_2, then the solutions of (1.17) are given by

$$x_n = c_1 w_1^{\ n} + c_2 w_2^{\ n}, \qquad n \geq 0$$

(b) If (1.18) has a double root w, then the solutions of (1.17) are given by

$$x_n = c_1 w^n + c_2 n w^{n-1}, \qquad n \geq 0$$

(In the special case when $w = 0$, we use the conventions that $0^0 = 1$ and $0 \cdot 0^{-1} = 0$.)

(c) [For students familiar with complex arithmetic.] If (1.18) has two nonreal complex roots $w_1 = re^{i\theta}$ and $w_2 = re^{-i\theta}$, then the solutions of (1.17) are given by

$$x_n = c_1 r^n \cos(n\theta) + c_2 r^n \sin(n\theta), \qquad n \geq 0$$

Proof The preceding discussion shows that we need only prove that in (a) $\{w_1^{\ n}\}$ and $\{w_2^{\ n}\}$ are solutions of (1.17) satisfying (1.20); in (b), $\{w^n\}$ and $\{nw^{n-1}\}$ are solutions of (1.17) satisfying (1.20); and in (c), $\{r^n \cos(n\theta)\}$ and $\{r^n \sin(n\theta)\}$ are solutions of (1.17) satisfying (1.20). The verification that each of these three pairs of solutions does satisfy the condition (1.20) is left to Exercise 23. Therefore, we have nothing left to prove for part (a), since we already observed that $\{w_1^{\ n}\}$ and $\{w_2^{\ n}\}$ are solutions. To complete the proof of (b), we need only verify that $\{nw^{n-1}\}$ is a solution. But, since w is a double root of the characteristic equation (1.18),

$$w = \frac{a}{2} \quad \text{and} \quad b = \frac{-a^2}{4} \tag{1.21}$$

Therefore, substituting $x_n = nw^{n-1}$ into the difference equation (1.17) and simplifying, we get

$$(n + 2)w^{n+1} = a(n + 1)w^n + bnw^{n-1}$$

or, after dividing by w^{n-1},

$$(n + 2)w^2 = a(n + 1)w + bn$$

or, after using (1.21),

$$\frac{(n + 2)a^2}{4} = \frac{a^2(n + 1)}{2} - \frac{a^2 n}{4}$$

or, after dividing by $a^2/4$,

$$n + 2 = 2(n + 1) - n$$

Since this last equation is true for all n, $\{nw^{n-1}\}$ is a solution of (1.17). To complete the proof of (c), note first that $\{w_1^n\}$ and $\{w_2^n\}$ are solutions of the difference equation (1.17) since w_1 and w_2 are solutions of (1.18). So, according to the remarks preceding the theorem, we need only show that the sequences $\{r^n \cos(n\theta)\}$ and $\{r^n \sin(n\theta)\}$ are linear combinations of $\{w_1^n\}$ and $\{w_2^n\}$. But

$$w_1^n = r^n e^{in\theta} = r^n \cos(n\theta) + ir^n \sin(n\theta)$$

$$w_2^n = r^n e^{-in\theta} = r^n \cos(n\theta) - ir^n \sin(n\theta)$$

So, if we add and then subtract these two equations, we get

$$r^n \cos(n\theta) = \left(\frac{1}{2}\right) w_1^n + \left(\frac{1}{2}\right) w_2^n$$

$$r^n \sin(n\theta) = \left(\frac{1}{2i}\right) w_1^n - \left(\frac{1}{2i}\right) w_2^n$$

This completes the proof of the theorem.

EXAMPLE 1.15   Solve for x_n in

$$x_{n+2} = 6x_{n+1} - 9x_n, \qquad x_0 = 1, \quad x_1 = 9$$

The characteristic equation is $w^2 = 6w - 9$, which has the double root $w = 3$. So, by Theorem 1.6b,

$$x_n = c_1 3^n + c_2 n 3^{n-1} \quad \text{for some constants } c_1, c_2$$

And, when the initial conditions are applied to this, we get

$$1 = x_0 = c_1(1) + c_2(0) \qquad \text{and so } c_1 = 1$$

$$9 = x_1 = c_1(3) + c_2 = 3 + c_2 \quad \text{and so } c_2 = 6$$

Therefore, $x_n = 3^n + 6n 3^{n-1}$.

EXAMPLE 1.16   [For students familiar with complex arithmetic]   Solve for x_n in

$$x_{n+2} = 2x_{n+1} - 2x_n, \qquad x_0 = 2, \quad x_1 = 5$$

The characteristic equation is $w^2 = 2w - 2$, which has the complex roots

$$w = (2 \pm \sqrt{4 - 8})/2 = 1 \pm i = \sqrt{2} e^{\pm i\pi/4}$$

So, by Theorem 1.6c,

$$x_n = c_1(\sqrt{2})^n \cos\left(\frac{n\pi}{4}\right) + c_2(\sqrt{2})^n \sin\left(\frac{n\pi}{4}\right) \quad \text{for some } c_1, c_2$$

And, when the initial conditions are applied to this, we get

$$2 = x_0 = c_1(1) + c_2(0) = c_1 \qquad \text{and so } c_1 = 2$$

$$5 = x_1 = c_1 \frac{\sqrt{2}}{\sqrt{2}} + c_2 \frac{\sqrt{2}}{\sqrt{2}} = 2 + c_2 \quad \text{and so } c_2 = 3$$

Therefore

$$x_n = 2(\sqrt{2})^n \cos\left(\frac{n\pi}{4}\right) + 3(\sqrt{2})^n \sin\left(\frac{n\pi}{4}\right)$$

It is also useful to be able to solve some first- and second-order linear, constant coefficient, *nonhomogeneous* difference equations:

$$x_{n+1} = a x_n + b_n \tag{1.22}$$

$$x_{n+2} = a x_{n+1} + b x_n + c_n \tag{1.23}$$

The following theorem describes how to find the general solution of such an equation by combining one particular solution of the equation with the general solution of the corresponding homogeneous equation. (Its proof is left to Exercise 24.) We have just learned how to find the general solution of the homogeneous equation. We will not, however, learn methods for finding a particular solution of the nonhomogeneous equation; instead, each example will contain a hint on how to find one. [See Goldberg (1958) for a more extensive discussion.]

Theorem 1.7 Suppose $\{z_n\}$ is a particular solution of the difference equation (1.23) and $\{y_n\}$ is the general solution of the corresponding homogeneous equation (1.17). Then $\{y_n + z_n\}$ is the general solution of (1.23). The same result holds for (1.22) and (1.16) in place of (1.23) and (1.17), respectively.

EXAMPLE 1.17 Find the general solution of $x_{n+1} = x_n/2 + 5$.

The general solution of the corresponding homogeneous equation $y_{n+1} = y_n/2$ is given by $y_n = c/2^n$, as in Example 1.13. We guess that the given nonhomogeneous equation has a solution of the form $z_n = d$, since this is the form of the nonhomogeneous term $c_n = 5$. (This simple reasoning works only some of the time; see exceptions in Exercises 13 and 14.) When we substitute the guess $z_n = d$ into the given difference equation, we get $d = d/2 + 5$ and so $d = 10$. Hence, a particular solution is $z_n = 10$. Therefore, by Theorem 1.7, the general solution of the nonhomogeneous equation is $x_n = c/2^n + 10$.

EXAMPLE 1.18 Find the general solution of

$$x_{n+2} = 8x_{n+1} - 15x_n + 6 \cdot 2^n$$

The characteristic equation of the corresponding homogeneous equation is

$$w^2 = 8w - 15 \quad \text{or} \quad (w - 3)(w - 5) = 0$$

which has the roots $w = 3$ and $w = 5$. Thus, the general solution of the homogeneous equation is given by

$$y_n = c_1 3^n + c_2 5^n$$

We guess that the nonhomogeneous equation has a solution of the form $z_n = d\, 2^n$. When we substitute this guess into the given difference equation, we get

$$d\, 2^{n+2} = 8\, d\, 2^{n+1} - 15\, d\, 2^n + 6 \cdot 2^n$$

or, after dividing by 2^n

$$4d = 16d - 15d + 6$$

which implies that $d = 2$. Hence, $z_n = 2 \cdot 2^n = 2^{n+1}$, and therefore

$$x_n = c_1 3^n + c_2 5^n + 2^{n+1}$$

Outside of mathematics, difference equations are especially useful in the social and life sciences. The next example is a typical application; several more appear in the exercises.

EXAMPLE 1.19   A loan of P dollars is made at an annual interest rate of $100r\%$ on the unpaid balance. Each year, Y dollars are repaid. What is the unpaid balance after n years, and how many years does it take to pay off the loan, with interest?

We let x_n stand for the unpaid balance after n years. Each year, the unpaid balance increases by the amount of interest due, rx_n, and decreases by the yearly payment, Y. Thus, we have the difference equation

$$x_{n+1} = x_n + rx_n - Y$$
$$= (1 + r)x_n - Y$$

The corresponding homogeneous equation $y_{n+1} = (1 + r)y_n$ has the general solution $y_n = c(1 + r)^n$. We guess at the particular solution $z_n = d$ and substitute it into the nonhomogeneous equation. This yields $d = (1 + r)d - Y$ or $d = Y/r$. Therefore

$$x_n = c(1 + r)^n + \frac{Y}{r}$$

Using the initial condition $x_0 = P$, we get

$$P = c + \frac{Y}{r}$$

and so

$$x_n = \left(P - \frac{Y}{r} \right)(1 + r)^n + \frac{Y}{r}$$

To find the number of years needed to pay off the loan, set $x_n = 0$ and solve for n in the last equation.

EXERCISES

In Exercises 1–16, solve for x_n. (In each exercise involving a nonhomogeneous equation, a hint is given for finding a particular solution z_n).

1. $x_{n+1} = (\frac{1}{3})x_n$, $\qquad x_0 = 2$

2. $x_{n+1} = 5x_n$, $\qquad x_0 = -3$

3. $x_{n+1} = (n+1)x_n$, $\qquad x_0 = 1$

4. $x_{n+2} = 7x_{n+1} - 10x_n$, $\qquad x_0 = 3$, $x_1 = 12$

5. $x_{n+2} = 10x_{n+1} - 25x_n$, $\qquad x_0 = 10$, $x_1 = 15$

6. $x_{n+2} = 16x_n$, $\qquad x_0 = 8$, $x_1 = 16$

7. $x_{n+2} = 2x_{n+1} - x_n$, $\qquad x_0 = 3$, $x_1 = 5$

8. $x_{n+2} = -x_n$, $\qquad x_0 = 2$, $x_1 = -3$

9. $x_{n+2} = -2x_{n+1} - 2x_n$, $\qquad x_0 = 5$, $x_1 = -2$

10. $x_{n+2} = 7x_{n+1} - 10x_n + 8$, $\qquad x_0 = 5$, $x_1 = 14$ $\;(z_n = c)$

11. $x_{n+2} = 10x_{n+1} - 25x_n + 16 \cdot 3^n$, $\qquad x_0 = 6$, $x_1 = 15$ $\;(z_n = c\,3^n)$

12. $x_{n+2} = 16x_n + 6\cos(n\pi/4)$, $\qquad x_0 = 3$, $x_1 = -1/\sqrt{2}$
$$(z_n = c_1 \cos(n\pi/4) + c_2 \sin(n\pi/4))$$

13. $x_{n+2} = 2x_{n+1} - x_n + 6$, $\qquad x_0 = 5$, $x_1 = -2$ $\;(z_n = cn^2)$

14. $x_{n+2} = 16x_n + 64 \cdot 4^n$, $\qquad x_0 = 2$, $x_1 = 32$ $\;(z_n = cn4^n)$

15. $x_{n+2} = 2(\cos\theta)x_{n+1} - x_n$, $\qquad x_0 = 0$, $x_1 = \sin\theta$

16. $x_{n+1} = x_n + (n+1)$, $\qquad x_0 = 0$

17. Solve the difference equation in Example 1.7 and then find the limit of the sequence.

18. Solve the difference equation $F_{n+2} = F_{n+1} + F_n$, $F_0 = 0$, $F_1 = 1$ (the Fibonacci sequence), and use your answer to find the limit of F_{n+1}/F_n.

19. Solve the system of difference equations
$$x_{n+1} = x_n + y_n$$
$$y_{n+1} = 2x_n - y_n$$
$\qquad x_0 = 2$, $y_0 = \sqrt{3} - 2$

    Hint: Eliminate y_n and y_{n+1} in the second equation by solving for y_n in the first equation.

20. If $x_{n+2} = 2tx_{n+1} - x_n$, $x_0 = 1$, $x_1 = t$, show that x_n is a polynomial of degree n in t with leading coefficient 2^{n-1}, for $n > 0$. (x_n is called the nth Tchebycheff polynomial.)

21. If $x_n = P(n)$, where P is a polynomial of degree k, show that $x_{n+1} - x_n = Q(n)$, where Q is a polynomial of degree $k - 1$.

22. Suppose $\{y_n\}$ and $\{z_n\}$ are solutions of the difference equation (1.17). (a) If c_1 and c_2 are any constants, prove that $\{c_1 y_n + c_2 z_n\}$ is a solution of (1.17). (b) If $y_0 z_1 \neq y_1 z_0$, prove that, for every solution $\{x_n\}$ of (1.17), there exist constants c_1 and c_2 such that $x_n = c_1 y_n + c_2 z_n$. *Hint:* First show there exist c_1 and c_2 so that the equation $x_n = c_1 y_n + c_2 z_n$ holds for $n = 0$ and $n = 1$; then explain why the equation holds for all $n \geq 0$.

23. Verify the condition $y_0 z_1 \neq y_1 z_0$ for each of the following pairs of sequences:

    (a) $y_n = w_1^n$, $\qquad z_n = w_2^n$ where $w_1 \neq w_2$

    (b) $y_n = w^n$, $\qquad z_n = nw^{n-1}$

    (c) $y_n = r^n \cos(n\theta)$, $\qquad z_n = r^n \sin(n\theta)$ where $r\sin\theta \neq 0$

24. Prove Theorem 1.7, except for the last sentence.

25. **(a)** Find necessary and sufficient conditions for all the solutions $\{x_n\}$ of the difference equation $x_{n+2} = ax_{n+1} + bx_n$ to converge. (These should be conditions on the roots of the associated characteristic equation.) What are the possible limits?

(b) Find necessary and sufficient conditions for the sequence of ratios $\{x_{n+1}/x_n\}$ to converge, where $\{x_n\}$ is any solution of the difference equation $x_{n+2} = ax_{n+1} + bx_n$. (These should be conditions on the roots of the associated characteristic equation.) What are the possible limits?

26. A family borrows $50,000 to buy a house. They will pay off the loan in equal yearly payments at an annual interest rate of 10%. If they wish the loan to be repaid in 25 years, how much money will they have to pay each year?

27. You earn $100r\%$ interest each month in your savings account, which you started with a deposit of P dollars.

(a) If you never add to or take from this account, what will its value be after n months?

(b) If you add M dollars to your account each month but never take from it, what will its value be after n months?

28. A pair of rabbits gives birth to a new pair of rabbits every month after the pair is 2 months old. Every new pair of rabbits does the same. If we begin with one pair of newborn rabbits, how many will we have after n months?

29. Certain shrubs have the property that every twig which is more than a year old sprouts one new twig each spring, but no younger twig can sprout a new twig. If a shrub begins with one new twig, how many twigs will it have after n years?

30. If a slant ray passing through two face-to-face glass plates is reflected exactly n times, in how many ways could this occur? (Figure 1.4 shows the three ways in which the ray could be reflected two times.) Omit rays that simply reflect off the top face.

31. (Dynamic equilibrium with lagged adjustment) Suppose demand, D_n, supply, S_n, and price, P_n, change at discrete times n and are related as follows:

$$D_n = a - bP_n$$
$$S_n = -c + dP_{n-1}$$

for some positive real constants a, b, c, d, where $d < b$. Assuming supply equals demand, find a difference equation satisfied by P_n and solve this equation for P_n; show that

$$P_n = (P_0 - P_e)\left(-\frac{d}{b}\right)^n + P_e$$

where $P_e = \lim_{n \to \infty} P_n$ (the equilibrium price).

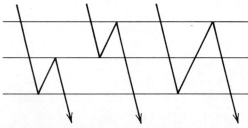

Figure 1.4 Slant rays that are reflected twice

32. The national income, Y_n, in any given year is the sum of the consumer expenditure, C_n, the investment in industry, I_n, and government expenditure, G_n, during that year. The following assumptions can be shown to be reasonable under certain conditions. Consumer expenditure is proportional to the national income in the preceding year; investment in industry is proportional to the change in consumer expenditure between the current year and the preceding year; and government expenditure is constant. Find a difference equation satisfied by Y_n; do not solve it.

33. In a population consisting of a predator and its prey, such as foxes and rabbits, their rates of growth are related to one another approximately as follows. (1) In the absence of foxes, the rabbit population would increase and the amount of increase in a given time period would be proportional to its current population. (2) In the absence of rabbits, the fox population would decrease and the amount of decrease in a given time period would be proportional to its current population. (3) The number of kills of rabbits by foxes in a given time period is proportional to the product of the rabbit and fox population. (4) For every kill, there is a proportional increase in the fox population; that is, this increase is proportional to the number of kills in the given time period.

(a) If r_n and f_n denote the number of rabbits and foxes, respectively, after n time periods, find a pair of difference equations which these sequences satisfy.

(b) Write a computer program that simulates the rabbit and fox populations; that is, write a program that computes the successive values of r_n and f_n. Describe how the populations vary over time when 150 is the initial number of rabbits, 100 is the initial number of foxes, and the four constants of proportionality referred to in conditions (1)–(4) are 0.1, 0.15, 0.001, 1, respectively. *Note:* $r_n = 150$ and $f_n = 100$ constitute a pair of stable populations for these four constants; change these initial populations slightly and then describe how the subsequent population fluctuations are related to the stable population values.

1.4 Taylor Polynomials

In this section we begin to study the problem of approximating a given function by a simpler function. The adjective "simpler" has no unique meaning in this context; but over the centuries mathematicians have found polynomials especially convenient since they are so easy to evaluate, differentiate, and integrate. Among the many classes of useful polynomials are those bearing the names of Taylor, Lagrange, Legendre, Tchebycheff, and Hermite. In this section and the next, we study methods for approximating functions by Taylor's polynomials and then Lagrange's.

Suppose we are given a function f and a point x_0 in the domain of f, and suppose we wish to find a polynomial that approximates f in such a way that the approximation is especially good at points near x_0. It is easier to find such a polynomial if we write it in the form

$$P_n(x) = c_0 + c_1(x - x_0) + c_2(x - x_0)^2 + \cdots + c_n(x - x_0)^n \qquad (1.24)$$

For example, this formula suggests that we should choose the coefficient c_0 to be $f(x_0)$ since then $P_n(x_0)$ and $f(x_0)$ will be equal. If, in addition, the first through nth derivatives of P_n at x_0 equal the first through nth derivatives of f at x_0, respectively, it is reasonable to guess that P_n is then a good approximation to f at points near x_0. Just how "good" the approximation is and how "near" the points x must be to x_0 will be discussed later. For now, let us try to determine the coefficients c_k so that the polynomial P_n satisfies the above criterion. By repeatedly differentiating (1.24) and substituting x_0 into the result, we get

$$f(x_0) = P_n(x_0) = c_0 + c_1(0) + \cdots + c_n(0) = c_0$$

$$P_n'(x) = c_1 + 2c_2(x - x_0) + 3c_3(x - x_0)^2 + \cdots + nc_n(x - x_0)^{n-1}$$

and so

$$f'(x_0) = P_n'(x_0) = c_1 + 2c_2(0) + \cdots + nc_n(0) = c_1$$

$$P_n''(x) = 2c_2 + 3\cdot2c_3(x - x_0) + \cdots + n(n-1)c_n(x - x_0)^{n-2}$$

and so

$$f''(x_0) = P_n''(x_0) = 2c_2 + 3\cdot2c_3(0) + \cdots + n(n-1)c_n(0) = 2c_2$$

and so on. The general pattern is that $f^{(k)}(x_0) = k!c_k$, where $f^{(k)}$ denotes the kth derivative of f (that is, $f^{(0)} = f$, $f^{(1)} = f'$, $f^{(2)} = f''$, and so on). Thus

$$c_k = \frac{f^{(k)}(x_0)}{k!} \quad \text{for} \quad k = 0, 1, \ldots, n \tag{1.25}$$

Thus, our reasoning suggests that the polynomial P_n given by (1.24) with coefficients c_k given by (1.25) ought to be a good approximation of f at points near x_0. We give these polynomials a name.

Definition 1.3 Suppose f is n times differentiable at a point x_0. Then

$$P_n(x) = \sum_{k=0}^{n} \frac{f^{(k)}(x_0)}{k!}(x - x_0)^k \tag{1.26}$$

is called the *nth Taylor polynomial of f at x_0*. Furthermore,

$$R_n(x) = f(x) - P_n(x) \tag{1.27}$$

is called the *nth remainder*.

The sequence of Taylor polynomials of f at x_0 may also be defined recursively:

$$P_0(x) = f(x_0)$$

$$P_{n+1}(x) = P_n(x) + \frac{f^{(n+1)}(x_0)(x - x_0)^{n+1}}{(n+1)!} \quad \text{for } n \geq 0 \tag{1.28}$$

EXAMPLE 1.20 If $f(x) = e^x$, find the nth Taylor polynomial of f at 0. For all $k \geq 0$, $f^{(k)}(x) = e^x$ and so $f^{(k)}(0) = 1$. Therefore

$$P_n(x) = 1 + x + \frac{x^2}{2!} + \frac{x^3}{3!} + \cdots + \frac{x^n}{n!} = \sum_{k=0}^{n} \frac{x^k}{k!}$$

EXAMPLE 1.21 If $f(x) = x^3 - 2x^2 + 5x + 3$, find the nth Taylor polynomial of f at 2 for $n \geq 3$.

$$f(x) = x^3 - 2x^2 + 5x + 3 \qquad f(2) = 13$$
$$f'(x) = 3x^2 - 4x + 5 \qquad f'(2) = 9$$
$$f^{(2)}(x) = 6x - 4 \qquad f^{(2)}(2) = 8$$
$$f^{(3)}(x) = 6 \qquad f^{(3)}(2) = 6$$
$$f^{(n)}(x) = 0 \quad \text{for } n \geq 4 \qquad f^{(n)}(2) = 0 \quad \text{for } n \geq 4$$

Therefore, for $n \geq 3$,

$$P_n(x) = 13 + 9(x - 2) + \frac{8}{2!}(x - 2)^2 + \frac{6}{3!}(x - 2)^3$$

Furthermore, $f(x)$ actually equals $P_3(x)$ at every point x since $f^{(4)}$ is the zero function (see Exercise 7).

EXAMPLE 1.22 If $f(x) = \sin x$, find the nth Taylor polynomial of f at 0.

$$f(x) = \sin x \qquad f(0) = 0$$
$$f'(x) = \cos x \qquad f'(0) = 1$$
$$f^{(2)}(x) = -\sin x \qquad f^{(2)}(0) = 0$$
$$f^{(3)}(x) = -\cos x \qquad f^{(3)}(0) = -1$$
$$f^{(4)}(x) = \sin x \qquad f^{(4)}(0) = 0$$
$$f^{(5)}(x) = \cos x \qquad f^{(5)}(0) = 1$$

and so on. Therefore

$$P_n(x) = x - \frac{x^3}{3!} + \frac{x^5}{5!} + \cdots + (-1)^{(p-1)/2} \frac{x^p}{p!} = \sum_{k=0}^{(p-1)/2} (-1)^k \frac{x^{2k+1}}{(2k+1)!}$$

where p is the largest odd integer not exceeding n.

EXAMPLE 1.23 If $f(x) = (1 - x)^{-1/2}$, find the nth Taylor polynomial of f at 0.

$$f(x) = (1 - x)^{-1/2} \qquad f(0) = 1$$
$$f'(x) = \frac{1}{2}(1 - x)^{-3/2} \qquad f'(0) = \frac{1}{2}$$
$$f^{(2)}(x) = \frac{1 \cdot 3}{2^2}(1 - x)^{-5/2} \qquad f^{(2)}(0) = \frac{1 \cdot 3}{2^2}$$
$$f^{(3)}(x) = \frac{1 \cdot 3 \cdot 5}{2^3}(1 - x)^{-7/2} \qquad f^{(3)}(0) = \frac{1 \cdot 3 \cdot 5}{2^3}$$

and so on. Therefore

$$P_n(x) = 1 + \frac{1}{2}x + \frac{1 \cdot 3}{2^2(2!)}x^2 + \cdots + \frac{1 \cdot 3 \cdots (2n - 1)}{2^n(n!)}x^n$$

If we want to prove that the sequence $\{P_n\}$ of Taylor polynomials of f at x_0 converges to f on some interval I, we might find Theorem 1.3b useful. That is, we might look for a sequence $\{y_n\}$ such that

$$|f(x) - P_n(x)| \leq y_n(x) \quad \text{and} \quad y_n(x) \to 0 \quad \text{for all } x \text{ in } I \qquad (1.29)$$

If, instead, we want to approximate f by a Taylor polynomial P_n, we proceed similarly. For a given interval I and a given positive number ε (which we think of as an acceptable bound for the error), we look for a single value of n such that

$$|f(x) - P_n(x)| \leq \varepsilon \quad \text{for all } x \text{ in } I \qquad (1.30)$$

This criterion guarantees that on the interval I, the values $P_n(x)$ of the approximation are within the distance ε of the values $f(x)$ of the given function. Neither of these two tasks is easy, neither finding a sequence $\{y_n\}$ such that (1.29) holds nor finding n such that (1.30) holds. Fortunately, Theorem 1.8 helps by giving us a way to rewrite $f(x) - P_n(x)$ more compactly and thus find an upper bound for $|f(x) - P_n(x)|$ which can be related to y_n or ε. For $n = 0$, Theorem 1.8 is just the mean value theorem (Corollary A.17), which says that

$$f(x) - P_0(x) = f(x) - f(x_0) = f'(z)(x - x_0)$$

for some number z between x and x_0. For $n = 1$, Theorem 1.8 says that

$$f(x) - P_1(x) = f(x) - [f(x_0) + f'(x_0)(x - x_0)] = \frac{f^{(2)}(z)}{2!}(x - x_0)^2$$

$$(1.31)$$

for some number z between x and x_0; this special case of the theorem occurs frequently.

Theorem 1.8 (Taylor's remainder formula) Suppose f is $n + 1$ times differentiable on some interval I. Then, for every pair of numbers x_0 and x in I, there exists a point z between x_0 and x such that

[handwritten: Use generated expressions to equate this to a sequence, & find limit.]

$$f(x) - P_n(x) = \frac{f^{(n+1)}(z)}{(n+1)!}(x - x_0)^{n+1}$$

where P_n denotes the nth Taylor polynomial of f at x_0.

Proof We will think of x and x_0 as constants—they will be fixed throughout. If we define c to be the unique number satisfying the equation

$$f(x) - P_n(x) = c(x - x_0)^{n+1}$$

we must show that $c = f^{(n+1)}(z)/(n + 1)!$ for some number z in $[x_0, x]$. (By writing $[x_0, x]$, we assumed that $x_0 < x$; a similar proof can be given when $x < x_0$.) To this end, we define a function g to which we can apply the mean value theorem. Let

$$g(t) = f(t) - P_n(t) - c(t - x_0)^{n+1} \quad \text{for all } t \text{ in } [x_0, x]$$

Then $g(x) = 0$ and

$$g^{(m)}(t) = f^{(m)}(t) - \sum_{k=0}^{n} \frac{f^{(k)}(x_0)}{k!} \frac{d^m}{dt^m}(t - x_0)^k - c\frac{d^m}{dt^m}(t - x_0)^{n+1}$$

$$= \begin{cases} f^{(n+1)}(t) - c(n+1)! & \text{if } m = n+1 \\ f^{(m)}(x_0) - f^{(m)}(x_0) = 0 & \text{if } m \le n \text{ and } t = x_0 \end{cases}$$

Therefore, by the mean value theorem applied successively to g, $g^{(1)}$, etc.,

$$g^{(1)}(z_1) = 0 \quad \text{for some } z_1 \text{ in } (x_0, x), \text{ since } g(x_0) = g(x) = 0$$

$$g^{(2)}(z_2) = 0 \quad \text{for some } z_2 \text{ in } (x_0, z_1), \text{ since } g^{(1)}(x_0) = g^{(1)}(z_1) = 0$$

$$\vdots$$

$$g^{(n+1)}(z_{n+1}) = 0 \quad \text{for some } z_{n+1} \text{ in } (x_0, z_n), \text{ since } g^{(n)}(x_0) = g^{(n)}(z_n) = 0$$

But we showed above that

$$g^{(n+1)}(z_{n+1}) = f^{n+1}(z_{n+1}) - c(n+1)!$$

So it follows that $c = f^{(n+1)}(z_{n+1})/(n+1)!$, which is what we needed to prove.

EXAMPLE 1.24 Let P_n be the nth Taylor polynomial of $f(x) = \sin x$ at 0. (a) Prove that $P_n(x) \to f(x)$ for every x. (b) Find a value of n such that P_n approximates f on the interval $[-\pi, \pi]$ with an error of at most 0.01.

By Taylor's remainder formula,

$$|f(x) - P_n(x)| = \frac{|f^{(n+1)}(z)|}{(n+1)!}|x|^{n+1}$$

for some number z between 0 and x. But, in Example 1.22 we saw that $f^{(k)}(x)$ equals $\pm\sin x$ or $\pm\cos x$; hence, $|f^{(n+1)}(z)| \le 1$. Therefore, $|f(x) - P_n(x)| \le |x|^{n+1}/(n+1)!$. Also, $|x|^{n+1}/(n+1)! \to 0$ (Example 1.5); hence, by Theorem 1.3b, $P_n(x) \to f(x)$ for every x. Furthermore, if $-\pi \le x \le \pi$, then $|x| \le \pi$ and so our preceding inequality implies that

$$|f(x) - P_n(x)| \le \frac{\pi^{n+1}}{(n+1)!} \quad \text{whenever } -\pi \le x \le \pi$$

By trial and error, we find that $\pi^{11}/11! = 0.0074 < 0.01$; so

$$|\sin x - P_{10}(x)| < 0.01 \quad \text{whenever } -\pi \le x \le \pi$$

The approximating polynomial is

$$P_{10}(x) = x - \frac{x^3}{3!} + \frac{x^5}{5!} - \frac{x^7}{7!} + \frac{x^9}{9!}$$

Figure 1.5 illustrates the typical behavior of Taylor polynomials: They usually approximate f quite well at points near x_0 and less well at points farther from x_0; and they usually approximate f more closely as n gets larger.

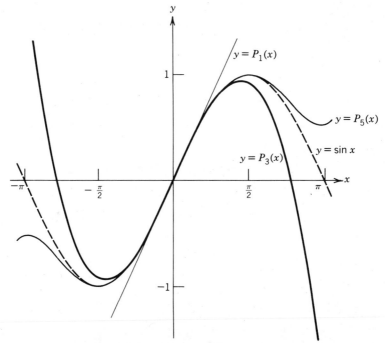

Figure 1.5 Taylor polynomials of sin x at 0

EXAMPLE 1.25 Let P_n be the nth Taylor polynomial of $f(x) = 1/\sqrt{x}$ at 1. Find a value of n such that P_n approximates f on the interval $[0.6, 1.5]$ with an error of at most 0.05.

By Taylor's remainder formula,

$$\left| f(x) - P_n(x) \right| = \left| \frac{f^{(n+1)}(z)}{(n+1)!} \right| |x - 1|^{n+1}$$

for some number z between 1 and x. But

$$f'(x) = -\frac{1}{2} x^{-3/2}$$

$$f^{(2)}(x) = \frac{1 \cdot 3}{2^2} x^{-5/2}$$

$$f^{(3)}(x) = -\frac{1 \cdot 3 \cdot 5}{2^3} x^{-7/2}$$

$$\vdots$$

$$f^{(k)}(x) = \frac{(-1)^k 1 \cdot 3 \cdot 5 \cdots (2k-1)}{2^k} x^{-(2k+1)/2} \quad \text{for } k \geq 1$$

Therefore

$$|f(x) - P_n(x)| = \frac{1 \cdot 3 \cdot 5 \cdots (2n + 1)|x - 1|^{n+1}}{2^{n+1}(n + 1)!|z|^{(2n+3)/2}}$$

$$< \frac{|x - 1|^{n+1}}{|z|^{(2n+3)/2}} = \frac{|x - 1|^{n+1}}{|z|^{n+3/2}}$$

The above inequality follows from the fact that

$$\frac{1 \cdot 3 \cdot 5 \cdots (2n + 1)}{2^{n+1}(n + 1)!} = \frac{1}{2 \cdot 1} \cdot \frac{3}{2 \cdot 2} \cdot \frac{5}{2 \cdot 3} \cdots \frac{2n + 1}{2(n + 1)}$$

which is less than 1 since each of the $n + 1$ factors is. We find an upper bound for $|x - 1|^{n+1}/|z|^{n+3/2}$ by considering separately the case $1 \le x \le 1.5$ and the case $0.6 \le x \le 1$. When $1 \le x \le 1.5$, then $z \ge 1$ and so

$$\frac{|x - 1|^{n+1}}{|z|^{n+3/2}} \le |x - 1|^{n+1} \le (1.5 - 1)^{n+1} = \frac{1}{2^{n+1}}$$

When $0.6 \le x \le 1$, then $z \ge x \ge 0.6$ and so

$$\frac{|x - 1|^{n+1}}{|z|^{n+3/2}} \le \frac{(1 - 0.6)^{n+1}}{(0.6)^{n+3/2}} = \frac{(0.4)^{n+1}}{(0.6)^{n+1}\sqrt{0.6}} = \left(\frac{2}{3}\right)^{n+1} \frac{1}{\sqrt{0.6}}$$

By trial and error, we find that both $1/2^{n+1}$ and $(2/3)^{n+1}/\sqrt{0.6}$ are less than 0.05 when $n = 8$; so

$$\left|\frac{1}{\sqrt{x}} - P_8(x)\right| \le 0.05 \quad \text{whenever } 0.6 \le x \le 1.5$$

We can also use Taylor's remainder formula to approximate an integral that annot be evaluated exactly.

EXAMPLE 1.26 Approximate $\int_{-1}^{1} \exp(x^2)\, dx$ with an error of at most 0.001.

Our strategy is to approximate $\exp(x^2)$ by a polynomial and then use the integral of this polynomial as an approximation of the integral of $\exp(x^2)$. But since $\exp(x)$ has much simpler derivatives than $\exp(x^2)$, we will compute Taylor polynomials for $\exp(x)$ and then replace x by x^2, rather than compute Taylor polynomials for $\exp(x^2)$. Furthermore, since we expect the Taylor polynomials of f at x_0 to be closer to f at points near x_0 and since we want a good approximation to f on the interval $[-1, 1]$, we choose x_0 to be the center, 0, of that interval. (Also the derivatives $f^{(k)}$ are especially simple at 0.)

So we let $f(x) = e^x$ and let P_n denote the nth Taylor polynomial of f at 0. Then, by Taylor's remainder formula, there exists a number z between 0 and x such that

$$|e^x - P_n(x)| = \frac{|f^{(n+1)}(z)||x|^{n+1}}{(n + 1)!} = \frac{e^z|x|^{n+1}}{(n + 1)!} \le \frac{e|x|^{n+1}}{(n + 1)!}$$

where $e^z \le e$ since $z \le 1$. Therefore, after replacing x by x^2, we get

$$\left| \exp(x^2) - P_n(x^2) \right| \le \frac{e|x^2|^{n+1}}{(n+1)!} = \frac{ex^{2n+2}}{(n+1)!}$$

and so

$$\left| \int_{-1}^{1} \exp(x^2)\, dx - \int_{-1}^{1} P_n(x^2)\, dx \right| = \left| \int_{-1}^{1} \left[\exp(x^2) - P_n(x^2) \right] dx \right|$$

$$\le \int_{-1}^{1} \left| \exp(x^2) - P_n(x^2) \right| dx \quad \text{(by Theorem A.24e)}$$

$$\le \int_{-1}^{1} \frac{ex^{2n+2}}{(n+1)!}\, dx$$

$$= \frac{2e}{(n+1)!(2n+3)}$$

By trial and error, we find that $2e/6!(13) = 0.00058 < 0.001$, and so $\int_{-1}^{1} P_5(x^2)\, dx$ is within 0.001 of $\int_{-1}^{1} \exp(x^2)\, dx$. Therefore, our desired approximation is

$$\int_{-1}^{1} P_5(x^2)\, dx = \int_{-1}^{1} \left(1 + x^2 + \frac{x^4}{2} + \frac{x^6}{6} + \frac{x^8}{24} + \frac{x^{10}}{120} \right) dx$$

$$= 2\left(1 + \frac{1}{3} + \frac{1}{10} + \frac{1}{42} + \frac{1}{216} + \frac{1}{1320} \right)$$

$$= 2.9251$$

In our final example, we use Taylor's remainder formula to derive a few inequalities of the type used in Example 1.10 and Exercises 14–16 of Section 1.2.

EXAMPLE 1.27   Show that

$$\left| \log(1+x) - x \right| \le \frac{x^2}{2} \quad \text{if } 0 \le x \tag{1.32}$$

$$e^x > \frac{x^n}{n!} \quad \text{if } x > 0 \text{ and } n = 0, 1, 2, \ldots \tag{1.33}$$

$$\log x < \frac{x^p}{p} \quad \text{if } x > 1 \text{ and } p > 0 \tag{1.34}$$

To verify inequality (1.32), we apply Taylor's remainder formula to $f(x) = \log(1+x)$ with $n=1$ and $x_0 = 0$. Then $f'(x) = (1+x)^{-1}$ and $f''(x) = -(1+x)^{-2}$; hence, there exists z between 0 and x such that

$$\left| f(x) - P_1(x) \right| = \frac{|f''(z)||x|^2}{2!}$$

or

$$\left| \log(1+x) - x \right| = \frac{|x|^2}{2(1+z)^2} \le \frac{|x|^2}{2}$$

To prove inequality (1.33), we apply Taylor's remainder formula to $f(x) = e^x$ with $x_0 = 0$. Then $f^{(k)}(x) = e^x$ and $P_n(x) = \sum_{k=0}^{n} x^k/k!$; so there exists z between 0 and x such that

$$f(x) - P_n(x) = \frac{f^{(n+1)}(z)x^{n+1}}{(n+1)!}$$

or

$$e^x = \sum_{k=0}^{n} \frac{x^k}{k!} + \frac{e^z x^{n+1}}{(n+1)!} > \frac{x^n}{n!}$$

The inequality is valid since a sum of positive numbers is greater than each term in the sum, and $x^n/n!$ is the last term in the sum $\sum_{k=0}^{n} x^k/k!$.

We derive inequality (1.34) from inequality (1.33). If we replace x by $p \log x$ and n by 1 in (1.33), we get

$$x^p = e^{p \log x} > p \log x \quad \text{if } x > 1$$

One use we have already made of such inequalities is to derive some limit formulas. In Example 1.10, we used inequality (1.32) to show that $(1 + 1/n)^n \to e$. From (1.33) and (1.34), we can derive the limits

$$\frac{n^p}{e^n} \to 0 \quad \text{for any } p > 0 \tag{1.35}$$

$$\frac{\log n}{n^p} \to 0 \quad \text{for any } p > 0 \tag{1.36}$$

(Exercises 22 and 23). Roughly speaking, these say that e^n tends toward infinity more rapidly than any power of n (no matter how large the power) and $\log n$ tends toward infinity more slowly than any positive power of n (no matter how small the power).

EXERCISES

In Exercises 1–5, (a) find P_n, the nth Taylor polynomial of f at x_0, and (b) use Taylor's remainder formula to show that $P_n(x) \to f(x)$ for all x in the indicated interval.

1. $f(x) = \cosh x = (e^x + e^{-x})/2$, $\quad x_0 = 0$, on $(-\infty, \infty)$
   Note: $\cosh x$ is called the "hyperbolic cosine of x."

2. $f(x) = \sinh x = (e^x - e^{-x})/2$, $\quad x_0 = 0$, on $(-\infty, \infty)$
   Note: $\sinh x$ is called the "hyperbolic sine of x."

3. $f(x) = \cos x$, $x_0 = 0$, on $(-\infty, \infty)$

4. $f(x) = \log(1 + x)$, $x_0 = 0$, on $[-0.5, 1]$

5. $f(x) = x^{-2}$, $x_0 = 1$, on $(0.5, 2)$

6. Find a Taylor polynomial of the form $a_0 + a_1(x + 1) + \cdots + a_n(x + 1)^n$ which equals $2x^3 - 5x + 7$ for all x.

7. If f is a polynomial of degree at most n, prove that f equals its nth Taylor polynomial at x_0 (for any x_0).

In Exercises 8–11, find a numerical bound for the error $|f(x) - P_n(x)|$ on the given interval, where P_n is the nth Taylor polynomial of $\cdot f$ at x_0.

8. $f(x) = \cos x$, $n = 3$, $x_0 = 0$, interval $= [-\pi/2, \pi/2]$
9. $f(x) = \sin x$, $n = 4$, $x_0 = 0$, interval $= [-\pi/2, \pi/2]$
10. $f(x) = \arctan x$, $n = 1$, $x_0 = 0$, interval $= [-0.5, 0.5]$
11. $f(x) = \sqrt{x}$, $n = 2$, $x_0 = 1$, interval $= [0.5, 1.5]$

In Exercises 12–15, find a Taylor polynomial that approximates f with an error of at most 0.01 on the indicated interval.

12. $f(x) = \cos x$, $[-\pi/2, \pi/2]$ 13. $f(x) = e^x$, $[-1, 1]$
14. $f(x) = \sqrt{x}$, $[0.5, 1.5]$ 15. $f(x) = \sin(x^2)$, $[-1.5, 1.5]$

16. Approximate $\int_{-\sqrt{\pi}}^{\sqrt{\pi}} \sin(x^2)\, dx$ with an error of at most 0.1.

17. Approximate $\int_0^{0.5} \log(1 + \sqrt{x})\, dx$ with an error of at most 0.02.

18. Approximate $\int_0^1 (\sin x)/x\, dx$ with an error of at most 0.0005.

19. Show that $|\sin x - x| \le |x|^3/6$ for all x.

20. Show that $|\cos x - (1 - x^2/2)| \le x^4/24$ for all x.

21. Show that $|\log(1 + x) - x + x^2/2| \le x^3/3$ when $0 \le x$.

22. Show that $n^p/e^n \to 0$ for any $p > 0$.

23. Show that $(\log n)/n^p \to 0$ for any $p > 0$.

1.5 Lagrange Polynomials

As we have seen, the Taylor polynomials of f at x_0 usually approximate f quite well near x_0 and less well farther from x_0. Lagrange interpolation polynomials, on the other hand, will enable us to approximate functions over larger domains.

Suppose we are given $n + 1$ distinct values $x_0, x_1, \ldots, x_n$ of the independent variable of some function and the $n + 1$ corresponding values $y_0, y_1, \ldots, y_n$ of the dependent variable. We want to find a polynomial which "interpolates" between such points, by which we mean a polynomial whose graph passes through the $n + 1$ points

$$(x_0, y_0), (x_1, y_1), \ldots, (x_n, y_n)$$

Our hope is that the graph of such a polynomial is reasonably close to the graph of the original function at intermediate points. The next theorem assures us that interpolating polynomials exist. It says, in effect, that through any two points we can draw a straight line, through three points a parabola, through four a cubic, and so on.

Theorem 1.9   Given any set of $n + 1$ points $(x_0, y_0), (x_1, y_1), \ldots, (x_n, y_n)$ with distinct x coordinates, there exists a unique polynomial p of degree $\le n$ such that $p(x_i) = y_i$ for $i = 0, 1, \ldots, n$.

Proof   First we prove that there cannot be more than one such polynomial. Suppose p and q are both polynomials of degree $\le n$ and that $p(x_i) = y_i$, $q(x_i) = y_i$ for $i = 0, 1, \ldots, n$. Then $p - q$ is a polynomial of degree $\le n$ which has $n + 1$ distinct zeros: the points x_i. But since a nonzero polynomial of degree n has at most n zeros, $p - q$ must be the zero polynomial. Therefore, p and q are the same polynomial.

To prove that p exists, we construct it. Let

$$p(x) = \sum_{k=0}^{n} y_k q_k(x)$$

where

$$q_k(x) = \prod_{\substack{i=0 \\ i \neq k}}^{n} \frac{(x - x_i)}{(x_k - x_i)} \quad \text{for } k = 0, 1, \ldots, n \quad (1.37)$$

This compact formula for q_k means that

$$q_0(x) = \frac{(x - x_1)(x - x_2) \cdots (x - x_n)}{(x_0 - x_1)(x_0 - x_2) \cdots (x_0 - x_n)}$$

$$q_1(x) = \frac{(x - x_0)(x - x_2) \cdots (x - x_n)}{(x_1 - x_0)(x_1 - x_2) \cdots (x_1 - x_n)}$$

and so on. Thus $q_k(x_i) = 0$ or 1, depending on whether $k \neq i$ or $k = i$, respectively. Therefore

$$p(x_i) = \sum_{k=0}^{n} y_k q_k(x_i) = y_i$$

which completes the proof of the theorem.

Definition 1.4 Given any $n + 1$ points $(x_0, y_0), (x_1, y_1), \ldots, (x_n, y_n)$ with distinct x coordinates, the unique polynomial p of degree $\leq n$ that satisfies $y_i = p(x_i)$ for $i = 0, 1, \ldots, n$ is called the *Lagrange polynomial* that interpolates between these $n + 1$ points.

The Lagrange polynomials we use most often are those that interpolate between three or fewer points:

$$p(x) = y_0 \tag{1.38}$$

$$p(x) = y_0 \frac{(x - x_1)}{(x_0 - x_1)} + y_1 \frac{(x - x_0)}{(x_1 - x_0)} \qquad \leftarrow degree\ 1 \qquad degree\ 2. \tag{1.39}$$

$$p(x) = y_0 \frac{(x - x_1)(x - x_2)}{(x_0 - x_1)(x_0 - x_2)} + y_1 \frac{(x - x_0)(x - x_2)}{(x_1 - x_0)(x_1 - x_2)} + y_2 \frac{(x - x_0)(x - x_1)}{(x_2 - x_0)(x_2 - x_1)}$$

$$\tag{1.40}$$

EXAMPLE 1.28 Find the Lagrange polynomial p that interpolates $\sin x$ between the points $x_0 = 0$, $x_1 = \pi/2$, $x_2 = \pi$.

Since $y_0 = 0$, $y_1 = 1$, $y_2 = 0$, we have, from (1.40)

$$p(x) = \frac{x(x - \pi)}{(\pi/2)(-\pi/2)}$$

Table 1.5 and Figure 1.6 help us compare $p(x)$ and $\sin x$.

Table 1.5 An Interpolation of sin x

x	p(x)	sin x
0.0	0.000	0.000
$\pi/6$	0.556	0.500
$\pi/4$	0.750	0.707
$\pi/3$	0.889	0.866
$\pi/2$	1.000	1.000

As in Section 1.4, if we approximate a function f by a Lagrange polynomial p, we would like an upper bound for the error $|f(x) - p(x)|$. The next theorem is very similar to Taylor's remainder formula.

Theorem 1.10   Suppose f is $n + 1$ times differentiable on some interval I containing the distinct points $x_0, x_1, \ldots, x_n$. If p is the Lagrange polynomial satisfying $p(x_i) = f(x_i)$ for $i = 0, 1, \ldots, n$, then, for every x in I, there exists a point z in I such that

$$f(x) - p(x) = \frac{f^{(n+1)}(z)}{(n+1)!} \prod_{i=0}^{n} (x - x_i) \tag{1.41}$$

Proof   The theorem is certainly true if x is one of the numbers $x_0, x_1, \ldots, x_n$. So, in the rest of the proof, we may assume x equals none of them. If we define c to be the unique number satisfying the equation

$$f(x) - p(x) = c \prod_{i=0}^{n} (x - x_i)$$

n is degree of p(x).
n+1 is number of points used.

we must show that $c = f^{(n+1)}(z)/(n + 1)!$ for some number z in I. To this end, we define a function g to which we can apply the mean value theorem. Let

$$g(t) = f(t) - p(t) - c \prod_{i=0}^{n} (t - x_i) \quad \text{for all } t \text{ in } I$$

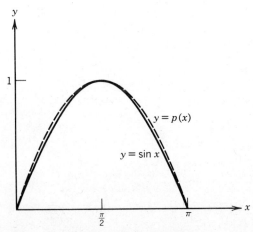

Figure 1.6   An interpolation of sin x

Then $g(x) = 0$ and $g(x_i) = 0$ for $i = 0, 1, \ldots, n$. Thus, g has at least $n + 2$ zeros in I. Therefore, since the mean value theorem assures us that between any two zeros of g there is at least one zero of the derivative g', g' must have at least $n + 1$ zeros in I. Similarly, $g^{(2)}$ must have at least n zeros, $\ldots$, and $g^{(n+1)}$ must have at least one zero; we let z denote a zero of $g^{(n+1)}$ in I. Furthermore

$$g^{(n+1)}(t) = f^{(n+1)}(t) - c(n + 1)!$$

since $p^{(n+1)}(t) = 0$ and since the term of $\prod_{i=0}^{n}(t - x_i)$ with the highest power of t is t^{n+1}. Therefore $0 = g^{(n+1)}(z) = f^{(n+1)}(z) - c(n + 1)!$, which is what we needed to prove.

In our applications of Theorem 1.10, n is usually small. For $n = 0, 1, 2$, equation (1.41) takes on the respective forms

$$f(x) - f(x_0) = f'(z)(x - x_0) \quad \text{(the mean value theorem)} \tag{1.42}$$

$$f(x) - \left[f(x_0) \frac{(x - x_1)}{(x_0 - x_1)} + f(x_1) \frac{(x - x_0)}{(x_1 - x_0)} \right]$$
$$= \frac{f^{(2)}(z)}{2}(x - x_0)(x - x_1) \tag{1.43}$$

$$f(x) - \left[f(x_0) \frac{(x - x_1)(x - x_2)}{(x_0 - x_1)(x_0 - x_2)} + f(x_1) \frac{(x - x_0)(x - x_2)}{(x_1 - x_0)(x_1 - x_2)} \right.$$
$$\left. + f(x_2) \frac{(x - x_0)(x - x_1)}{(x_2 - x_0)(x_2 - x_1)} \right]$$
$$= \frac{f^{(3)}(z)}{6}(x - x_0)(x - x_1)(x - x_2) \tag{1.44}$$

EXAMPLE 1.29 Find the Lagrange polynomial p that interpolates the function x^{-1} between the points $x_0 = 2$ and $x_1 = 4$; find a bound for the error $|x^{-1} - p(x)|$, where $2 \leq x \leq 4$.

$y_0 = \frac{1}{2}$ and $y_1 = \frac{1}{4}$; so, from (1.39)

$$p(x) = \frac{1}{2} \frac{(x - 4)}{(-2)} + \frac{1}{4} \frac{(x - 2)}{(2)}$$
$$= \frac{(4 - x)}{4} + \frac{(x - 2)}{8} = \frac{(6 - x)}{8}$$

By (1.43) with $f(x) = x^{-1}$, there exists a point z between 2 and 4 such that

$$|x^{-1} - p(x)| = \left| \frac{f^{(2)}(z)}{2}(x - 2)(x - 4) \right|$$
$$= \frac{2z^{-3}}{2}(x - 2)(4 - x)$$
$$\leq (0.125)(x - 2)(4 - x) \quad \text{since } z \geq 2$$
$$\leq (0.125)(3 - 2)(4 - 3) = 0.125$$

for all x in $[2, 4]$. The last inequality comes from the fact that the quadratic function $(x - 2)(4 - x)$ has its maximum value midway between its two zeros $x = 2$ and $x = 4$.

Finding a bound for the expression $|(x - x_0)(x - x_1) \cdots (x - x_n)|$ when $n > 1$ can be much more difficult than finding a bound when $n = 1$, which we did in the above example. But, if the points x_i are equally spaced and if n is not too large, we can derive the following bounds. Let

$$x_i = x_0 + ih \quad \text{for } i = 0, 1, \ldots, n$$

(where h is the distance between successive points x_i) and assume x is in the interval $[x_0, x_n]$. Then, for $n = 1$,

$$|(x - x_0)(x - x_1)| \le \frac{h^2}{4} \tag{1.45}$$

since, as we saw in Example 1.29, the quadratic function $(x - x_0)(x_1 - x)$ has its maximum value at the point $x = (x_0 + x_1)/2$ which is halfway between its zeros x_0 and x_1. For $n = 2$ and $n = 3$, we have

$$|(x - x_0)(x - x_1)(x - x_2)| \le \frac{2h^3}{3\sqrt{3}} \tag{1.46}$$

$$|(x - x_0)(x - x_1)(x - x_2)(x - x_3)| \le h^4 \tag{1.47}$$

whose proofs are left to the exercises.

EXAMPLE 1.30 If $f(x) = \sin x$ and $p(x)$ is the Lagrange polynomial in Example 1.28, find a bound for the error $|f(x) - p(x)|$, where $0 \le x \le \pi$.

For x in $[0, \pi]$, there exists z such that

$$|\sin x - p(x)| = \left| \frac{f^{(3)}(z)}{3!}(x - x_0)(x - x_1)(x - x_2) \right|$$

$$= \left| \frac{(\cos z)(x - x_0)(x - x_1)(x - x_2)}{6} \right|$$

$$\le \frac{2h^3}{6 \cdot 3\sqrt{3}} \quad \text{by inequality (1.46)}$$

$$= \frac{(\pi/2)^3}{9\sqrt{3}} = 0.249$$

Judging by Table 1.5 and Figure 1.6, however, this error bound is overly conservative.

The formulas (1.37) are not always the most convenient ones to use for computing Lagrange interpolation polynomials. Suppose, for example, we have formulas for computing Lagrange polynomials recursively, just as we have formulas (1.28) for computing Taylor polynomials recursively. Then, if we find that a particular Lagrange polynomial of degree n does not provide a sufficiently good approximation, we may be able to add just one more term to it to compute a Lagrange polynomial of degree $n + 1$ and thus obtain a better approximation. Formulas (1.37) do not enable us to compute a Lagrange polynomial of degree $n + 1$ from one of degree n.

The trick to finding a recursion formula for Lagrange interpolation polynomials is to try the following analog of a Taylor polynomial:

$$Q_n = c_0 + c_1(x - x_0) + c_2(x - x_0)(x - x_1)$$
$$+ \cdots + c_n(x - x_0)(x - x_1) \cdots (x - x_{n-1}) \qquad (1.48)$$

where the x_i are distinct values of the independent variable. It can be shown that, for any set of corresponding values y_i of the dependent variable, there exist coefficients $c_0, c_1, \ldots, c_n$ such that $Q_n(x_i) = y_i$, for $i = 0, 1, \ldots, n$. Therefore, by Theorem 1.9, Q_n must be the unique Lagrange polynomial that interpolates between the points $(x_0, y_0), \ldots, (x_n, y_n)$. Furthermore, it turns out that each coefficient c_i depends only on the points $(x_0, y_0), \ldots, (x_i, y_i)$. Therefore, it makes sense to rewrite (1.48) so that Q_n is defined recursively:

$$Q_0(x) = c_0$$
$$Q_n(x) = Q_{n-1}(x) + c_n(x - x_0) \cdots (x - x_{n-1}) \quad \text{for } n > 0 \qquad (1.49)$$

EXAMPLE 1.31 Approximate $\sqrt{2}$ by using formula (1.49) to interpolate between the points $(2.25, \sqrt{2.25}\,)$, $(1, \sqrt{1}\,)$, and $(4, \sqrt{4}\,)$, in that order.

Thus, $x_0 = 2.25$, $y_0 = 1.5$, $x_1 = 1$, $y_1 = 1$, $x_2 = 4$, $y_2 = 2$. We compute Q_0, Q_1, Q_2 recursively.

$$Q_0(x) = c_0$$

and so

$$y_0 = Q_0(x_0) = c_0 \quad \text{or} \quad c_0 = 1.5$$

Therefore

$$Q_0(x) = 1.5$$

and so

$$Q_0(2) = 1.5$$

Next

$$Q_1(x) = Q_0(x) + c_1(x - x_0)$$

and so

$$y_1 = Q_1(x_1) = Q_0(x_1) + c_1(x_1 - x_0)$$

or

$$1 = 1.5 + c_1(-1.25) \quad \text{or} \quad c_1 = 0.4$$

Therefore

$$Q_1(x) = Q_0(x) + 0.4(x - x_0)$$

and so

$$Q_1(2) = 1.5 + 0.4(-0.25) = 1.4$$

Next

$$Q_2(x) = Q_1(x) + c_2(x - x_0)(x - x_1)$$

and so

$$y_2 = Q_1(x_2) + c_2(x_2 - x_0)(x_2 - x_1)$$

or

$$2 = 1.5 + 0.4(1.75) + c_2(1.75)(3) \quad \text{or} \quad c_2 = -0.038$$

Therefore

$$Q_2(x) = Q_1(x) - 0.038(x - x_0)(x - x_1)$$

and so

$$Q_2(2) = 1.4 - 0.038(-0.25)(1) = 1.4095$$

There are much more efficient ways to organize a computation such as the one above, but we will go no further with this discussion. [See the discussion of Newton's divided difference formula in Ralston and Rabinowitz (1978).]

EXERCISES

In Exercises 1–5, (a) find the Lagrange polynomial p that interpolates the given function between the indicated points, and (b) find a numerical bound for the error $|f(x) - p(x)|$, where $x_0 \leq x \leq x_n$.

1. $f(x) = x^2, \quad x_0 = 0, \quad x_1 = 1$
2. $f(x) = \cos x, \quad x_0 = 0, \quad x_1 = \pi/4, \quad x_2 = \pi/2$
3. $f(x) = 1/x, \quad x_0 = 1, \quad x_1 = \frac{3}{2}, \quad x_2 = 2$
4. $f(x) = \exp(-x^2), \quad x_0 = -1, \quad x_1 = 0, \quad x_2 = 1$
5. $f(x) = \log x, \quad x_0 = 1, \quad x_1 = \frac{4}{3}, \quad x_2 = \frac{5}{3}, \quad x_3 = 2$

In Exercises 6–8, find a Lagrange polynomial that interpolates the given function on the given interval with an error of at most 0.06.

6. $f(x) = \sin x \quad \text{on } [0, \pi]$
7. $f(x) = e^{-x} \quad \text{on } [-1, 1]$
8. $f(x) = \dfrac{1}{x} \quad \text{on } [1, 2]$
9. Approximate $\int_1^2 1/x\, dx$ by the integral of an appropriate Lagrange polynomial, with an error of at most 0.05.
10. According to the U. S. Bureau of the Census' *Historical Statistics of the United States, Colonial Times to 1970, Bicentennial Edition*, the number of cases of measles in the United States was 337.9, 245.4, 135.1, and 23.2 (per 100,000 population) in the years 1955, 1960, 1965, and 1970, respectively. Interpolate between the first three points and use the interpolation polynomial to predict the fourth point. (*Suggestion:* Number the years 0, 1, 2, and 3, respectively.) Discuss the reliability of the answer predicted by the interpolation polynomial.

11. If $x_i = x_0 + ih$ for $i = 0, 1, 2$, show that

$$\left|(x - x_0)(x - x_1)(x - x_2)\right| \le \frac{2h^3}{3\sqrt{3}}$$

for all x in $[x_0, x_2]$. *Hint:* Replace x by $u + x_1$ in the function $g(x) = (x - x_0)(x - x_1)(x - x_2)$, and then use calculus to show that $g(u + x_1)$ has critical points at $u = \pm h/\sqrt{3}$.

12. If $x_i = x_0 + ih$ for $i = 0, 1, 2, 3$, show that

$$\left|(x - x_0)(x - x_1)(x - x_2)(x - x_3)\right| \le h^4$$

for all x in $[x_0, x_3]$. *Hint:* Replace x by $u + (x_1 + x_2)/2$.

13. Find the cubic polynomial $y = a + bx + cx^2 + dx^3$ such that

$$y(0) = y_0, \qquad y'(0) = z_0, \qquad y(1) = y_1, \qquad y'(1) = z_1$$

where y_0, z_0, y_1, and z_1 are given. (This is an example of a *Hermite* interpolation polynomial, a polynomial that agrees with a function and with its derivative at points $x_0, x_1, \ldots, x_n$ and that has degree at most $2n + 1$.)

14. Does the polynomial p in Theorem 1.9 always have degree exactly n? Explain.

15. If f is a polynomial of degree $\le n$ and p is a Lagrange polynomial that interpolates f between $n + 1$ points, prove that $f = p$.

16. If $q_k(x)$ is defined as in (1.37) and if $w(x) = \prod_{i=0}^{n}(x - x_i)$, prove that

$$q_k(x) = \frac{w(x)}{(x - x_k)w'(x_k)}$$

17. Prove that $\sum_{k=0}^{n} q_k(x) = 1$ for all x, where the q_k are defined as in (1.37).

18. Show that if

$$Q_1(x) = y_0 + \frac{y_1 - y_0}{x_1 - x_0}(x - x_0)$$

then $Q_1(x_i) = y_i$ for $i = 0, 1$.

19. Show that if

$$Q_2(x) = y_0 + \frac{y_1 - y_0}{x_1 - x_0}(x - x_0)$$

$$+ \left(\frac{y_2 - y_1}{(x_2 - x_1)(x_2 - x_0)} - \frac{y_1 - y_0}{(x_1 - x_0)(x_2 - x_0)} \right)(x - x_0)(x - x_1)$$

then $Q_2(x_i) = y_i$ for $i = 0, 1, 2$.

1.6 Uniform Convergence

In Sections 1.4 and 1.5 we studied two methods for approximating functions by sequences of polynomials. But we skirted several delicate issues. For example, given a function f with domain D, can we find a sequence $\{P_n\}$ of Taylor polynomials or Lagrange polynomials which converges to f at every point of D? If we can, does it follow (as Example 1.26 and Exercise 9 of Section 1.5 might suggest) that $\{ \int P_n(x) \, dx \}$

converges to $\int f(x)\,dx$? Also, does the rate of speed at which $\{P_n\}$ converges to f depend on the domain D? For example, for a fixed value of n might $|f(x) - P_n(x)|$ be small for some values of x and much larger for others?

These questions have no simple answers. But they and others like them will impede our progress if we leave them unresolved. So we use this section to analyze a few such questions. The discussion will be in the more general context of arbitrary sequences of functions (not just polynomials) so that the conclusions are more widely applicable.

A *sequence of functions with domain D* is an infinite list

$$f_1, f_2, \ldots, f_n, \ldots$$

in which each f_n is a function (usually real valued) with domain D (usually an interval). The following definition has been implicit in all our discussions of convergent sequences of functions.

Definition 1.5 Suppose $\{f_n\}$ is a sequence of functions with domain D. We say that $\{f_n\}$ *converges pointwise* to a function f on D if $\{f_n(x)\}$ converges to $f(x)$ for every x in D; that is, if for each x in D and every $\varepsilon > 0$ there exists N such that

$$|f_n(x) - f(x)| < \varepsilon \quad \text{for all } n \geq N$$

Then we write

$$f_n \rightarrow f \text{ pointwise on } D$$

The first question we ask about such a sequence is one suggested in the opening paragraph. If $f_n \rightarrow f$ pointwise on an interval $[a, b]$, is it always true that $\int_a^b f_n(x)\,dx \rightarrow \int_a^b f(x)\,dx$? As the first example shows, the answer is "no."

EXAMPLE 1.32 Let f_n be the function with domain $[0,1]$ whose graph is the jagged dashed line in Figure 1.7. Show that $\{f_n\}$ converges to the zero function pointwise on $[0,1]$, but $\{\int_0^1 f_n(x)\,dx\}$ does not converge to $\int_0^1 0\,dx = 0$.

Although it is possible to write out a formula for f_n, it is easier to work with its graph in Figure 1.7. Certainly $f_n(0) = 0 \rightarrow 0$. Now suppose $0 < x \leq 1$. Then $f_n(x) = 0$ when $n \geq 1/x$, and so $f_n(x) \rightarrow 0$. On the other hand, $\int_0^1 f_n(x)\,dx$ is just the area of the triangle whose base is the segment of the x axis from 0 to $1/n$ and whose height is n. Therefore, $\int_0^1 f_n(x)\,dx = \frac{1}{2} \rightarrow \frac{1}{2}$, which is not 0.

The behavior of the functions f_n in the above example is neither highly unusual nor extreme. In particular, the functions could have been chosen to be polynomials, even though the ones we selected were not (see Exercise 1). Also, they could have been chosen so their integrals diverge (see Exercise 2).

Our second example sheds light on the question in the opening paragraph about rates of speed.

EXAMPLE 1.33 Let $f_n(x) = x^n$, where $0 \leq x \leq 1$. Then $f_n(x) \rightarrow 0$ if $0 \leq x < 1$ and $f_n(1) = 1 \rightarrow 1$ (see Figure 1.8). So even though the functions f_n are simple polynomials (and are therefore continuous), the limit function f is not continuous

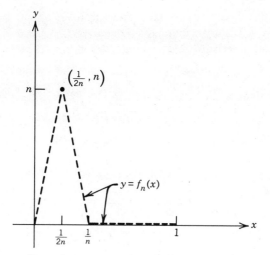

Figure 1.7   Area under the graph of f_n is $\frac{1}{2}$

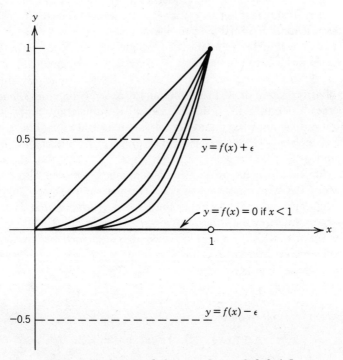

Figure 1.8 $f_n(x) = x^n$, $0 \le x \le 1$, $n = 1, 2, 3, 4, 5$

at $x = 1$; instead, it has a sudden jump from $f(x) = 0$ for x less than 1 to $f(x) = 1$ at $x = 1$. Furthermore, the sequence $\{ f_n(x) \}$ converges more and more slowly as x gets closer to 1. For example, if $0 \leq x \leq 0.9$, then

$$| f_n(x) - f(x) | = | x^n - 0 | = x^n \leq (0.9)^n$$

which is less than 0.1, for example, when $n \geq 22$. But if $x = 0.99$, then

$$| f_n(x) - f(x) | = (0.99)^n$$

which is much larger than $(0.9)^n$. For example, $(0.99)^{22} > 0.8$.

In contrast, we will see that sequences satisfying the following more restrictive definition behave more predictably.

Definition 1.6 Suppose $\{ f_n \}$ is a sequence of functions with domain D. We say that $\{ f_n \}$ *converges uniformly* to a function f on D if for every $\varepsilon > 0$ there exists N such that

$$| f_n(x) - f(x) | < \varepsilon \quad \text{for all } n \geq N \text{ and all } x \text{ in } D.$$

Then we write

$$f_n \to f \text{ uniformly on } D$$

As we will see, neither of the two sequences in Examples 1.32 and 1.33 converges uniformly on $[0, 1]$. But first, let us compare Definitions 1.5 and 1.6. Although they may seem almost indistinguishable from one another, there is a crucial difference in the order of their words. In Definition 1.5, the phrase "for every x in D and every $\varepsilon > 0$" precedes the phrase "there exists N"; we will understand this to signify that the value of N can depend on both x and ε. By the same reasoning, we will understand that the value of N in Definition 1.6 depends only on ε, not x. Thus, if $\{ f_n \}$ converges uniformly to f on D and n is sufficiently large, the distance $| f_n(x) - f(x) |$ can be made less than ε for all values of x at once. This is certainly not the case in Example 1.32; in that example, $| f_n(x) - f(x) |$ has the value n when $x = 1/2n$, and so there is no value of n for which $| f(x_n) - f(x) |$ can be made less than $\varepsilon = 1$ (for example) for all values of x at once. We can also draw a useful conclusion from the similarities between the two definitions. Since any N given by Definition 1.6 is an acceptable N for Definition 1.5, it follows that a sequence $\{ f_n \}$ which converges uniformly on D also converges pointwise on D.

Another way in which we can perceive the difference between uniform and pointwise convergence is from a picture. In Figure 1.9 we see the general idea of uniform convergence: If $n \geq N$, then

$$| f_n(x) - f(x) | < \varepsilon \quad \text{for all } x \text{ in } D$$

or, by property (1.13) of absolute values,

$$-\varepsilon < f_n(x) - f(x) < \varepsilon \quad \text{for all } x \text{ in } D$$

or

$$f(x) - \varepsilon < f_n(x) < f(x) + \varepsilon \quad \text{for all } x \text{ in } D$$

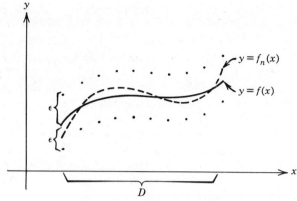

Figure 1.9 Uniform convergence

This last line says that the graph of f_n must lie within a band that extends from ε units below the graph of f to ε units above the graph of f. Such a picture helps us see why the sequence $\{x^n\}$ does not converge to 0 uniformly on $[0, 1)$. In Figure 1.8, we have indicated the band $f(x) - \varepsilon < y < f(x) + \varepsilon$, where $\varepsilon = 0.5$ and $f(x) = 0$ $(0 \le x < 1)$. Clearly, no matter how large n is, we see that part of the graph of $y = x^n$ lies above this band.

To help us work with the concept of uniform convergence without always going back to the definition, we seek a simpler criterion, one analogous to that given in Theorem 1.3b. The next theorem provides one. The reason it guarantees uniform convergence, not just pointwise convergence, is that the bounds y_n are independent of x.

Theorem 1.11 If $|f_n(x) - f(x)| \le y_n$ for all x in the domain D and if $y_n \to 0$ (where each y_n is independent of x), then $f_n \to f$ uniformly on D.

Proof Let ε be an arbitrary, given positive number. Then, since $y_n \to 0$, there exists N such that $|y_n - 0| < \varepsilon$ for all $n \ge N$. (Note that this N is independent of x since the terms y_n are.) Hence, for all $n \ge N$ and for all x in D,

$$|f_n(x) - f(x)| \le y_n = |y_n - 0| < \varepsilon$$

Therefore, $f_n \to f$ uniformly on D.

The inequality $|f_n(x) - f(x)| \le y_n$ is familiar not only because it is analogous to the one in Theorem 1.3b we have used so often; we actually wrote several inequalities of this form in our computations of error bounds in Sections 1.4 and 1.5. Those error bounds were almost always independent of x. This suggests that we perhaps have techniques at our disposal for proving that sequences of Taylor polynomials and Lagrange polynomials converge uniformly. Unfortunately, there are many examples of Taylor polynomials and Lagrange polynomials that do not converge or do not converge uniformly, even under conditions that seem, at first examination, sufficient to guarantee convergence. Although we will not take up this problem in detail, more information is provided in Exercise 19 and Section 3.9.

EXAMPLE 1.34 Show that $x^n \to 0$ uniformly on $[-c, c]$ for every number c satisfying $0 \leq c < 1$.

Note that $|x^n - 0| = |x|^n \leq c^n$ since $-c \leq x \leq c$, and $c^n \to 0$ since $0 \leq c < 1$. Therefore, by Theorem 1.11, $x^n \to 0$ uniformly on $[-c, c]$.

Certain aspects of the behavior of the sequence $\{x^n\}$ occur for a great many other sequences. Namely, the sequence converges pointwise, but not uniformly, on some open interval I (in the above example, $I = (-1, 1)$); yet it does converge uniformly on every closed and bounded subinterval of I. As we will see in Section 3.9, sequences of Taylor polynomials often behave like this.

For more complicated sequences than that in Example 1.34, the following more systematic approach to the problem of checking for uniform convergence may help. After finding the limit f of a given sequence $\{f_n\}$, compute the maximum (if it exists) of $|f_n(x) - f(x)|$ over all x in the domain D; denote this maximum by y_n. If $y_n \to 0$, then $f_n \to f$ uniformly on D, by Theorem 1.11. But if it is not true that $y_n \to 0$, then the convergence is not uniform on D (Exercise 4). Even if the maximum of $|f_n(x) - f(x)|$ does not exist, its least upper bound (Definition 3.3) might. That number will work as well for y_n.

EXAMPLE 1.35 Let $f_n(x) = n/(1 + n^2 x)$ and let $D = (0, \infty)$. Find the limit f of $\{f_n\}$ on D, and find the subintervals of D on which $\{f_n\}$ converges uniformly.

$$f_n(x) = \frac{n}{1 + n^2 x} = \frac{1/n}{1/n^2 + x} \to 0$$

Then

$$|f_n(x) - f(x)| = \left| \frac{n}{1 + n^2 x} - 0 \right| = \frac{n}{1 + n^2 x}$$

If we hold n fixed and let x vary throughout D, then the expression $n/(1 + n^2 x)$ is always less than n but gets arbitrarily close to n as x gets close to 0. We let $y_n = n$, which is the least upper bound, not the maximum, of $|f_n(x) - f(x)|$. Since $\{n\}$ does not have the limit 0, $\{f_n\}$ does not converge to f uniformly on D. In fact, the convergence cannot be uniform on any interval of the form $(0, c]$, $c > 0$, since $y_n = n$ on every such interval. However, $n/(1 + n^2 x)$ is a decreasing function of x, and so

$$\frac{n}{1 + n^2 x} \leq \frac{n}{1 + n^2 c} \quad \text{if } x \geq c$$

Therefore, by letting $y_n = n/(1 + n^2 c)$ and noting that $y_n \to 0$ if $c > 0$, we deduce that $f_n \to f$ uniformly on $[c, \infty)$ for every $c > 0$. Furthermore, every interval on which $\{f_n\}$ converges uniformly is contained in an interval of this type.

EXAMPLE 1.36 Let $f_n(x) = (nx)e^{-nx}$ and let $D = [0, \infty)$. Find the limit f of $\{f_n\}$ on D, and find the subintervals of D on which $\{f_n\}$ converges uniformly.

Clearly, $f_n(0) = 0 \to 0$. And, if $x > 0$,

$$f_n(x) = (nx)e^{-nx} = x\left[n(e^{-x})^n\right] \to 0$$

since $e^{-x} < 1$ (see Example 1.4, which says that $na^n \to 0$ when $0 \leq a < 1$). Next,

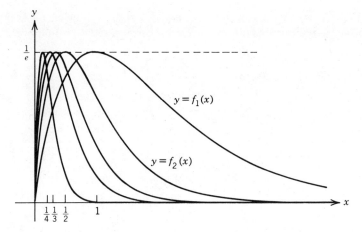

Figure 1.10 $f_n(x) = (nx)e^{-nx}$, $n = 1, 2, 3, 4, 8$

for each fixed n, we seek an upper bound for

$$| f_n(x) - f(x)| = f_n(x) = (nx)e^{-nx}$$

We use calculus to find the maximum of f_n, which is the upper bound we seek.

$$f_n'(x) = ne^{-nx} - (n^2x)e^{-nx}$$

$$= ne^{-nx}(1 - nx)$$

$$= 0 \quad \text{when } x = 1/n$$

Since $f_n(1/n) = 1/e$, the maximum of $| f_n(x) - f(x)|$ on $[0, \infty)$ is $1/e$. Therefore, by Exercise 4, $\{ f_n\}$ does not converge uniformly to 0 on $[0, \infty)$. Or, more geometrically (see Figure 1.10), if $\varepsilon < 1/e$ then the graph of f_n lies partly outside the band $f(x) - \varepsilon < y < f(x) + \varepsilon$ no matter how large n is. However, $\{ f_n\}$ does converge uniformly to 0 on $[c, \infty)$ for every $c > 0$, as we now show. From the above formula for f_n', we see that $f_n'(x) < 0$ for $x > 1/n$. So $f_n'(x) < 0$ for $x \geq c$ if $n > 1/c$. Hence, f_n is a decreasing function on $[c, \infty)$ if $n > 1/c$. Therefore, if $n > 1/c$ and $x \geq c$,

$$| f_n(x) - f(x)| = f_n(x) \leq f_n(c) = nce^{-nc}$$

Since $nce^{-nc} \to 0$, it follows from Theorem 1.11 that $f_n \to 0$ uniformly on $[c, \infty)$.

The two major theorems of this section, Theorems 1.12 and 1.13, show that, for uniformly convergent sequences, we can give more predictable answers to the kinds of questions raised earlier.

Theorem 1.12   Suppose $\{ f_n\}$ is a sequence of continuous functions on an interval I. If, for some point x_0 in I, $f_n \to f$ uniformly on $I \setminus \{ x_0\}$ (the interval I with x_0 removed) and if $f_n(x_0) \to y_0$, then $\lim_{x \to x_0} f(x) = y_0$. In particular, if $f_n \to f$ uniformly on all of I, then f is continuous on I.

Proof   We first recall the definition of $\lim_{x \to x_0} f(x) = y_0$ given in Appendix A. It states: For every positive real number ε, there exists a positive real number δ such that

$$| f(x) - y_0 | < \varepsilon$$

whenever x is a point in I satisfying

$$0 < |x - x_0| < \delta$$

So we let ε be an arbitrary, given positive number and seek such a number δ. Since $f_n \to f$ uniformly on $I \setminus \{ x_0 \}$, there exists N_1 such that

$$| f_n(x) - f(x) | < \frac{\varepsilon}{3} \quad \text{for all } n \geq N_1 \text{ and all } x \text{ in } I \setminus \{ x_0 \}$$

Also, since $f_n(x_0) \to y_0$, there exists N_2 such that

$$| f_n(x_0) - y_0 | < \frac{\varepsilon}{3} \quad \text{for all } n \geq N_2$$

Then we choose any fixed value of n such that $n \geq N_1$ and $n \geq N_2$. But since f_n is continuous at x_0, $\lim_{x \to x_0} f_n(x) = f_n(x_0)$; hence, there exists $\delta > 0$ such that

$$| f_n(x) - f_n(x_0) | < \frac{\varepsilon}{3} \quad \text{whenever } |x - x_0| < \delta$$

Therefore, by the triangle inequality,

$$| f(x) - y_0 | = |(f(x) - f_n(x)) + (f_n(x) - f_n(x_0)) + (f_n(x_0) - y_0)|$$
$$\leq | f(x) - f_n(x) | + | f_n(x) - f_n(x_0) | + | f_n(x_0) - y_0 |$$
$$< \frac{\varepsilon}{3} + \frac{\varepsilon}{3} + \frac{\varepsilon}{3} = \varepsilon$$

whenever $0 < |x - x_0| < \delta$.

Furthermore, if $f_n \to f$ uniformly on I, then, for each point x_0 in I, $f_n \to f$ uniformly on $I \setminus \{ x_0 \}$ and $f_n(x_0) \to f(x_0)$. Therefore, by the first part of the theorem, $\lim_{x \to x_0} f(x) = f(x_0)$, which means that f is continuous at x_0.

Theorem 1.13   Suppose $\{ f_n \}$ is a sequence of continuous functions on an interval of the form $[a, b]$. If $f_n \to f$ uniformly on $[a, b]$, then $\int_a^b f_n(x)\, dx \to \int_a^b f(x)\, dx$.

Proof   For every $\varepsilon > 0$ there exists N such that

$$| f_n(x) - f(x) | < \frac{\varepsilon}{2(b - a)}$$

for all x in $[a, b]$ and all $n \geq N$. Then

$$\left| \int_a^b f_n(x)\, dx - \int_a^b f(x)\, dx \right| = \left| \int_a^b (f_n(x) - f(x))\, dx \right|$$
$$\leq \int_a^b | f_n(x) - f(x) |\, dx \quad \text{(Theorem A.24e)}$$
$$\leq \int_a^b \frac{\varepsilon}{2(b - a)}\, dx \quad \text{(Theorem A.24d)}$$
$$= \frac{\varepsilon}{2} < \varepsilon$$

for all $n \geq N$. Therefore, $\int_a^b f_n(x)\, dx \to \int_a^b f(x)\, dx$.

EXERCISES

1. Let $f_n(x) = nx(1 - x^2)^n$ for $0 \le x \le 1$. Show that $\{ f_n \}$ converges to the zero function pointwise on $[0, 1]$ but that $\{ \int_0^1 f_n(x)\, dx \}$ does not converge to $\int_0^1 0\, dx$.

2. Modify the sequence $\{ f_n \}$ in Example 1.32 so that $\{ f_n \}$ still converges to the zero function pointwise on $[0, 1]$ but $\{ \int_0^1 f_n(x)\, dx \}$ diverges. *Suggestion:* Just change the second coordinate of the vertex $(1/2n, n)$ in Figure 1.7.

3. Let $f_n(x) = 1/(1 + nx)$ and let $D = [0, \infty)$.

    (a) Find f such that $f_n \to f$ pointwise on D, and show that $\lim_{n \to \infty} f_n(0) \ne \lim_{x \to 0} f(x)$.

    (b) Sketch graphs of f and several of the functions f_n, and draw in a band of the form $f(x) - \varepsilon < y < f(x) + \varepsilon$ for a particular value of ε. Use your picture and your value of ε to explain why $\{ f_n \}$ does not converge to f uniformly on $(0, \infty)$.

4. Suppose that, for each fixed value of n, the maximum y_n of $|f_n(x) - f(x)|$, where x varies over the domain D, exists. If $\{ y_n \}$ does not converge to 0, prove that $\{ f_n \}$ does not converge uniformly to f on D.

In Exercises 5–15, (a) Find the limit $f(x)$ of $\{ f_n(x) \}$ for all x in D. (b) Find out whether $\{ f_n \}$ converges uniformly to f on D. (c) If $\{ f_n \}$ does not converge uniformly on D, find the subintervals of D on which $\{ f_n \}$ does converge uniformly.

5. $f_n(x) = x^n/n, \qquad D = [-1, 1]$

6. $f_n(x) = 1/(1 + nx), \qquad D = [0, \infty)$

7. $f_n(x) = \sin^n(x), \qquad D = [0, \pi/2]$

8. $f_n(x) = \exp(-nx^2), \qquad D = [0, \infty)$

9. $f_n(x) = (1 - x)^n x^n, \qquad D = [0, 1]$

10. $f_n(x) = x^n/(1 + nx), \qquad D = [0, 1]$

11. $f_n(x) = x^n/(1 + x^n), \qquad D = [0, \infty)$

12. $f_n(x) = x/(1 + nx^2), \qquad D = [0, \infty)$

13. $f_n(x) = nx/(1 + n^2 x^2), \qquad D = [0, \infty)$

14. $f_n(x) = (nx)\exp(-nx^2), \qquad D = [0, \infty)$

15. $f_n(x) = nx(1 - x)^n, \qquad D = [0, 1]$

16. If $f_n \to f$ uniformly on D and if $g_n \to g$ uniformly on D, prove that $f_n + g_n \to f + g$ uniformly on D.

17. Let $f_n(x) = x + 1/n$ and $f(x) = x$. Check that $f_n \to f$ uniformly on $[0, \infty)$ but that $\{ f_n^2 \}$ does not converge uniformly to f^2 on $[0, \infty)$.

18. **(a)** Verify that $\int_0^1 x^n\, dx \to 0 = \int_0^1 0\, dx$ even though $\{ x^n \}$ does not converge uniformly to 0 on $[0, 1]$.

    (b) Why does this not violate Theorem 1.13?

19. Suppose f has derivatives of all orders on some interval I, and suppose there is a constant M such that $|f^{(n)}(x)| \le M$ for all non-negative integers n and all x in I. If I is bounded, prove that $P_n \to f$ uniformly on I, where P_n is the nth Taylor polynomial of f at x_0 and x_0 is any point in I. *Hint:* Use Taylor's remainder formula.

20. Suppose $f_n \to f$ uniformly on I and $f_n \to f$ uniformly on J. Prove that $f_n \to f$ uniformly on $I \cup J$ (the union of I and J).

21. Suppose $\{f_n\}$ is a sequence of continuous functions on an interval I and f is also continuous on I. If, for some point x_0 in I, $f_n \to f$ uniformly on $I \setminus \{x_0\}$, prove that $f_n \to f$ uniformly on I.

22. (Approximation of a continuous function by piecewise linear functions) Suppose f is continuous at every point of the interval $[a, b]$. Prove that there exists a sequence of piecewise linear functions $\{f_n\}$ such that $f_n \to f$ uniformly on $[a, b]$. (A function g is "piecewise linear" on $[a, b]$ if g is continuous and if $[a, b]$ can be partitioned into finitely many closed subintervals in such a way that, on each of these subintervals, the graph of g is a straight line.) *Hint:* Use Theorem A.10.

CHAPTER 2

Numerical approximation

In this chapter, we undertake a systematic though modest study of methods of numerical approximation for the following three important types of problems:

1. Approximate a solution of the functional equation $f(x) = 0$.

2. Approximate the value of the definite integral $\int_a^b f(x)\, dx$.

3. Approximate a solution of the differential equation $\dfrac{dy}{dx} = f(x, y)$.

These three types of computational problems occur frequently in applications of mathematics. In real-world problems, there are often no known mathematical methods for finding exact answers. Or, if there is an exact method, it may take much longer to carry out than an approximate method, and the solution it produces may be no more satisfactory than an approximate one.

The first of the three problems is studied in Sections 2.1, 2.2, and 2.4, the second in 2.5 and 2.6, and the third in 2.7 and 2.8. Section 2.3 discusses the difficulties of assessing the kind and amount of error in any approximation. Many of the methods to be studied build on the work we did with Taylor polynomials and Lagrange polynomials in Sections 1.4 and 1.5.

The approach used in the first four sections is adapted from "Algorithms for Finding Zeros of Functions" (UMAP Unit 264) by Werner C. Rheinboldt (Newton, Mass: COMAP, 1978, 1980). Some of the examples are also adapted from that pamphlet.

2.1 The Bisection Method

A *root* of the equation $f(x) = 0$ (also called a *zero* of the function f) is a number z such that $f(z) = 0$. We will look only for real, not complex, roots; these are usually the ones needed in applications. The first difficulty we observe is that different equations can have different numbers of roots (see Table 2.1).

Table 2.1 Number of Roots of an Equation

$x^2 + 1 = 0$	No real roots
$1/x = 0$	No roots
$3x + 4 = 0$	One root: $-\frac{4}{3}$
$(x - 1)^2 = 0$	One double root: 1
$x^2 - 3x + 2 = 0$	Two roots: 1, 2
$(x - x_1)(x - x_2) \cdots (x - x_n) = 0$	Finitely many roots: $x_1, \ldots, x_n$
$\sin x = 0$	Infinitely many roots: $n\pi$, n any integer

The usual procedure for overcoming this difficulty is to make a table of values or a graph in the hope of finding a pair of numbers a and b for which $f(a)$ and $f(b)$ have opposite signs (see Table 2.2 and Figure 2.1). Then, according to the inter- mediate value theorem (Theorem A.5 and its corollary), the equation $f(x) = 0$ must have a root between a and b, provided f is continuous. Furthermore, if a and b are fairly close together, we will suspect that there is only one root between them. Thus, we know that there is at least one root of $x^3 - 3x - 5 = 0$ in the interval $[2,3]$ and we suspect there is only one.

As the example $(x - 1)^2$ in Table 2.1 shows, however, we might not isolate any roots by such an application of the intermediate value theorem; that is, $[a, b]$ could contain one or more roots of $f(x) = 0$ even if $f(a)$ and $f(b)$ have the same sign. The source of difficulty in this example is that $z = 1$ is a root of "multiplicity two" (or "double root"). In general, if

$$f(x) = g(x)(x - z)^m$$

for some positive integer m, where g is continuous and $g(z) \neq 0$, then z is called a *root of multiplicity m*. It can be shown that if the sum of the multiplicities of the roots of $f(x) = 0$ in the interval $[a, b]$ is even, then $f(a)$ and $f(b)$ have the same sign.

But let us suppose that $f(a)$ and $f(b)$ do have opposite signs and let us continue to assume that f is continuous. How can we find a root of $f(x) = 0$ between a and b?

There are few general methods for finding exact roots. For example, high school students learn how to solve all polynomial equations of degrees 1 and 2, but only special equations of higher degree. There are also methods, much less well known, for

Table 2.2

x	$f(x) = x^3 - 3x - 5$
-2	-7
-1	-3
0	-5
1	-7
2	-3
3	13

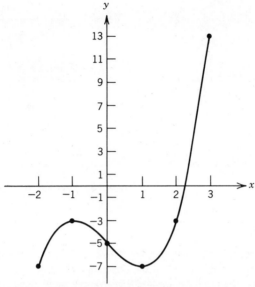

Figure 2.1 $y = x^3 - 3x - 5$

solving all polynomial equations of degrees 3 and 4 [see Clark (1971)]. However, for polynomial equations of degree greater than 4, there are no general methods for finding exact roots. And methods for solving nonpolynomial equations, such as $2x - 2 - \sin x = 0$, are nearly nonexistant.

In contrast, for approximating roots of equations there are quite a few general methods, although of varying efficiency and reliability. If possible, we want a method that tells us how to generate a sequence of approximations $\{x_n\}$ and a theorem that assures us that $\{x_n\}$ converges to a root z of $f(x) = 0$. Furthermore, we want the convergence to be sufficiently rapid that, for a given error bound ε, we can find a moderately small value of n for which the error $|x_n - z|$ is no greater than ε. The reason we want n to be small is that sequences of approximations are usually defined recursively; so it may take 99 applications of a recursion formula to compute the 100th term in a sequence of approximations but only 9 applications to compute the 10th term. Certainly, we prefer a method that requires less computation for the desired accuracy.

Our first method, the *bisection method*, is reliable but slow. Roughly speaking, we bisect the given interval $[a, b]$ repeatedly, each time keeping the half that contains the root. More precisely, we construct the midpoint $m = (a + b)/2$ of the interval $[a, b]$ and then replace $[a, b]$ by $[a, m]$ if $f(a)$ and $f(m)$ have opposite signs and by $[m, b]$ otherwise. Thus, the new interval $[a, b]$ is half the length of the previous interval $[a, b]$, and it still contains a root of $f(x) = 0$ (see Figures 2.2 and 2.3). We continue this process until $b - a \le 2\varepsilon$. If we define our sequence of approximations $\{x_n\}$ to be the successive midpoints m, then

$$|x_n - z| = |m - z| \le \frac{(b - a)}{2} \le \varepsilon$$

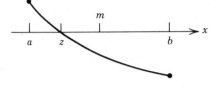

Figure 2.2 Bisection method: new $a = a$; new $b = m$

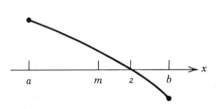

Figure 2.3 Bisection method: new $a = m$; new $b = b$

EXAMPLE 2.1 Use the bisection method to approximate a root of $x^3 - 3x - 5 = 0$ with an error of at most 0.05.

We start with the information in Table 2.2—we let the initial values of a and b be 2 and 3, respectively. Then we construct Table 2.3; note that the last row of the table is the first row for which $b - a \le 0.1 = 2\varepsilon$. Therefore, if z denotes the root,

$$|2.28125 - z| \le \frac{0.0625}{2} = 0.03125$$

Furthermore, Table 2.3 shows that $z < 2.28125$, since $f(a)$ and $f(m)$ have opposite signs in the last row.

The proof of the intermediate value theorem (Theorem A.5) is interesting to consider now, for two reasons. First, the proof simply consists in showing that the sequence generated by the bisection method converges to a root of $f(x) = 0$; thus, it is an example of a proof technique suggested by a computational technique. Furthermore, since the proof shows that this sequence does converge to a root, we have a proof of the reliability of the bisection method.

We conclude this section with an algorithm for the bisection method. By an "algorithm" we mean a precise description of a mechanical step-by-step method for carrying out a type of task. The method should be so mechanical that a computer can carry out the task, and the description of the method should be so precise that the process of converting the algorithm into a computer program is straightforward. Every numerical method in this chapter is accompanied by an algorithm. And, at the ends of

Table 2.3 The Bisection Method for $x^3 - 3x - 5 = 0$

n	a	b	$b - a$	$x_n = m$	$f(a)$	$f(x_n) = f(m)$
1	2	3	1.0	2.5	-3	3.125
2	2	2.5	0.5	2.25	-3	-0.359
3	2.25	2.5	0.25	2.375	-0.359	1.271
4	2.25	2.375	0.125	2.3125	-0.359	0.429
5	2.25	2.3125	0.0625	2.28125	-0.359	0.028

the sections, there are exercises that ask the reader to convert the algorithms into computer programs and then to use those programs to generate highly accurate approximations.

Algorithm 2.1   *The bisection method*

    1. **Input** *a, b, epsilon*
    2. $n \Leftarrow 0$
    3. **Repeat steps** 4–7 **until** $b - a \leq epsilon$
        4. $n \Leftarrow n + 1$
        5. $m \Leftarrow (a + b)/2$
        6. **Print** $n, m, f(m)$
        7. **If** $sign(f(a)) = sign(f(m))$
                then $a \Leftarrow m$ **else** $b \Leftarrow m$
    8. **End**

There is no standard format for describing algorithms. We have chosen a rather stylized one that makes algorithms look like computer programs; this should make it easier to translate the algorithms into programs. The words in boldface are instructions to the computer, whereas most of the remaining strings of letters (such as "*a*," "*epsilon*," and "*n*") are names of variables. Each numbered step of an algorithm should translate into one or more lines of a program. Steps with an arrow (such as 2, 4, and 5) indicate the usual assignment statements of a computer program—the value of the expression on the right is to be stored in the location named by the variable on the left of the arrow. Steps 1 and 6 are the usual input and output statements, respectively —the values of the listed variables are to be input to the program and output from the program, respectively. In some languages, "input" must be translated as "read" and "print" as "write." Step 7 is a typical conditional statement. Step 8, which consists of the word "End," indicates where the algorithm logically ends, not necessarily the last physical line of the program; some algorithms will have more than one "End" statement. Loops will always be indicated by the words "repeat steps," followed by the numbers of the steps constituting the loop. Also, the steps of the loop will be indented. Conditions describing exist tests for loops will be placed at the top of the loop (as is "until $b - a < epsilon$"), but in a computer program it is sometimes better to place the test at the bottom of the loop.

Interpreting the results of a computer program based on an algorithm such as the one above is not a simple matter. But trying to do so can be quite worthwhile; one can gain a much better understanding of what the algorithm accomplishes and what its weaknesses are. An especially appropriate format for such an interpretation is a report written in clear English. Imagine that the report is addressed to a scientist who has requested an approximate solution of a problem but who might not understand the algorithm that produced the approximation. The report should include:

1. A clear statement of the problem and an approximate solution.

2. A bound or estimate of the error.

3. A discussion of the reliability of the error bound or error estimate.

EXERCISES

In Exercises 1–7, use the intermediate value theorem to find an interval containing a root of $f(x) = 0$.

1. $f(x) = x^3 - 2x + 2$ 2. $f(x) = \log x + x^2 - 8$

3. $f(x) = 2x - 2 - \sin x$ 4. $f(x) = x - e^{-x}$

5. $f(x) = 1 + 9(\log x)/x^2$

6. $f(x) = x^5 - 5x^4 + 10x^3 - 10x^2 + 5x - 1$

7. $f(x) = 1 - 57(\log x)/x^3, \quad x < 4$

In Exercises 8–14, use the bisection method to approximate a root of the equation $f(x) = 0$ with an error of at most 0.05. (A calculator will be helpful.)

8. $f(x) = x^3 - 2x + 2$ 9. $f(x) = \log x + x^2 - 8$

10. $f(x) = 2x - 2 - \sin x$ 11. $f(x) = x - e^{-x}$

12. $f(x) = 1 + 9(\log x)/x^2$

13. $f(x) = x^5 - 5x^4 + 10x^3 - 10x^2 + 5x - 1$

14. $f(x) = 1 - 57(\log x)/x^3, \quad x < 4$

For Exercises 15–21, first write a computer program that implements Algorithm 2.1 for an arbitrary function f. Then use the program to approximate a root of the equation $f(x) = 0$, for each assigned function f, with an error of at most 0.000001. For each assigned function, describe your conclusion in words, following the recommendations at the end of Section 2.1.

15. $f(x) = x^3 - 2x + 2$ 16. $f(x) = \log x + x^2 - 8$

17. $f(x) = 2x - 2 - \sin x$ 18. $f(x) = x - e^{-x}$

19. $f(x) = 1 + 9(\log x)/x^2$

20. $f(x) = x^5 - 5x^4 + 10x^3 - 10x^2 + 5x - 1$

21. $f(x) = 1 - 57(\log x)/x^3, \quad x < 4$

22. **(a)** In the bisection method, find out how large n must be in order to guarantee that $|x_n - z| \leq 0.01$, if initially $b - a = 1.0$.

    (b) Find a general formula for how large n must be in order to guarantee that $|x_n - z| \leq \varepsilon$, if initially $b - a = D$.

There is a simple method, called *Picard iteration*, for approximating roots of equations of the form $f(x) = x$:

$$x_{n+1} = f(x_n)$$

where x_0 is any initial approximation. This method may fail completely or may converge extremely slowly. It works well when $|f'|$ is substantially less than 1 on an interval containing both the root and x_0 (see Exercise 19 of Section 3.3). In Exercises 23–26, use Picard iteration to approximate a root of the equation $f(x) = x$. Stop the iteration when two successive terms differ by at most 0.01. (A calculator will be helpful.)

23. $f(x) = x^{-2} - 0.5 + x, \quad x_0 = 1$ (to approximate $\sqrt{2}$)

24. $f(x) = e^{-x}, \quad x_0 = 0$

25. $f(x) = 2 + \log x, \quad x_0 = 1$

26. $f(x) = \pi + \arctan x, \quad x_0 = 0$

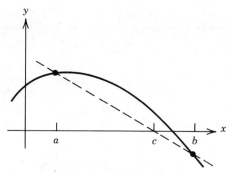

Figure 2.4 The secant method

2.2 The Secant Method and Newton's Method

Although the bisection method for solving the equation $f(x) = 0$ is very reliable, it is slow. We can often get faster convergence by using a "linearizing" method, in which we approximate f by a sequence of linear functions p and find the root of each of the linear equations $p(x) = 0$. We consider two such methods.

In the *secant method*, we approximate f by a Lagrange polynomial of degree ≤ 1; that is, we join two points on the graph of f by a secant line. Then we find the point where this secant line crosses the x axis (Figure 2.4). If a and b are the x coordinates of the two given points, the approximate root c may be computed as follows. The secant line has the equation

$$y - f(b) = \frac{(f(b) - f(a))}{b - a}(x - b)$$

Thus, setting $y = 0$ and solving for x, we get

$$c = x = b - \frac{f(b)(b - a)}{f(b) - f(a)} \tag{2.1}$$

We define our sequence of approximations so that the point c is the current term, b is the previous term, and a is the term before that; that is, we convert (2.1) into the recursion formula

$$x_{n+2} = x_{n+1} - \frac{f(x_{n+1})(x_{n+1} - x_n)}{f(x_{n+1}) - f(x_n)} \tag{2.2}$$

EXAMPLE 2.2   Use the secant method to approximate a root of $x^3 - 3x - 5 = 0$ with an error of at most 0.05.

As in Example 2.1, we start with $a = 2$ and $b = 3$. Then we construct Table 2.4; note that the last row of the table is the first row for which $|c - b| \leq 0.05$. Unlike the bisection method, however, there might be no roots of $f(x) = 0$ between two consecutive terms of a secant method approximation. Fortunately, the last column of Table 2.4 shows that the given function has opposite signs at 2.2506 and 2.2805; therefore, there is a root z between these two approximations,

Table 2.4 The Secant Method for $x^3 - 3x - 5 = 0$

n	a	b	$x_n = c$	$f(x_n) = f(c)$
1	2	3	2.1875	-1.095
2	3	2.1875	2.2506	-0.352
3	2.1875	2.2506	2.2805	0.019

and

$$|2.2805 - z| < |2.2805 - 2.2506| = 0.0299$$

Algorithm 2.2   *The secant method*

    1. **Input** *a, b, epsilon, maxn*
    2. $n \Leftarrow 0$
    3. **Repeat steps** 4–10 **until** $|difference| \leq epsilon$
       4. $n \Leftarrow n + 1$
       5. **If** $f(a) = f(b)$ **then**
            5a. Print "Algorithm fails: division by zero"
            5b. End
       6. **If** $n > maxn$ **then**
            6b. Print "Algorithm fails: no convergence"
            6b. End
       7. *difference* $\Leftarrow f(b)(b - a)/(f(b) - f(a))$
       8. $a \Leftarrow b$
       9. $b \Leftarrow b - difference$
      10. **Print** $n, b, f(b)$
  11. **End**

Note: The value of the variable named "difference" is undefined before the first execution of the loop. Any computer program based on Algorithm 2.2 must, of course, avoid this defect.

In practice, the sequences produced by the secant method usually converge much more rapidly than those produced by the bisection method. However, the method is unreliable. If, for example, the secant line at some stage is horizontal, then the next approximation cannot be computed (step 5). Even if it can be computed, it could fall outside the interval $[a, b]$, as in Figure 2.5; then the sequence $\{x_n\}$ might fail to converge or might converge to another root of $f(x) = 0$. Step 6 is designed to detect the case when the sequence of approximations fails to converge. For example, if we have given *maxn* the value 10, then the algorithm will declare that the sequence of approximations has failed to converge whenever the 10th and 11th terms of the sequence and all earlier pairs of successive terms are more than *epsilon* units apart. (Experience helps one learn to choose an appropriate value for *maxn*.) Furthermore, even if the approximations seem to converge, it may be difficult to judge how close the current approximation is to a root. The loop's exit test in step 3 of the algorithm seems

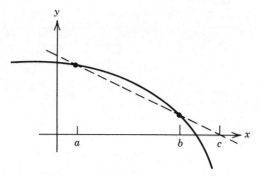

Figure 2.5 The current secant approximation is not between the preceding two

to suggest that whenever the distance between the two most recent approximations is less than *epsilon*, the same will be true for the root and the most recent approximation. But this is false. As Figure 2.6 shows, the root might not lie between the two most recent approximations; in that case, we can make no reliable prediction about the distance between the root and the most recent approximation. In Table 2.4, we saw that the root of $x^3 - 3x - 5 = 0$ is indeed between x_2 and x_3, since $f(x_2)$ and $f(x_3)$ have opposite signs. But .if we had ended the table at the second line, we could have made no reliable statement about the distance between the root and x_2, since $f(x_1)$ and $f(x_2)$ have the same sign. Some additional reliability may be gained if we put in an extra error test of the form $|f(c)| < \varepsilon_1$, where ε_1 is another error bound.

In the *Newton* (or *Newton–Raphson*) *method*, we approximate f by a Taylor polynomial of degree ≤ 1; that is, we draw a line tangent to the graph of f at a given point and find the point where this line crosses the x axis (Figure 2.7). If a is the x coordinate of the given point, the approximate root b may be computed as follows. The tangent line has the equation

$$y - f(a) = f'(a)(x - a)$$

Thus, setting $y = 0$ and solving for x, we get

$$b = x = a - \frac{f(a)}{f'(a)} \tag{2.3}$$

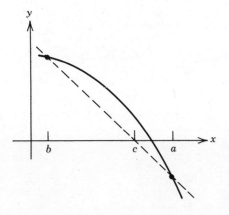

Figure 2.6 The root is not between the two most recent secant approximations

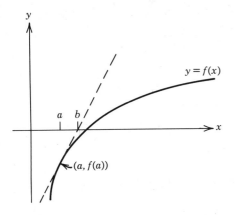

Figure 2.7   Newton's method

We define our sequence of approximations so that the point b is the current term and a is the previous term; that is, we convert (2.3) into the recursion formula

$$x_{n+1} = x_n - \frac{f(x_n)}{f'(x_n)} \tag{2.4}$$

EXAMPLE 2.3   Use Newton's method to approximate a root of $x^3 - 3x - 5 = 0$ with an error of at most 0.05.

In the bisection and secant methods, we needed two initial terms before we could compute the remaining terms recursively. In Newton's method, only one is needed. If we again take our cue from Table 2.2, $a = 2$ appears to be a better choice of initial term than $a = 3$, since $f(2)$ is closer to zero than $f(3)$ (and therefore 2 might be closer to the root than 3). Then we construct Table 2.5, whose last row is the first row for which $|b - a| \leq 0.05$. This time $f(x_n)$ never changes sign. Evidently, the terms x_n are all on the right of the root and decreasing toward it. Therefore, the distance between the root z and its approximation 2.27902 could be larger than $|2.27902 - 2.28056| = 0.00154$. But it is almost certainly less than 0.05, especially since the terms seem to be converging so rapidly.

The advantages and disadvantages of Newton's method are much like those of the secant method; Newton's method usually converges a bit faster, but it requires that we know the derivative of f. It also leads to greater uncertainty about the error. In our discussion of the bisection method, we saw that the root z satisfies the inequality $|x_n - z| \leq |x_n - x_{n-1}|$; so we refer to $|x_n - x_{n-1}|$ as an *error bound*. The same in-

Table 2.5   Newton's Method for $x^3 - 3x - 5 = 0$

n	a	$x_n = b$	$f(x_n) = f(b)$
1	2	2.33333	0.70370
2	2.33333	2.28056	0.01935
3	2.28056	2.27902	0.00002

equality is satisfied in the secant method, provided $f(x_n)$ and $f(x_{n-1})$ have opposite signs; and, in practice, this often does occur. In Newton's method, however, the approximating terms are almost always on just one side of the root, as we will see in the next theorem; hence, it can happen that $|x_n - z| > |x_n - x_{n-1}|$. In practice, however, $|x_n - z|$ and $|x_n - x_{n-1}|$ are often quite close to one another, and so we refer to $|x_n - x_{n-1}|$ as an *error estimate*, which can be larger or smaller than the actual error.

Algorithm 2.3 *Newton's method*

    1. ***Input*** *a, epsilon, maxn*
    2. $n \Leftarrow 0$
    3. ***Repeat steps*** 4–9 ***until*** $|difference| \leq epsilon$
        4. $n \Leftarrow n + 1$
        5. ***If*** $f'(a) = 0$ ***then***
            5a. ***Print*** "Algorithm fails: division by zero"
            5b. ***End***
        6. ***If*** $n > maxn$ ***then***
            6a. ***Print*** "Algorithm fails: no convergence"
            6b. ***End***
        7. *difference* $\Leftarrow f(a)/f'(a)$
        8. $a \Leftarrow a - $ *difference*
        9. ***Print*** *n, a,* $f(a)$
  10. ***End***

Note: *The value of the variable named "difference" is undefined before the first execution of the loop.*

As our last topic of this section, we discuss the convergence of these linearizing methods. The following theorem concerns Newton's method.

Theorem 2.1 Suppose the function f satisfies the following conditions:

    1. f' and f'' exist and have constant sign on $[a, b]$. *Slope & concavity dont change sign.*
    2. There exists $m > 0$ such that $|f'(x)| \geq m$ on $[a, b]$.
    3. There exists $M > 0$ such that $|f''(x)| \leq M$ on $[a, b]$.
    4. $f(a)$ and $f(b)$ have opposite signs.

Then the sequence $\{x_n\}$ defined by the formula (2.4), where $x_1 = b$ if $f'f'' > 0$ and $x_1 = a$ if $f'f'' < 0$, converges to the unique root z of $f(x) = 0$ in $[a, b]$. Furthermore

$$|z - x_{n+1}| \leq M(z - x_n)^2/2m \tag{2.5}$$

Proof We assume that $f' > 0$ and $f'' > 0$ on $[a, b]$. (The proofs for the other three combinations of signs possible in hypothesis (1) are similar.) Since $f(a)$ and $f(b)$ have opposite signs, the intermediate value theorem guarantees that the equation $f(x) = 0$ has a root in $[a, b]$. Furthermore, since $f' > 0$, it follows that f is strictly increasing; hence, there is no other root in $[a, b]$. Let z denote that unique root. We will prove, as

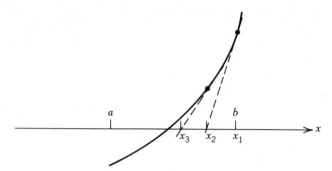

Figure 2.8   Convergence of Newton's method

Figure 2.8 suggests, that the terms x_n are all to the right of z and decreasing; that is,

$$x_{n+1} < x_n \quad \text{and} \quad z < x_n \leq b \quad \text{for all } n \geq 1 \tag{2.6}$$

First, we prove that $z < x_n \leq b$ for all $n \geq 1$, by using the principle of mathematical induction (Appendix E). Since $f'f'' > 0$, our definition of x_1 in the statement of the theorem says that $x_1 = b$; thus, $z < x_1 \leq b$. Now we suppose that $z < x_n \leq b$ for some value of $n \geq 1$ and prove it for the next value of n. To do so, we approximate f by its first Taylor polynomial at x_n and apply Taylor's remainder formula:

$$0 = f(z) = f(x_n) + f'(x_n)(z - x_n) + \frac{f''(c)}{2}(z - x_n)^2 \tag{2.7}$$

for some c between x_n and z. Then, since $f'' > 0$,

$$f(x_n) + f'(x_n)(z - x_n) < 0$$

or

$$f'(x_n)(z - x_n) < -f(x_n)$$

or

$$z - x_n < -\frac{f(x_n)}{f'(x_n)}$$

or

$$z < x_n - \frac{f(x_n)}{f'(x_n)} = x_{n+1} \quad \text{by formula (2.4)}$$

Furthermore, since $z < x_n$ and f is increasing, $0 = f(z) < f(x_n)$. Thus, formula (2.4) also shows that $x_{n+1} < x_n$. But since $x_n \leq b$, it follows that $x_{n+1} \leq b$. This completes the proof that $z < x_n \leq b$ for all $n \geq 1$ and also proves that the other half of statement (2.6) is true.

In Theorem 3.6 we show that any sequence $\{x_n\}$ satisfying the inequalities (2.6) must converge. Thus, assuming that $\{x_n\}$ has a limit c and letting n go to infinity in formula (2.4), we conclude that $c = c - f(c)/f'(c)$, since f and f' are continuous. Therefore, $f(c) = 0$, and so c must be the unique root z.

Finally, we derive inequality (2.5). If we rewrite formula (2.4) in the form $f(x_n) = f'(x_n)(x_n - x_{n+1})$, we can replace $f(x_n)$ in (2.7) by this expression to get

$$0 = f'(x_n)(x_n - x_{n+1}) + f'(x_n)(z - x_n) + \frac{f''(c)}{2}(z - x_n)^2$$

$$= f'(x_n)(z - x_{n+1}) + \frac{f''(c)}{2}(z - x_n)^2$$

and so

$$|z - x_{n+1}| = \left| \frac{f''(c)(z - x_n)^2}{2f'(x_n)} \right| \leq \frac{M(z - x_n)^2}{2m}$$

This completes the proof.

Inequality (2.5) contains useful information about the rate of convergence of the sequences produced by Newton's method. We can read it as telling us that, at each stage of the approximation process, the error $|z - x_{n+1}|$ is roughly proportional to the square of the preceding error. If, for example, the error at some stage is proportional to 10^{-d}, then the next is roughly proportional to 10^{-2d}. In other words, we get roughly twice as many decimal places of accuracy with each approximation. This highly desirable phenomenon is referred to as "quadratic convergence." A similar theorem can be proved for the secant method, but the analog of formula (2.5) has the exponent $(1 + \sqrt{5})/2 = 1.618$ in place of the exponent 2. Thus, the convergence of this method is likely to be somewhat slower than quadratic. [See Ralston and Rabinowiz (1978) for details and further discussion; the number $(1 + \sqrt{5})/2$ is related to the Fibonacci sequence.]

EXERCISES

In Exercises 1–3, use the secant method to approximate a root of the equation $f(x) = 0$; stop when two successive approximations differ by no more than 0.05. (A calculator will be helpful.)

1. $f(x) = x^2 - 2$ 2. $f(x) = \log x - 1$ 3. $f(x) = x - e^{-x}$

For Exercises 4–10, first write a computer program that implements Algorithm 2.2 for an arbitrary function f. Then use the program to approximate a root of the equation $f(x) = 0$, for each assigned function f, with *epsilon* = 0.000001 and *maxn* = 20. For each assigned function, describe your conclusion in words, following the recommendations at the end of Section 2.1.

4. $f(x) = x^3 - 2x + 2$ 5. $f(x) = \log x + x^2 - 8$

6. $f(x) = 2x - 2 - \sin x$ 7. $f(x) = x - e^{-x}$

8. $f(x) = 1 + 9(\log x)/x^2$

9. $f(x) = x^5 - 5x^4 + 10x^3 - 10x^2 + 5x - 1$

10. $f(x) = 1 - 57(\log x)/x^3$, $x < 4$

In Exercises 11–13, use Newton's method to approximate a root of the equation $f(x) = 0$; stop when two successive approximations differ by no more than 0.05. (A calculator will be helpful.)

11. $f(x) = x^2 - 2$ 12. $f(x) = \log x - 1$ 13. $f(x) = x - e^{-x}$

For Exercises 14–20, first write a computer program that implements Algorithm 2.3 for an arbitrary function f. Then use the program to approximate a root of the equation $f(x) = 0$, for each assigned function f, with *epsilon* = 0.000001 and *maxn* = 20. For each assigned function, describe your conclusion in words, following the recommendations at the end of Section 2.1.

14. $f(x) = x^3 - 2x + 2$ **15.** $f(x) = \log x + x^2 - 8$

16. $f(x) = 2x - 2 - \sin x$ **17.** $f(x) = x - e^{-x}$

18. $f(x) = 1 + 9(\log x)/x^2$

19. $f(x) = x^5 - 5x^4 + 10x^3 - 10x^2 + 5x - 1$

20. $f(x) = 1 - 57(\log x)/x^3$, $x < 4$

21. (Approximation of square roots)   Let $f(x) = x^2 - c$, where $c > 0$.

   (a) Show that (2.2) simplifies to

$$x_{n+2} = \frac{x_n x_{n+1} + c}{x_{n+1} + x_n}$$

   (b) Show that (2.4) simplifies to

$$x_{n+1} = \frac{x_n^2 + c}{2x_n} = \frac{1}{2}\left(x_n + \frac{c}{x_n}\right)$$

22. (Approximation of reciprocals by products)   Let $f(x) = 1/x - c$, where $c \neq 0$.

   (a) Show that (2.2) simplifies to

$$x_{n+2} = x_{n+1} + x_n - cx_{n+1}x_n$$

   (b) Show that (2.4) simplifies to

$$x_{n+1} = 2x_n - cx_n^2 = x_n(2 - cx_n)$$

In Exercises 23–27, you will find equations that cannot be solved by hand. Use one of our numerical methods to compute a reasonable approximation.

23. A family borrows $6000 to buy a new car. They will pay off the loan, with interest, in monthly installments of $150 over a period of 4 years. What is their monthly interest rate? (See the formula developed in Example 1.19.)

24. (See Figure 2.9)   It can be shown that a cable suspended from its two ends hangs in the shape of a "catenary"; that is, it is in the shape of the graph of an equation of the form $y = a\cosh(x/a)$ for some a. (Recall that $\cosh x = (e^x + e^{-x})/2$.) Suppose a telephone wire is suspended between two poles 100 meters apart and that the wire sags 5 meters at its lowest point. Find the value of a.

25. [From Eberhart and Sweet (1960)]   A radioactive nuclide, A, decays into another radioactive nuclide, B, which further decays. If, initially, only A nuclides are present, then the number of B nuclides present after t days is

$$N_B = \frac{r_A n_0}{r_B - r_A}\left(\exp(-r_A t) - \exp(-r_B t)\right)$$

where n_0 is the initial number of A nuclides and r_A and r_B are the rate constants for A and B, respectively. If $r_A = 0.0866$ and $r_B = 0.3466$, at what time(s) will there be 10^{19} B nuclides if there are 10^{20} A nuclides initially?

26. An object moves along a parabolic track described by the equation $y = x^2$. At what point of the track is the object closest to the point $(1, 2)$? (Use calculus to set up a critical point equation, and solve that equation numerically.)

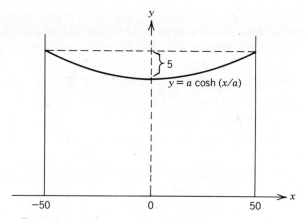

Figure 2.9   Telephone wire in the shape of a catenary

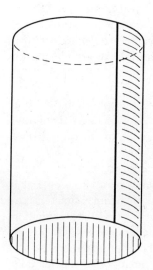

Figure 2.10   Can with two seams

27. A cylindrical can with a bottom, but no top, is to be made from sheet metal which costs 60 cents per square foot. The side seam and the base seam (see Figure 2.10) each cost 10 cents per linear foot. Assuming there is no waste when the bottom disk is cut, find the dimensions which minimize the total cost of the can if its volume is to be 1 cubic foot. (Use calculus to set up a critical point equation, and solve that equation numerically.)

2.3 Sources of Error

As we saw in Section 2.2, the secant method and Newton's method have the advantage that the sequences they produce usually converge faster than those produced by the bisection method and have the disadvantage of involving a greater risk of serious error or even failure. But the risk of error is present, to some degree, in every computation.

In this section we discuss the nonhuman errors that are created or magnified when a computation is carried out on a computer or calculator. The discussion assumes that a computer is being used, but most of the remarks apply equally to calculator computations.

The errors that we discuss fall into the following categories.

1. *Representation errors* caused by the limitations on how a machine can represent the mathematical system of real numbers and operations on those numbers.

2. *Propagation errors*, in which existing errors are magnified by a sometimes lengthy computation.

3. *Sensitivity errors*, in which a particular problem or a particular method of solving that problem tends to magnify existing errors.

An inherent limitation of any computer is the finite size of its memory. As a consequence of this limitation, a computer can store only finitely many real numbers and can store each number to only finitely many places of precision. In practice, each computer may be thought of as having a fixed, finite set of real numbers which it is capable of representing; we refer to these as the computer's "internal numbers." Also, each computer has a method for converting the numbers we feed it into its own internal numbers. This method for representing real numbers involves making approximations, and so it creates errors.

Next, once numbers have been stored in the computer, we want to perform on them the usual arithmetic operations: addition, subtraction, multiplication, and division. But the true sum, difference, product, and quotient of two internal numbers might itself not be an internal number. So, to represent the result, another approximation is made by the machine; that is, another error is introduced.

We also want to apply the elementary functions of calculus (sin, log, and so on) to the computer's internal numbers. But, since the four arithmetic operations are the only mathematical operations built into a computer's circuits, the elementary functions can only be approximated. The natural types of functions to use for these approximations are polynomials and quotients of polynomials since such functions can be evaluated by means of just the four arithmetic operations. Of course, the approximations used in modern computers are more sophisticated than our approximations by Taylor polynomials and Lagrange polynomials in Sections 1.4 and 1.5; still, they too are inexact and provide a source of error.

To better understand the consequences of these representation errors, we simulate a computer on paper by making up a set of internal numbers, a method for converting real numbers into these internal numbers, a method for performing arithmetic on these numbers, and a method for approximating the elementary functions on these numbers. The internal numbers of our simulated computer are all the real numbers that can be represented in base 10 using no more than 4 digits. For example, 9999, -32.46, and .0026 are internal numbers in our computer, but 99999, -32.465, and 0.00026 are not. To convert a real number to an internal number, our computer will round it to 4 digits if that is possible and give an error signal if it is not. For example, if it is fed 9999.5 or any larger real number, our simulated computer will signal an error. But it will round -32.465 to -32.47, 0.00026 to .0003, and -0.00004 to zero. (Rounding is done by dropping the sign of the number, adding 5 to the fifth digit, dropping all digits after the fourth, and then restoring the sign.) To add two numbers, our computer will use as many digits as necessary to perform the addition and then round the result to 4 digits, if possible; it will do likewise for subtraction, multiplication, and division. Finally, it will approximate values of the elementary functions by rounding the exact values to 4 digits, if possible.

This simulation of a computer is quite inaccurate in some ways. For example, real computers usually represent numbers in base 2 or some power of 2, not base 10. But the pattern of errors revealed in the examples below is much like that for a real computer.

As our initial example, we subtract 0.9996 from 4.0014. The number 0.9996 is fed into our simulated computer without error, but 4.0014 is rounded to 4.001, which introduces an error of 0.0004. Such "round-off" errors are unavoidable but can be no worse than 5 units in the fifth digit of a number. To perform the subtraction, we subtract the rounded numbers exactly and then round again:

$$4.0010 - 0.9996 = 3.0014 \rightarrow 3.001$$

(where the arrow indicates a rounding off). So subtracting causes an additional error of 0.0004, and the total error is 0.0008.

Subtracting nearly equal numbers causes errors that can be much more severe since they constitute a larger percentage of the correct result. For example, let us subtract 3.9996 from 4.0014. Then

$$4.0014 \rightarrow 4.001 \qquad 3.9996 \rightarrow 4.000$$

and

$$4.001 - 4.000 = .0010$$

as opposed to the exact difference of 0.0018. Thus the error, 0.0008, is almost 50% of the correct answer. The error caused by subtracting nearly equal numbers is referred to as a "cancellation error."

Severe cancellation errors can easily occur in the secant method of the previous section. In that method, the difference quotient $(b - a)/(f(b) - f(a))$ is computed repeatedly. Thus, when b and a are close, the computation of both $b - a$ and $f(b) - f(a)$ can lead to cancellation errors, and computing their quotient can magnify those errors. For example, if $b = 4.0014$, $a = 3.9006$ and $f(x) = \sqrt{x}$, then, in our simulated computer,

$$\frac{b - a}{f(b) - f(a)} \rightarrow \frac{4.001 - 3.901}{2.000 - 1.975} = \frac{.1000}{.0250} = 4.000$$

The numerator .1000 is only off by 0.0008 from the exact value of $b - a$, and the denominator .0250 is off by less than 0.0004. But even though these cancellation errors might not seem severe, the quotient is off by almost 0.025.

Cancellation errors can even affect the supposedly reliable bisection method. Its reliability depends on the fact that $f(x) = 0$ is guaranteed to have a root between two points at which f has opposite signs. But the computation of the sign can be unreliable. We will compute the sign of $f(x) = (x - 1)^3$ for several values of x on

Table 2.6 Values of $f(x) = (x - 1)^3$

x	f(x)	x	f(x)
0.95	−0.0006	1.01	−0.0001
0.96	0.0000	1.02	−0.0004
0.97	0.0000	1.03	0.0000
0.98	−0.0004	1.04	0.0000
0.99	−0.0001	1.05	−0.0004

either side of the root $x = 1$. But first we rewrite $(x - 1)^3$ in a form in which cancellation errors are more likely.

$$f(x) = (x - 1)^3 = x^3 - 3x^2 + 3x - 1$$
$$= (x^2 - 3x + 3)x - 1$$
$$= ((x - 3)x + 3)x - 1$$

This process of factoring out x repeatedly until we are left with no power of x higher than the first is referred to as "Horner's method" for representing polynomials. However, even though this method is often recommended for use in machine computations since it reduces the number of multiplications needed to evaluate a polynomial, it leads to severe cancellation errors in this example. Table 2.6 shows the result of evaluating $((x - 3)x + 3)x - 1$ on our simulated computer. Note the incorrect signs at $x = 1.01, 1.02,$ and 1.05.

Even if there are no cancellation errors, a lengthy computation can magnify small errors into large ones. For example, if we add together 2000 numbers on our simulated computer and if each term in the sum has a round-off error of 0.0005, then, unless these errors cancel, the sum could be off by $2000(0.0005) = 1.0$. This is a typical example of "accumulated round-off error," which is a type of propagation error. The algorithms in Sections 2.5 through 2.8 are all susceptible to errors of this type.

The computations in Sections 1.4 and 1.5 are also susceptible to propagation errors. Suppose we want to approximate e^x on the interval $[-5, 0]$ by Taylor polynomials of e^x at 0. According to Taylor's remainder formula, for each value of x there exists a point z between x and 0 such that

$$|e^x - P_n(x)| = \frac{|e^z x^{n+1}|}{(n+1)!} \le \frac{5^{n+1}}{(n+1)!}$$

where P_n denotes the nth Taylor polynomial of e^x at 0. Therefore, since $5^{20}/20!$ equals zero to four places, $P_{19}(x)$ ought to equal e^x on the interval $[-5, 0]$, as far as our simulated computer can judge. However, suppose $P_{19}(-5)$ is computed as follows. We write

$$P_{19}(-5) = 1 - 5 + \frac{5^2}{2!} - \frac{5^3}{3!} + \cdots - \frac{5^{19}}{19!}$$

Then we evaluate each term of the sum exactly, round each term to 4 digits, and add

Table 2.7 Values of $\sum_{k=0}^{n}(-5)^k/k! = P_n(-5)$

n	nth term	$P_n(-5)$	n	nth term	$P_n(-5)$
0	1.000	1.000	10	2.691	0.8670
1	−5.000	−4.000	11	−1.223	−0.3560
2	12.50	8.500	12	0.5097	0.1537
3	−20.83	−12.33	13	−0.1960	−0.0423
4	26.04	13.71	14	0.0700	0.0277
5	−26.04	−12.33	15	−0.0233	0.0044
6	21.70	9.370	16	0.0073	0.0117
7	−15.50	−5.130	17	−0.0021	0.0096
8	9.688	3.558	18	0.0006	0.0102
9	−5.382	−1.824	19	−0.0002	0.0100

the terms up from left to right. Table 2.7 shows the intermediate computations and the final result: The approximate value of e^{-5} is .0100. The exact value of e^{-5}, however, is .0067 (rounded to 4 digits).

There is another way to look at the source of error in the last example, other than simply to attribute it to accumulated small cancellation errors. We can blame ourselves. Note that although the exact value of e^{-5} is rather small, the computation involved terms as large as 26.04, in which digits that would have contributed significantly to the correct answer of .0067 were rounded off. As we will soon see, this poorly designed computation can be replaced by one that introduces far less error.

Of course, not all errors can be avoided. Representation errors, such as those caused by the approximations that are made when real numbers are converted to internal numbers, when arithmetic operations are performed on internal numbers, and when the elementary functions are approximated, can rarely be avoided. One can learn to estimate these errors and evaluate their impact on the results of a particular computation on a particular computer, but we will not aspire to this level of expertise.

Propagation errors, in contrast, are often magnified by a poorly designed computation. In some of the above examples, we can greatly reduce the propagation errors by using a small amount of algebra. For instance, since $e^{-5} = 1/e^5$, the value of e^{-5} can be computed by using a Taylor polynomial to approximate e^5 and then taking the reciprocal. There will be no cancellation errors since the terms of the Taylor polynomial are all positive when $x = 5$. In the example in which we showed that computing the fraction $(b - a)/(f(b) - f(a))$ can magnify a cancellation error, we could have tried to factor out and cancel the troublesome terms:

$$\frac{b - a}{\sqrt{b} - \sqrt{a}} = \frac{(b - a)(\sqrt{b} + \sqrt{a})}{(\sqrt{b} - \sqrt{a})(\sqrt{b} + \sqrt{a})} = \frac{(b - a)(\sqrt{b} + \sqrt{a})}{b - a} = \sqrt{b} + \sqrt{a}$$

There will be no cancellation error in computing $\sqrt{b} + \sqrt{a}$.

Sometimes there is no way to reduce a large error by redesigning the computation. It may be that our algorithm itself is magnifying errors. Although the algorithms in this chapter were all selected because of their good features, no general-purpose algorithm for computing approximations works equally well for all problems. The secant method, for example, involves cancellation errors for points close to the root if we cannot factor and cancel as we did above.

At other times, it is not the algorithm but the problem that is overly sensitive to errors, as the next example shows. Suppose we are to solve the quadratic equation $x^2 - 4x + 3.98 = 0$. Its exact roots are 2.141 and 1.859 (to four digits). But if an error causes the coefficient 3.98 to be computed as 3.95, then the roots would be 2.224 and 1.776. Thus, an error of 0.03 in a coefficient has been magnified, through no error in a computation or weakness of an algorithm, to an error of more than 0.08 in the roots.

There is a fourth category of errors that we should not overlook: data errors. Quite often, the numbers we feed to the computer are based on real data that have been collected in the laboratory or field. But since measurements and counts are almost never exact, neither are these numbers. Thus, even if there were no representation errors, many computations would still suffer from data errors that become magnified by propagation and sensitivity errors.

The moral of this troubling tale is that we should exercise great care in interpreting the result of a computation, and, if possible, we should design the computation to minimize propagation errors. Even if we lack the expertise to make a reliable estimate of the amount of error, we should take steps to check the result. For example, we might compute the result by more than one method; we might carry out the computation much further than necessary, say by choosing smaller values of *epsilon* and thus larger values of n in Algorithms 2.1–2.3; we might alter the values of certain parameters (such as the initial guesses in the secant method or the step size in later algorithms); we might use double-precision arithmetic, if it is available; we might print out some of the intermediate computations to see if any unfavorable patterns appear (such as the wild swings that a poorly designed computation might produce, as in Table 2.7); we might draw a picture, or have a computer graphics device draw a picture, of a graph of a function near its root or (in connection with later algorithms) of a direction field of a differential equation. Finally, we might even try to find a rough approximation by means of a hand calculation.

EXERCISES

1. Find the results of the following computations on our 4-digit simulated computer.
    (a) $1.3926 + 52.103$;
    (b) $(0.0025)(0.016)$;
    (c) $(78.2)(127.85)$.

2. (Nonassociativity of addition) On our 4-digit simulated computer, show that $(a + b) + c \neq a + (b + c)$, if $a = 762$, $b = 0.0430$, and $c = 0.0083$.

3. (Nonassociativity of rounding) Find a 6-digit number which gives one result when first rounded to 5 digits and then to 4 and a different result when it is rounded in one step to 4 digits.

4. In a typical computer there is a fixed, finite set of internal numbers and a fixed method for representing every real number either as an internal number or as an error. It follows that there must be at least one pair of real numbers with the same representation. Why?

5. The solutions of a quadratic equation $ax^2 + bx + c = 0$ are given by the familiar formula

$$x = \left(-b \pm \sqrt{b^2 - 4ac}\right)/2a$$

If $4ac$ is small compared to b^2, then this formula could lead to substantial cancellation error in the computation of one of the two roots. Show how to rewrite the formula so that the computation of this root avoids cancellation error.

6. The computation of $1 - \cos x$ could lead to substantial cancellation error if x is close to zero. Show how to rewrite $1 - \cos x$ to avoid cancellation error.

7. The computation of $\left(\sqrt[3]{a} - \sqrt[3]{b}\right)/(a - b)$ could lead to substantial cancellation error if a is close to b. Show how to rewrite this fraction to avoid cancellation error.

8. Recall that $\sinh x = (e^x - e^{-x})/2$. It would seem that the computation of $\sinh x$ for x near 0 could lead to substantial cancellation error. However, if $\sinh x$ is approximated by one of its Taylor polynomials at 0, cancellation error is avoided. Why?

9. Add the numbers $1, 1/3, 1/3^2, \ldots, 1/3^9$ in that order and then in the reverse order, on our 4-digit simulated computer. Explain the discrepancy between your two answers. Is one method of adding better than the other?

10. Solve the system of equations

$$x + y = 1$$

$$x + 1.01y = 3$$

Then change the coefficient of y in the second equation to 1.02 and solve again. By how much did the solutions change? Draw the graphs of the first pair of equations and use this picture to explain how it could happen that a small change in a coefficient could lead to a large change in the solution.

11. For any collection of data points $x_1, \ldots, x_n$, their *mean*, μ_n, and *standard derivation*, σ_n, can be computed by any of the following three equivalent sets of formulas:

(a) $\mu_n = \dfrac{x_1 + \cdots + x_n}{n}$, $\quad \sigma_n^2 = \dfrac{1}{n} \sum_{k=1}^{n} (x_k - \mu_n)^2$

(b) $\mu_n = \dfrac{x_1 + \cdots + x_n}{n}$, $\quad \sigma_n^2 = \dfrac{1}{n} \sum_{k=1}^{n} x_k^2 - \mu_n^2$

(c) $\mu_1 = x_1, \ \sigma_1 = 0$

and, if $n \geq 1$,

$$\mu_{n+1} = \frac{n\mu_n}{n+1} + \frac{x_{n+1}}{n+1}, \qquad \sigma_{n+1}^2 = \frac{n}{n+1}\sigma_n^2 + \frac{\left(x_{n+1} - \mu_{n+1}\right)^2}{n}$$

Compute μ_3 and σ_3^2 for each of the following two sets of data, using each of the above three formulas. Do the computations on our 4-digit simulated computer. Does any one of the three sets of formulas seem less sensitive to errors than the others?

$$x_1 = 1.006 \qquad x_1 = 1.001$$

$$x_2 = 1.006 \qquad x_2 = 1.001$$

$$x_3 = 0.988 \qquad x_3 = 0.9976$$

2.4 A Hybrid Method

We noted in Section 2.2 that the bisection method for finding a root of an equation $f(x) = 0$ is more reliable than the secant method but that the latter usually converges more rapidly. However, we do not need to choose between the two methods; we can combine them to get the advantages of both in a single method. Such a combination is referred to as a "hybrid" method.

At each stage of the particular hybrid method to be described here, we keep track of three points: a, b, and c. Point a is the current approximation to the root; point b is an earlier approximation that, together with a, is used to generate the next bisection method approximation; and point c is an earlier approximation that, together with a, is used to generate the next secant method approximation. As in the bisection method, b is chosen so that $f(a)$ and $f(b)$ have opposite signs; and, as in the secant method, c is usually chosen so that a and c are the two most recent approximations.

The most crucial step of the algorithm is the one in which a choice is made between using the bisection and secant methods for the next approximation to the root. If we let a_b and a_s denote these respective approximations, recall that

$$a_b = a + \frac{b - a}{2} \tag{2.8}$$

$$a_s = a + \frac{(a - c)f(a)}{f(c) - f(a)} \tag{2.9}$$

We reject a_s in favor of a_b if a_s does not fall between a and b. This gives our hybrid method the reliability of the bisection method. If a_s does fall between and a and b, we choose a_s over a_b provided a_s seems to be closer to the root than a_b. But how can we decide which approximation is likely to be closer to the root? We guess that the root is closer to a than to b if $|f(a)| \le |f(b)|$. (If this inequality is not true, we interchange the values of a and b to make it true.) So, if a_s is between a and a_b, we guess that a_s is closer to the root than a_b is (see Figure 2.11).

Algorithm 2.4   *A hybrid method*

1. ***Input*** *a, b, epsilon, maxn*
2. $c \Leftarrow b$
3. $n \Leftarrow 0$

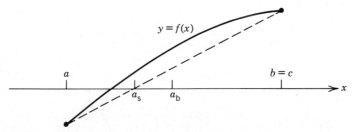

Figure 2.11   A typical step of the hybrid method in which the secant approximation is chosen over the bisection approximation

   4. Repeat steps 5–13 **until** $|a - b| \leq epsilon$
      5. If $|f(a)| > |f(b)|$ **then**
         5a. $c \Leftarrow a$
         5b. $a \Leftarrow b$
         5c. $b \Leftarrow c$
      6. $n \Leftarrow n + 1$
      7. If $n > maxn$ **then**
      7a. Print "Algorithm fails: no convergence"
      7b. End
   8. $bdiff \Leftarrow (b - a)/2$
   9. $sdiff \Leftarrow (a - c)f(a)/(f(c) - f(a))$
   10. $c \Leftarrow a$
   11. If $(0 < sdiff < bdiff)$ **or** $(0 > sdiff > bdiff)$
            then $a \Leftarrow a + sdiff$
            else $a \Leftarrow (a + b)/2$
   12. If $sign f(a) = sign f(b)$ **then** $b \Leftarrow c$
   13. Print $n, a, f(a)$
14. End

In steps 8 and 9 of the algorithm, the differences $a_b - a$ and $a_s - a$ are computed using formulas (2.8) and (2.9), respectively; the crucial choice between a_b and a_s is made in step 11. Point c is updated to the next most recent approximation in step 10, and b is updated in step 12. Step 12 is designed to guarantee that the updated values of $f(a)$ and $f(b)$ have opposite signs. In step 5, we insure that $|f(a)| \leq |f(b)|$ is always true by interchanging the values of a and b when it is not.

Algorithm 2.4 still has some significant defects. As in the secant method, we need to be concerned about division by zero in step 9. Even if $f(c) - f(a)$ is only close to zero, not equal to zero, the division in step 9 could produce too large a quotient for a computer to calculate and store. But in such a case, we surely want to choose the bisection method over the secant method. Thus, to avoid division by zero or by an overly small number, we compute the numerator and denominator of $(a - c)f(a)/(f(c) - f(a))$ separately and then cross-multiply the inequalities in step 11. For example, the inequality $sdiff < bdiff$ is equivalent to

$$(a - c)f(a) < (f(c) - f(a))(bdiff) \tag{2.10}$$

if $f(c) - f(a)$ is positive. In fact, if we force the numerator of $sdiff$ to be positive (by changing the sign of its numerator and denominator, if necessary), then the four inequalities in step 11 can be replaced by inequality (2.10) alone.

Another defect in Algorithm 2.4 is that it sometimes chooses the secant method when the bisection method is faster. For example, near a multiple root the secant method usually converges more slowly than the bisection method, and yet Algorithm 2.4 chooses the secant method over the bisection method in most such problems. Table 2.8 shows the results of both the hybrid method and the bisection method applied to the equation $(x - 1)^3 = 0$, which has a root of multiplicity 3 at $x = 1$. The hybrid method used the secant method for every approximation.

Table 2.8 Two Methods Applied to $(x - 1)^3 = 0$, with $a = 0.5$ and $b = 2.0$

n	Bisection	Hybrid
1	1.2500	0.6667
2	0.8750	0.7368
3	1.0625	0.8048
4	0.9688	0.8517
5	1.0156	0.8883
6	0.9923	0.9156
7	1.0039	0.9363
8	0.9980	0.9519

To avoid a long string of slowly converging secant approximations, we will force an occasional bisection approximation whenever the convergence threatens to become slower than the bisection method. Our technique for insuring this is to force a bisection approximation whenever three successive approximations have been made that have not reduced the distance between a and b by seven-eighths (which is what three bisections would have accomplished). Algorithm 2.5 incorporates these two improvements; the first occurs in steps 11, 12, 13, and 18; the second in steps 4, 5, 15, 16, and 17.

Algorithm 2.5   *An improved hybrid method*

1. **Input** a, b, *epsilon*, *maxn*
2. $c \Leftarrow b$
3. $n \Leftarrow 0$
4. *bcount* $\Leftarrow 0$
5. *oldbdiff* $\Leftarrow (b - a)/2$
6. **Repeat steps** 7–20 **until** $|a - b| \le$ *epsilon*
   7. **If** $|f(a)| > |f(b)|$ **then**
      7a. $c \Leftarrow a$
      7b. $a \Leftarrow b$
      7c. $b \Leftarrow c$
   8. $n \Leftarrow n + 1$
   9. **If** $n >$ *maxn* **then**
      9a. **Print** "Algorithm fails: no convergence"
      9b. **End**
   10. *bdiff* $\Leftarrow (b - a)/2$
   11. *num* $\Leftarrow (a - c)f(a)$
   12. *denom* $\Leftarrow f(c) - f(a)$
   13. **If** *num* < 0 **then**
      13a. *num* $\Leftarrow - num$
      13b. *denom* $\Leftarrow - denom$
   14. $c \Leftarrow a$

15. $bcount \Leftarrow bcount + 1$

16. If $bcount \leq 3$ **then go to** 18

17. If $8|bdiff| \geq |oldbdiff|$

　　　　then $a \Leftarrow (a + b)/2;$ **go to** 19

　　　　else $bcount \Leftarrow 0;$ $oldbdiff \Leftarrow bdiff$

18. If $num < (denom)(bdiff)$

　　　　then $a \Leftarrow a + num/denom$

　　　　else $a \Leftarrow (a + b)/2$

19. If $sign f(a) = sign f(b)$ **then** $b \Leftarrow c$

20. Print n, a, $f(a)$

　21. End

Algorithm 2.5 is generally regarded as an excellent method for finding the roots of a great variety of equations. No one method, however, is best for every equation. Standard textbooks on numerical analysis, such as those in the list of references at the end of this book, should be consulted for a more extensive discussion of the problem.

EXERCISES

For Exercises 1–7, first write a computer program that implements Algorithm 2.5 for an arbitrary function f. Then use the program to approximate a root of the equation $f(x) = 0$, for each assigned function f, with $epsilon$ = 0.000001 and $maxn$ = 20. For each assigned function, describe your conclusion in words, following the recommendations at the end of Section 2.1.

1.　$f(x) = x^3 - 2x + 2$

2.　$f(x) = \log x + x^2 - 8$

3.　$f(x) = 2x - 2 - \sin x$

4.　$f(x) = x - e^{-x}$

5.　$f(x) = 1 + 9(\log x)/x^2$

6.　$f(x) = x^5 - 5x^4 + 10x^3 - 10x^2 + 5x - 1$

7.　$f(x) = 1 - 57(\log x)/x^3, \quad x < 4$

8.　Find a recursion formula for the terms x_n of the sequence produced by Newton's method for $f(x) = (x - 1)^3$. Use this formula to show that

$$\left| \frac{x_{n+1} - x_n}{x_n - x_{n-1}} \right| = \frac{2}{3} \quad \text{for all } n$$

Note: This shows that Newton's method, when applied to the equation $(x - 1)^3 = 0$, converges more slowly than the bisection method, since this ratio has the value $\frac{1}{2}$ for the latter method.

2.5 The Trapezoid Rule and Simpson's Rule

Now we take up the second of the three types of problems listed in the introduction to this chapter:

Approximate the value of the integral $\int_a^b f(x)\, dx$.

In a typical calculus course, it is shown that if f is a continuous function, then f has an antiderivative F and this antiderivative can be used to evaluate the integral as follows:

$$\int_a^b f(x)\,dx = F(b) - F(a)$$

(see Theorem A.23). Then a variety of techniques for finding antiderivatives are taught. It would be incorrect to infer, however, that this theorem and these techniques are all one needs to evaluate any integral that is likely to arise in practice. The proof that continuous functions have antiderivatives does not provide a practical method for constructing antiderivatives. In fact, it can be shown that the antiderivatives of some quite uncomplicated continuous functions, such as $\exp(-x^2)$, cannot be expressed in terms of the elementary functions of calculus. By this we mean that the antiderivatives cannot be expressed in terms of trigonometric, logarithmic, exponential, and power functions using only the operations of addition, subtraction, multiplication, division, composition, and inversion of functions [see Rosenlicht (1972)]. Furthermore, even if f does have an antiderivative that can be expressed in terms of the elementary functions of calculus, it may be unreasonably difficult to find. Thus, integrals that arise in real-world problems often cannot be evaluated exactly by the techniques of calculus but must be approximated instead.

Since a definite integral is a limit of Riemann sums (Theorem A.21), we could approximate an integral by such a sum. But Riemann sums converge to the integral much more slowly than certain other sums which are almost as easy to compute. As a first step toward finding these sums, we consider the relative merits of approximating f by a Taylor or Lagrange polynomial P and then approximating the integral of f by the integral of P. Theorem 1.13 suggests that if $|f(x) - P(x)|$ has an error bound that is independent of x and small, then the integral of P is close to the integral of f. Furthermore, we may be able to find such error bounds from the remainder formulas in Theorems 1.8 and 1.10. Taylor polynomials, however, turn out to be less suitable than Lagrange polynomials for approximating most integrals. In part, we reject Taylor polynomials because they can be difficult to evaluate; to find their coefficients we would need to differentiate f repeatedly. But also recall that, although the Taylor polynomials of f at x_0 approximate f very well near x_0, they are often poor approximations of f on a large interval. In particular, the interval of integration $[a, b]$ is often too large for Taylor polynomials to be the best choice when approximating an integral. (However, see Exercises 35 and 36, in which Taylor polynomials of degree 1 are used.)

The simplest way of using Lagrange polynomials to generate approximations to $\int_a^b f(x)\,dx$ is by subdividing the interval $[a, b]$ into n congruent subintervals and interpolating f between the $n + 1$ endpoints. Thus, let

$$x_i = a + ih, \quad \text{where } h = \frac{b - a}{n} \quad \text{and} \quad i = 0, 1, \ldots, n \qquad (2.11)$$

and let P denote the Lagrange polynomial that interpolates f between the $n + 1$ points $x_0, x_1, \ldots, x_n$. Then $\int_a^b P(x)\,dx$ is called a *Newton–Cotes* approximation to $\int_a^b f(x)\,dx$.

For $n = 1$, the Newton–Cotes approximation to $\int_a^b f(x)\,dx$ is [see (1.39)]

$$\int_{x_0}^{x_1}\left[f(x_0)\frac{x - x_1}{x_0 - x_1} + f(x_1)\frac{x - x_0}{x_1 - x_0}\right] dx$$

$$= \int_0^h\left[f(x_0)\frac{u - h}{-h} + f(x_1)\frac{u}{h}\right] du$$

$$\text{(where } u = x - x_0 \quad \text{and} \quad h = x_1 - x_0)$$

$$= \left. -f(x_0)\frac{(u - h)^2}{2h} + f(x_1)\frac{u^2}{2h}\right|_0^h$$

$$= f(x_0)\frac{h}{2} + f(x_1)\frac{h}{2}$$

$$= \frac{h}{2}[f(x_0) + f(x_1)] \tag{2.12}$$

Formula (2.12) is called the *trapezoid rule* for approximating $\int_{x_0}^{x_1} f(x)\,dx$. Its name is suggested by Figure 2.12; if f is positive on the interval $[x_0, x_1]$, then the area under the graph of f and above the interval $[x_0, x_1]$ can be approximated by the area of the shaded trapezoid, and the latter area is given by the formula $h[f(x_0) + f(x_1)]/2$.

For $n = 2$, the Newton–Cotes approximation to $\int_a^b f(x)\,dx$ is [see (1.40)]

$$\int_{x_0}^{x_2}\left[f(x_0)\frac{(x - x_1)(x - x_2)}{(x_0 - x_1)(x_0 - x_2)} \right.$$

$$\left. +f(x_1)\frac{(x - x_0)(x - x_2)}{(x_1 - x_0)(x_1 - x_2)} + f(x_2)\frac{(x - x_0)(x - x_1)}{(x_2 - x_0)(x_2 - x_1)}\right] dx$$

$$= \int_{-h}^h\left[f(x_0)\frac{u(u - h)}{(-h)(-2h)} + f(x_1)\frac{(u + h)(u - h)}{h(-h)} + f(x_2)\frac{(u + h)u}{(2h)h}\right] du$$

$$\text{(where } u = x - x_1 \quad \text{and} \quad h = x_2 - x_1 = x_1 - x_0)$$

$$= \int_{-h}^h\left[f(x_0)\frac{(u^2 - hu)}{2h^2} - f(x_1)\frac{(u^2 - h^2)}{h^2} + f(x_2)\frac{(u^2 + uh)}{2h^2}\right] du$$

$$= \frac{1}{h^2}\int_0^h\left[f(x_0)u^2 - 2f(x_1)(u^2 - h^2) + f(x_2)u^2\right] du$$

$$\text{(by Exercise 28, since } u^2 \text{ and 1 are even and } u \text{ is odd)}$$

$$= \frac{1}{h^2}\left[f(x_0)\frac{u^3}{3} - 2f(x_1)\left(\frac{u^3}{3} - h^2 u\right) + f(x_2)\frac{u^3}{3}\right]\Bigg|_0^h$$

$$= \frac{h}{3}[f(x_0) + 4f(x_1) + f(x_2)] \tag{2.13}$$

Formula (2.13) is called *Simpson's rule* for approximating $\int_{x_0}^{x_2} f(x)\,dx$.

These two Newton–Cotes formulas usually provide quite poor approximations to $\int_a^b f(x)\,dx$, especially if the interval $[a, b]$ is large. After all, we are merely approximat-

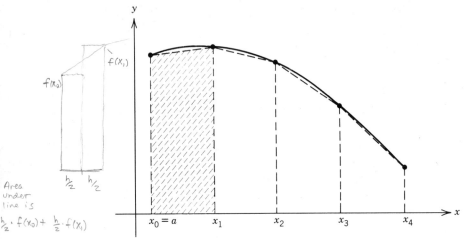

Figure 2.12 The composite trapezoid rule

ing the graph of f by a single straight line or a single parabola; so, if the domain is large, the fit is likely to be poor on much of that domain. But it is not usually advisable to seek a better approximation by increasing the value of n. Not only do the Newton–Cotes formulas become unwieldy for large n, but the remainder, $f^{(n+1)}(z)\prod_{i=0}^{n}(x - x_i)/(n + 1)!$, can be substantial for large n. Even though the factor $1/(n + 1)!$ gets small rapidly, the remaining factors can get large equally rapidly. Instead, we subdivide the interval $[a, b]$ into several small subintervals and apply the trapezoid rule or Simpson's rule on each subinterval.

If we apply the trapezoid rule on each of the subintervals

$$[x_0, x_1], [x_1, x_2], \ldots, [x_{n-1}, x_n]$$

and add the approximations, the resulting formula is called the *composite trapezoid rule*. Thus

$$\int_a^b f(x)\, dx = \int_{x_0}^{x_1} f(x)\, dx + \int_{x_1}^{x_2} f(x)\, dx + \cdots + \int_{x_{n-1}}^{x_n} f(x)\, dx$$

which is approximately equal to

$$\frac{h}{2}\left[f(x_0) + f(x_1)\right] + \frac{h}{2}\left[f(x_1) + f(x_2)\right] + \cdots + \frac{h}{2}\left[f(x_{n-1}) + f(x_n)\right]$$

$$= \frac{h}{2}\left[f(x_0) + 2f(x_1) + 2f(x_2) + \cdots + 2f(x_{n-1}) + f(x_n)\right] \tag{2.14}$$

Formula (2.14), the composite trapezoid rule, is denoted by $T_n(f)$.

Algorithm 2.6   *The composite trapezoid rule*

1. ***Input*** *a, b, n*
2. $h \Leftarrow (b - a)/n$
3. $sum \Leftarrow f(a) + f(b)$
4. $sumtwo \Leftarrow f(a + h) + f(a + 2h) + \cdots + f(a + (n - 1)h)$
5. $sum \Leftarrow h[sum + 2\ sumtwo]/2$
6. ***Print*** *sum*
7. ***End***

Note: *In most programming languages, the sum in step 4 must be evaluated by means of a loop that calculates a running sum.*

If we apply Simpson's rule on each of the subintervals

$$[x_0, x_2], [x_2, x_4], \ldots, [x_{n-2}, x_n]$$

(where n has to be even) and add the approximations, the resulting formula is called the *composite Simpson's rule.* Thus

$$\int_a^b f(x)\, dx = \int_{x_0}^{x_2} f(x)\, dx + \int_{x_2}^{x_4} f(x)\, dx + \cdots + \int_{x_{n-2}}^{x_n} f(x)\, dx$$

which is approximately equal to

$$\frac{h}{3}\left[f(x_0) + 4f(x_1) + f(x_2)\right] + \frac{h}{3}\left[f(x_2) + 4f(x_3) + f(x_4)\right]$$

$$+ \cdots + \frac{h}{3}\left[f(x_{n-2}) + 4f(x_{n-1}) + f(x_n)\right]$$

$$= \frac{h}{3}\left[f(x_0) + 4f(x_1) + 2f(x_2) + 4f(x_3) + \cdots + 2f(x_{n-2}) + 4f(x_{n-1}) + f(x_n)\right]$$

$$(2.15)$$

Formula (2.15), the composite Simpson's rule, is denoted by $S_n(f)$.

Algorithm 2.7   *The composite Simpson's rule*

1. ***Input*** *a, b, n*
2. $h \Leftarrow (b - a)/n$
3. $sum \Leftarrow f(a) + f(b)$
4. $sumtwo \Leftarrow f(a + 2h) + f(a + 4h) + \cdots + f(a + (n - 2)h)$
5. $sumfour \Leftarrow f(a + h) + f(a + 3h) + \cdots + f(a + (n - 1)h)$
6. $sum \Leftarrow h[sum + 2\ sumtwo + 4\ sumfour]/3$
7. ***Print*** *sum*
8. ***End***

Note: *In most programming languages, the sums in steps 4 and 5 must be evaluated by means of a loop that calculates a running sum.*

From the remainder formula in Theorem 1.10, we can derive bounds or estimates for the errors $|\int_a^b f(x)\, dx - T_n(f)|$ and $|\int_a^b f(x)\, dx - S_n(f)|$.

Theorem 2.2   Suppose the second derivative of f exists at every point of the interval $[a, b]$, and suppose there exists a constant M such that $|f''(x)| \le M$ for all x in

$[a, b]$. Then, for every positive integer n, if we let h denote $(b - a)/n$ (the length of each one of the n subintervals)

$$\left| \int_a^b f(x)\, dx - T_n(f) \right| \le \frac{M(b-a)h^2}{12} = \frac{M(b-a)^3}{12n^2} \qquad (2.16)$$

TRAP?

Proof   We let $x_i = a + ih$ and let p_i denote the Lagrange polynomial that interpolates f between x_{i-1} and x_i. Then

$$\left| \int_a^b f(x)\, dx - T_n(f) \right|$$

$$= \left| \sum_{i=1}^n \int_{x_{i-1}}^{x_i} f(x)\, dx - \sum_{i=1}^n \int_{x_{i-1}}^{x_i} p_i(x)\, dx \right|, \; = \left| \sum_{i=1}^n \int_{x_{i-1}}^{x_i} [f(x) - p_i(x)]\, dx \right|$$

$$\le \sum_{i=1}^n \left| \int_{x_{i-1}}^{x_i} [f(x) - p_i(x)]\, dx \right| \qquad \text{(by the triangle inequality)}$$

$$= \sum_{i=1}^n \left| \int_{x_{i-1}}^{x_i} \frac{1}{2} [f''(z(x))(x - x_{i-1})(x - x_i)]\, dx \right| \text{(by Theorem 1.10 with } n = 1)$$

$$\le \sum_{i=1}^n \int_{x_{i-1}}^{x_i} \frac{M}{2} (x - x_{i-1})(x_i - x)\, dx \qquad \text{(by Theorem A.24e)}$$

$$= \sum_{i=1}^n \frac{M}{2} \int_0^h u(h - u)\, du = nM\left(\frac{hu^2}{4} - \frac{u^3}{6} \right)\Big|_0^h$$

$$\text{(where } u = x - x_{i-1} \quad \text{and} \quad h = (b-a)/n = x_i - x_{i-1})$$

$$= nM\left(\frac{h^3}{4} - \frac{h^3}{6} \right) = \frac{Mnh^3}{12} = \frac{M(b-a)h^2}{12}$$

EXAMPLE 2.4   Use the composite trapezoid rule to approximate $\int_{-1}^1 \exp(-x^2)\, dx$ with an error of at most 0.05.

Let $f(x) = \exp(-x^2)$. Then

$$f'(x) = -2x \exp(-x^2)$$

$$f''(x) = -2\exp(-x^2) + 4x^2 \exp(-x^2) = 2\exp(-x^2)(2x^2 - 1)$$

Therefore, if $-1 \le x \le 1$, then $|f''(x)| \le 2$ since $\exp(-x^2) \le 1$ and $|2x^2 - 1| \le 1$ on $[-1, 1]$. So, by Theorem 2.2,

$$\left| \int_{-1}^1 \exp(-x^2)\, dx - T_n(f) \right| \le \frac{2(2^3)}{12n^2} = \frac{4}{3n^2}$$

which is less than 0.05 if $n = 6$. Then $h = \frac{1}{3}$, and so our approximation to $\int_{-1}^1 \exp(-x^2)\, dx$ is $T_6(f)$, which equals

$$\tfrac{1}{6}\left[f(-1) + 2f(-\tfrac{2}{3}) + 2f(-\tfrac{1}{3}) + 2f(0) + 2f(\tfrac{1}{3}) + 2f(\tfrac{2}{3}) + f(1) \right] = 1.480$$

An error bound similar to (2.16) can be proved for the composite Simpson's rule:

$$\left| \int_a^b f(x)\, dx - S_n(f) \right| \le \frac{M(b-a)h^4}{180} = \frac{M(b-a)^5}{180n^4} \tag{2.17}$$

if $|f^{(4)}(x)| \le M$ for all x in $[a, b]$. The proof is a good bit messier than that of Theorem 2.2, and so we omit it [see Ralston and Rabinowitz (1978)]. Note, however, the surprising appearance of h^4 in formula (2.17) where we would have expected h^3, as in inequality (1.46). This is caused by the fortuitous circumstance that $\int_{x_0}^{x_2} (x - x_0)(x - x_1)(x - x_2)\, dx = 0$ (Exercise 27). It is fortuitous since h^4 is much smaller than h^2 when h is small, and so we expect the composite Simpson's rule approximations to converge much more rapidly than the composite trapezoid rule approximations. However, since the constant M in inequality (2.17) is sometimes much larger than the constant M in inequality (2.16), it can happen that, for a particular value of n, $T_n(f)$ is a better approximation to the integral than $S_n(f)$.

EXAMPLE 2.5 Use the composite Simpson's rule to approximate $\int_0^1 \sqrt[3]{1 + x}\, dx$ with an error of at most 0.0001.

Let $f(x) = (1 + x)^{1/3}$. Then

$$f^{(4)}(x) = -\frac{80}{81}(1 + x)^{-11/3}$$

Therefore, $|f^{(4)}(x)| \le 1$ when $0 \le x \le 1$ since $1 + x \ge 1$. So, by inequality (2.17),

$$\left| \int_0^1 \sqrt[3]{1 + x}\, dx - S_n(f) \right| \le \frac{(1)(1)^5}{180n^4}$$

which is less than 0.0001 if $n = 4$. (Recall that n has to be even for Simpson's rule.) Then $h = \frac{1}{4}$, and so our approximation to $\int_0^1 \sqrt[3]{1 + x}\, dx$ is $S_4(f)$, which equals

$$\tfrac{1}{12}\left[f(0) + 4f(0.25) + 2f(0.5) + 4f(0.75) + f(1) \right] = 1.13988$$

Examples 2.4 and 2.5 are, in a sense, unrepresentative. In practice, the number M in formulas (2.16) and (2.17) is often too hard to find. Or, even if we can find a value for M, the resulting bound in formulas (2.16) and (2.17) is often too conservative. That is, the errors $|\int_a^b f(x)\, dx - T_n(f)|$ and $|\int_a^b f(x)\, dx - S_n(f)|$ are often much less than their respective bounds $M(b-a)^3/12n^2$ and $M(b-a)^5/180n^4$. Instead, it is common to use the following rather imprecise error estimates in their place.

$$\left| \int_a^b f(x)\, dx - T_{2n}(f) \right| = \text{(approximately)} \; |T_n - T_{2n}| \tag{2.18}$$

$$\left| \int_a^b f(x)\, dx - S_{2n}(f) \right| = \text{(approximately)} \; |S_n - S_{2n}| \tag{2.19}$$

That is, we double n repeatedly until two successive approximations differ by the given error tolerance. These error estimates, however, can be far from reliable.

EXERCISES

In Exercises 1–6, use the composite trapezoid rule to approximate the given integral with an error of at most 0.05. (A calculator will be useful.)

1. $\int_1^2 1/x \, dx$

2. $\int_0^1 1/(1 + x^2) \, dx$

3. $\int_1^2 e^x/x \, dx$

4. $\int_0^{\pi/2} \cos^2 x \, dx$

5. $\int_0^1 (e^x - 1)/\sqrt{x} \, dx$

   Hint: $\int_0^1 (e^x - 1)/\sqrt{x} \, dx = \int_0^{0.5} (e^x - 1)/\sqrt{x} \, dx + \int_{0.5}^1 (e^x - 1)/\sqrt{x} \, dx$

   Approximate the first integral in the above sum with an error of at most 0.025 by using a Taylor polynomial to approximate $e^x - 1$. Approximate the second integral in the sum with an error of at most 0.025 by using the composite trapezoid rule. You may use the fact $|f''(x)| \le 7.3$ when $0.5 \le x \le 1$.

6. $\int_0^1 e^x/\sqrt{x} \, dx$

   Hint: $e^x/\sqrt{x} = (e^x - 1)/\sqrt{x} + 1/\sqrt{x}$. Use your answer to Exercise 5 to approximate the integral of the first function in this sum, and evaluate the integral of the second function exactly.

In Exercises 7–12, use the composite Simpson's rule to approximate the given integral with an error of at most 0.02. (A calculator will be useful.)

7. $\int_1^2 1/x \, dx$

8. $\int_0^1 1/(1 + x^2) \, dx$

9. $\int_1^2 e^x/x \, dx$

10. $\int_0^{\pi/2} \cos^2 x \, dx$

11. $\int_0^1 (e^x - 1)/\sqrt{x} \, dx$ (See hint for Exercise 5 and use $|f^{(4)}(x)| \le 170$)

12. $\int_0^1 e^x/\sqrt{x} \, dx$ (See hint for Exercise 6)

For Exercises 13–19, first write a computer program that implements Algorithm 2.6 for an arbitrary function f. When you use the program to approximate an integral, apply the error estimate (2.18) with an error tolerance of 0.0001; that is, double n repeatedly until two successive approximations differ by no more than 0.0001. (Either have the program double n and make this test for convergence, or run the program repeatedly, inputting twice the previous value of n each time.) For each assigned integral, describe your conclusion in words, following the recommendations at the end of Section 2.1.

13. $\int_1^2 1/x \, dx$

14. $\int_0^1 1/(1 + x^2) \, dx$

15. $\int_0^2 \sqrt{4 - x^2} \, dx$

16. $\int_0^1 \exp(-x^2) \, dx$

17. $\int_0^1 (e^x - 1)/\sqrt{x} \, dx$ (See hint for Exercise 5)

18. $\int_0^1 x \sin(1/x) \, dx$

    Hint: Although the integrand is undefined at $x = 0$, its limit at $x = 0$ is 0. Therefore, you might alter your program to force $f(0)$ to have the value 0.

19. $\int_0^\pi (\sin x)/x \, dx$

Hint: Although the integrand is undefined at $x = 0$, its limit at $x = 0$ is 1. Therefore, you might alter your program to force $f(0)$ to have the value 1.

For Exercises 20–26, first write a computer program that implements Algorithm 2.7 for an arbitrary function f. When you use the program to approximate an integral, apply the error estimate (2.19) with an error tolerance of 0.0001; that is, double n repeatedly until two successive approximations differ by no more than 0.0001. (Either have the program double n and make this test for convergence, or run the program repeatedly, inputting twice the previous value of n each time.) For each assigned integral, describe your conclusion in words, following the recommendations at the end of Section 2.1.

20. $\int_1^2 1/x \, dx$

21. $\int_0^1 1/(1 + x^2) \, dx \approx .78539952$

22. $\int_0^2 \sqrt{4 - x^2} \, dx$

23. $\int_0^1 \exp(-x^2) \, dx$

24. $\int_0^1 (e^x - 1)/\sqrt{x} \, dx$ (See hint for Exercise 5)

25. $\int_0^1 x \sin(1/x) \, dx$ (See hint for Exercise 18)

26. $\int_0^\pi (\sin x)/x \, dx$ (See hint for Exercise 19)

27. If $x_i = x_0 + ih$ for $i = 0, 1, 2$, show that

$$\int_{x_0}^{x_2} (x - x_0)(x - x_1)(x - x_2) \, dx = 0$$

Hint: Make the substitution $u = x - x_1$.

28. Suppose f is a continuous function on $[-h, h]$.

(a) If f is *even* (that is, $f(-x) = f(x)$ for all x in $[-h, h]$), show that $\int_{-h}^h f(x) \, dx = 2 \int_0^h f(x) \, dx$.

(b) If f is *odd* (that is, $f(-x) = -f(x)$ for all x in $[-h, h]$), show that $\int_{-h}^h f(x) \, dx = 0$.

29. Write a computer program that computes approximate values of the natural logarithm function. Use the fact that $\log x = \int_1^x 1/u \, du$ and use Algorithm 2.6 or 2.7 to approximate the integral. Your program should permit the user to enter initial and final values for x and an increment for x.

30. Write a computer program that computes approximate values of the "error function," which is defined by

$$\text{erf}(x) = \frac{2}{\sqrt{\pi}} \int_0^x \exp(-u^2) \, du, \qquad x \ge 0$$

Your program should permit the user to enter initial and final values of x and an increment for x.

31. Write a computer program that computes approximate values of the "complete elliptic integral of the second kind," which is defined by

$$E(x) = \int_0^{\pi/2} \sqrt{1 - x^2 \sin^2 u} \, du, \qquad 0 \le x \le 1$$

Your program should permit the user to enter initial and final values of x and an increment for x.

32. Write a computer program that computes approximate values of "Bessel's function of order n," which is defined by

$$J_n(x) = \frac{1}{\pi} \int_0^\pi \cos(nu - x \sin u)\, du, \qquad n = 0, 1, 2, \ldots$$

Your program should permit the user to enter initial and final values of x and an increment for x.

33. Approximate the value of $\int_0^\infty \exp(-x^2)\, dx$ with an error of at most 0.0001. *Hint:*

$$\int_0^\infty \exp(-x^2)\, dx = \int_0^m \exp(-x^2)\, dx + \int_m^\infty \exp(-x^2)\, dx$$

and

$$\int_m^\infty \exp(-x^2)\, dx \le \int_m^\infty \exp(-mx)\, dx = \exp(-m^2)/m$$

Thus, if m is chosen so that $\exp(-m^2)/m < 0.00005$, then $\int_0^m \exp(-x^2)\, dx$ approximates $\int_0^\infty \exp(-x^2)\, dx$ with an error of at most 0.00005. It then remains to approximate $\int_0^m \exp(-x^2)\, dx$ with an error of at most 0.00005.

34. Write a computer program to approximate the roots of the equation $J_n(x) = 0$, where J_n is Bessel's function of order n, defined in Exercise 32. Your program should permit the user to enter the order, n, and two approximations that are on opposite sides of a root.

35. Derive a formula analogous to formulas (2.14) and (2.15) for the "composite midpoint rule." In this method for approximating $\int_a^b f(x)\, dx$, the function f is approximated on each subinterval $[x_{i-1}, x_i]$ by its value at the midpoint, $(x_{i-1} + x_i)/2$.

36. Suppose $|f''(x)| \le M$ for all x in $[a, b]$. Prove that

$$\left| \int_a^b f(x)\, dx - M_n(f) \right| \le \frac{M(b-a)h^2}{24} = \frac{M(b-a)^3}{24\, n^2}$$

where $M_n(f)$ denotes the composite midpoint rule approximation (described in Exercise 35), and $h = (b-a)/n$. *Hint:* On each subinterval $[x_{i-1}, x_i]$, approximate f by its Taylor polynomial of degree 1 at the midpoint of that interval, and apply Theorem 1.8.

37. Approximate $\int_0^\pi (\sin x)/x\, dx$ by means of the composite midpoint rule (described in Exercise 35).

38. Find, approximately, the arc length of the curve described by $y = x^3$, $0 \le x \le 1$.

39. At temperature T, the specific heat capacity of certain chemical substances may be computed from Debye's law:

$$C = \frac{17.88}{a^3} \int_0^a \frac{x^4 e^x}{(e^x - 1)^2}\, dx$$

where T is measured in degrees Kelvin, C is measured in calories per mole, $a = \theta_D/T$, and θ_D is the "Debye temperature" of the substance. If silver has the Debye temperature of 215 K, compute its specific heat capacity at 298 K. (*Note:* The denominator of the integrand is zero at $x = 0$. So change the lower limit of integration from 0 to 0.0001; this will not throw the answer off significantly.)

Table 2.9 Annual Precipitation

Year	Inches	Year	Inches
1960	46.7	1966	41.1
1961	50.7	1967	54.1
1962	51.6	1968	49.9
1963	41.6	1969	58.4
1964	40.2	1970	48.3
1965	27.0		

Source: The U.S. Bureau of the Census, *Historical Statistics of the United States, Colonial Times to 1970, Bicentennial Edition.*

40. The numbers in Table 2.9 give the total annual precipitation, in inches, at Blue Hill Observatory, Massachusetts, in the years 1960–1970. Compute the average annual precipitation as follows. If f is a continuous function on the interval $[a, b]$, the *average value* of f is defined to be $[1/(b - a)] \int_a^b f(x)\, dx$. Imagine the numbers in the second column of Table 2.9 to be the values of some function f at the points $x_i = i$ for $i = 0, 1, 2, \ldots, 10$. Approximate $\int_0^{10} f(x)\, dx$ by the composite Simpson's rule $S_{10}(f)$.

2.6 The Romberg Method

Although the composite trapezoid rule and composite Simpson's rule are efficient and reliable methods for evaluating a great many integrals, they are not always sufficiently efficient and reliable. Sometimes n must be quite large for $T_n(f)$ or $S_n(f)$ to be acceptably close to the exact value of the integral, and this can cause two difficulties. For one, the time needed to compute $T_n(f)$ or $S_n(f)$ could be unacceptably long, even if a computer is being used to perform the computations. This is an especially significant difficulty when f is so complicated that the evaluation of $f(x)$ requires a lengthy computation. For another, the computation of $T_n(f)$ or $S_n(f)$ could introduce a substantial amount of accumulated round-off error. This is so because our approximations are sums that, as n gets larger, have more terms and therefore have a greater accumulation of the round-off error inherent in the computation of each term.

Thus, at times we need methods that require as few function evaluations and as few additions as possible for a given error bound. In this sense, the composite Simpson's rule is usually superior to the composite trapezoid rule, and the *Romberg method*, which we describe next, is usually better yet.

The idea behind the Romberg method is one that is often used to improve the performance of a method of approximation. If we can estimate the error of the nth term of a sequence of approximations, we may be able to use this estimate to modify the sequence so that it converges more rapidly. The Romberg method results from applying this idea to the composite trapezoid rule as follows.

First, we claim (see Exercise 11) that the proof of Theorem 2.2 can be modified so that, instead of producing an error bound, it produces the following exact expression for the error:

$$\int_a^b f(x)\,dx - T_n(f) = -\frac{f''(z_n)(b-a)^3}{12n^2} \tag{2.20}$$

for some number z_n between a and b. Of course, we have no way of knowing the value of z_n. Still, equation (2.20) at least gives us the form of the error. Furthermore, if f'' does not vary too much in the interval $[a, b]$, we should be able to replace z_n by any point in $[a, b]$ without changing the error term very much and thus obtain a good error estimate. Next, we use this error estimate to modify the trapezoid rule approximations so they converge more rapidly. Specifically, we combine two different approximations, $T_n(f)$ and $T_m(f)$, as follows. Since

$$\int_a^b f(x)\,dx - T_m(f) = -\frac{f''(z_m)(b-a)^3}{12m^2}$$

then, when we multiply this equation by m^2 and equation (2.20) by n^2 and subtract, we get

$$(m^2 - n^2)\int_a^b f(x)\,dx - \left(m^2 T_m(f) - n^2 T_n(f)\right) = \frac{\left(f''(z_n) - f''(z_m)\right)(b-a)^3}{12}$$

The right side of this equation will be quite small if $f''(z_n)$ and $f''(z_m)$ are nearly equal. Thus, a more promising approximation of $\int_a^b f(x)\,dx$ than either $T_n(f)$ or $T_m(f)$ is

$$\frac{m^2 T_m(f) - n^2 T_n(f)}{m^2 - n^2}$$

The usual choice for m is $2n$, in part because, as we will see below, $T_{2n}(f)$ can be computed from $T_n(f)$ with only n additions instead of $2n$ additions. This yields the approximation

$$\frac{4n^2 T_{2n}(f) - n^2 T_n(f)}{3n^2} = \frac{1}{3}\left(4T_{2n}(f) - T_n(f)\right) \tag{2.21}$$

But $\frac{1}{3}(4T_{2n}(f) - T_n(f))$ approximates $\int_a^b f(x)\,dx$ with an error that can again be estimated. In fact, the approximation (2.21) is exactly $S_{2n}(f)$ (Exercise 12), and the following analog of inequality (2.17) holds:

$$\int_a^b f(x)\,dx - S_n(f) = -\frac{f^{(4)}(z_n)(b-a)^5}{180n^4}$$

for some number z_n between a and b. Therefore, by an argument much like the one above, we deduce that the approximations $S_n(f)$ and $S_m(f)$ can be combined to produce the more promising approximation

$$\frac{m^4 S_m(f) - n^4 S_n(f)}{m^4 - n^4}$$

which equals, when $m = 2n$,

$$\frac{16n^4 S_{2n}(f) - n^4 S_n(f)}{15n^4} = \frac{1}{15}\left(16 S_{2n}(f) - S_n(f)\right) \tag{2.22}$$

Table 2.10 The Romberg Method

T_1^0

T_2^0 T_2^1

T_4^0 T_4^1 T_4^2

T_8^0 $T_8^1 \longrightarrow T_8^2$ T_8^3

. . . .

Furthermore, the approximation (2.22) differs from $\int_a^b f(x)\,dx$ by an error that can again be estimated; in fact, the approximations begun in formulas (2.21) and (2.22) can be continued indefinitely. The result is called the Romberg method; it is usually described by the formulas

$$T_n^0 = T_n(f)$$

$$T_{2n}^k = \frac{4^k T_{2n}^{k-1} - T_n^{k-1}}{4^k - 1} \quad \text{for } k > 0 \tag{2.23}$$

Table 2.10 displays more clearly the order in which the Romberg approximations, T_n^k, are usually computed.

According to formula (2.23), every entry of the table, other than those in the first column, is computed from the entry directly to its left and the entry above that, as the arrows in the table indicate. The entries in the first column of the table are simply composite trapezoid rule approximations. However, we can cut in half the number of additions needed to compute these sums by using T_n^0 in the computation of T_{2n}^0 as follows. If we let $h = (b - a)/n$, $n' = 2n$, and $h' = (b - a)/n' = h/2$, then, according to (2.14),

$$T_{2n}^0 = \frac{h'}{2}\left[f(a) + 2f(a + h') + 2f(a + 2h') + 2f(a + 3h') + \cdots \right.$$

$$\left. + 2f(a + (n' - 1)h') + f(b)\right]$$

$$= \frac{h'}{2}\left[f(a) + 2f(a + 2h') + \cdots + 2f(a + (n' - 2)h') + f(b)\right]$$

$$+ \frac{h'}{2}\left[2f(a + h') + 2f(a + 3h') + \cdots + 2f(a + (n' - 1)h')\right]$$

$$= \frac{T_n^0}{2} + h'\left[f(a + h') + f(a + 3h') + \cdots + f(a + (n' - 1)h')\right] \tag{2.24}$$

EXAMPLE 2.6 Use the Romberg method through the third row of Table 2.10 to approximate $\int_0^\pi \sin x\,dx$.

$$T_1^0 = \frac{\pi}{2}(0 + 0) = 0$$

$$T_2^0 = \frac{T_1^0}{2} + \frac{\pi}{2}\sin\left(\frac{\pi}{2}\right) = 1.5708$$

Table 2.11 Approximations to $\int_0^\pi \sin x\, dx$

n	$T_n(f)$	n	$S_n(f)$
1	0.00000	2	2.09439
2	1.57080	4	2.00456
4	1.89612	8	2.00027
8	1.97423	16	2.00002
16	1.99357	32	2.00000
32	1.99839		
64	1.99960		
128	1.99990		
256	1.99997		
512	1.99999		
1024	2.00000		

n	T_n^0	T_n^1	T_n^2	T_n^3	T_n^4
1	0.00000				
2	1.57080	2.09439			
4	1.89612	2.00456	1.99857		
8	1.97423	2.00027	1.99998	2.00001	
16	1.99357	2.00002	2.00000	2.00000	2.00000

$$T_2^1 = \frac{4T_2^0 - T_1^0}{3} = 2.09439$$

$$T_4^0 = \frac{T_2^0}{2} + \frac{\pi}{4}\left(\sin\left(\frac{\pi}{4}\right) + \sin\left(\frac{3\pi}{4}\right)\right) = 1.89612$$

$$T_4^1 = \frac{4T_4^0 - T_2^0}{3} = 2.00456$$

$$T_4^2 = \frac{16T_4^1 - T_2^1}{15} = 1.99857$$

Table 2.11 continues this computation through T_{16}^4 and compares the results with those of the composite trapezoid and Simpson's rules. The numbers were generated by computer programs that implement Algorithms 2.6, 2.7, and 2.8. Note that the last composite trapezoid rule approximation in the table required 1024 additions and 1025 function evaluations; the last composite Simpson's rule approximation required only 32 additions and 33 function evaluations; and the entire Romberg method table required merely 16 additions and 17 function evaluations for the first column and no further function evaluations for the other 10 table entries.

Algorithm 2.8 *The Romberg method*

  1. **Input** a, b, *maxcol*
  2. $T^0 \Leftarrow [\,f(a) + f(b)](b - a)/2$
  3. $n \Leftarrow 1$

4. **Repeat steps** 5–15 **with** $col \Leftarrow 1, \ldots, maxcol$
   5. $n \Leftarrow 2n$
   6. $h \Leftarrow (b - a)/n$
   7. $aboveT \Leftarrow T^0$
   8. $T^0 \Leftarrow T^0/2 + h[\,f(a + h) + f(a + 3h) + \cdots + f(a + (n - 1)h)]$
   9. $fourpower \Leftarrow 1$
  10. **Repeat steps** 11–14 **with** $k \Leftarrow 1, \ldots, col$
     11. $fourpower \Leftarrow 4\ fourpower$
     12. $savedT \Leftarrow T^k$
     13. $T^k \Leftarrow (T^{k-1}\ fourpower - aboveT)/(fourpower - 1)$
     14. $aboveT \Leftarrow savedT$
  15. **Print** n, T^0, $T^1, \ldots, T^{col}$
16. **End**

Note: In most programming languages, the sum in step 8 must be computed in a loop, and the numbers T^k must be saved in an array with subscript k.

The composite trapezoid rule approximations, T_n^0, are computed at step 8, using (2.24). The remaining approximations in the nth row of Table 2.10 (that is, T_n^k for $k \geq 1$) are computed in steps 11–14.

EXERCISES

In Exercises 1–5, use the Romberg method through the third row of Table 2.10 to approximate the given integral. (A calculator will be useful.)

1. $\int_1^2 1/x\, dx$
      2. $\int_0^1 1/(1 + x^2)\, dx$
3. $\int_0^1 \exp(-x^2)\, dx$
      4. $\int_0^2 \sqrt{4 - x^2}\, dx$
5. $\int_0^\pi (\sin x)/x\, dx$ (*Note:* Alter your program so that $f(0)$ is given the value 1, since
     $\lim_{x \to 0} (\sin x)/x = 1$.)

For Exercises 6–10, first write a computer program that implements Algorithm 2.8 for an arbitrary function f. By experimenting with small values of *maxcol*, approximate each integral to within about 0.0001. For each assigned integral, describe your conclusion in words, following the recommendations at the end of Section 2.1.

6. $\int_1^2 1/x\, dx$
    7. $\int_0^1 \exp(-x^2)\, dx$
    8. $\int_0^2 \sqrt{4 - x^2}\, dx$
9. $\int_0^\pi (\sin x)/x\, dx$ (*Note:* Alter your program so that $f(0)$ is given the value 1.)
10. $\int_0^1 x \sin(1/x)\, dx$ (*Note:* Alter your program so that $f(0)$ is given the value of 0.)
11. Suppose f'' exists and is continuous at every point of $[a, b]$. Then, for every positive integer n, show there exists a point z_n in $[a, b]$ such that

$$\int_a^b f(x)\, dx - T_n(f) = -\frac{f''(z_n)(b - a)^3}{12n^2}$$

Hint: Modify the proof of Theorem 2.2. Instead of using inequalities, use Theorem A.27 to show that, for each i,

$$\int_{x_{i-1}}^{x_i} \frac{f''(z_i(x))}{2}(x - x_{i-1})(x - x_i)\, dx = \frac{f''(w_i)}{2}\int_{x_{i-1}}^{x_i}(x - x_{i-1})(x - x_i)\, dx$$

for some w_i in $[x_{i-1}, x_i]$. Then use the intermediate value theorem to show that $\sum_{i=1}^{n} f''(w_i)/n$ can be written as $f''(z_n)$.

12. Show that $S_{2n}(f) = (4T_{2n}(f) - T_n(f))/3$.

2.7 The Euler Method

The third problem listed in the introduction of the chapter

Approximate a solution of the differential equation $\dfrac{dy}{dx} = f(x, y)$

is the most challenging and important of the three. It is an important problem because differential equations arise so frequently in the descriptions of scientific problems and because so many differential equations cannot be solved exactly. The reasons that some differential equations cannot be solved exactly include all the reasons that some integrals cannot be evaluated exactly and more. Some differential equations are simply unreasonably difficult to solve exactly; others can be solved readily using elementary techniques such as those in Appendix D but lead to integrals that cannot be evaluated exactly; and still others cannot be solved by any of the known elementary techniques. For example, $y' = 1 - 2xy$ is a linear differential equation and so can be solved by a technique described in Section D.2. However, this technique produces the solutions

$$y = \exp\left(-x^2\right)\left[\int_0^x \exp\left(t^2\right) dt + c\right]$$

which are of little use since the antiderivative of $\exp(t^2)$ cannot be expressed in terms of the elementary functions of calculus. The equation $y' = x^2 + y^2$ exemplifies a different phenomenon: It is one of many differential equations that cannot be solved by any of the usual textbook techniques. Therefore, for many differential equations, numerical methods provide the only practical means for getting a useful solution.

The differential equation problems that we consider in this chapter are initial-value problems of the form

$$\frac{dy}{dx} = f(x, y), \qquad y(a) = y_0, \qquad a \le x \le b \tag{2.25}$$

We will not consider higher-order differential equations or systems of differential equations. Higher-order differential equations, however, can be expressed as first-order systems, and first-order systems can be solved numerically by methods much like those we will develop for a single equation.

One way to see how we might approximate the solutions of an initial-value problem of the form (2.25) is to examine their geometric properties. Figure 2.13

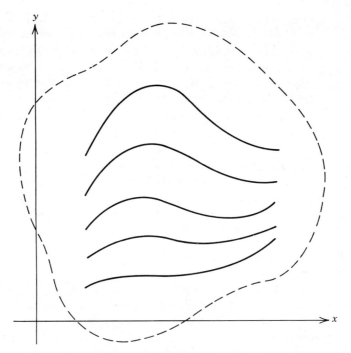

Figure 2.13   Solution curves of a differential equation

displays the graphs of a few solutions of a typical first-order differential equation $y' = f(x, y)$. From existence–uniqueness theorems such as Theorems 3.23 and D.1, we can infer the following properties of these graphs (provided f satisfies certain hypotheses that we do not discuss here). First, for every point (u, v) in the domain of f, there exists a solution $y = y(x)$ of the initial-value problem

$$\frac{dy}{dx} = f(x, y), \qquad y(u) = v \tag{2.26}$$

Thus, since the graph of this solution passes through the point (u, v) and since the point (u, v) is arbitrary, we see that the graphs of all the solutions taken together must completely fill the domain of f. In particular, if the dashed curve in Figure 2.13 indicates the boundary enclosing the domain of f, the set of all solution curves of $y' = f(x, y)$ must completely fill that enclosed region. Theorems 3.23 and D.1 also tell us that the solution satisfying the initial condition $y(u) = v$ is unique, which implies that no two distinct solution curves can intersect. Finally, it can be shown that the domain of each solution can be chosen so that the graph of the solution extends all the way to the boundary of the domain of f.

Another piece of graphical information can be read directly from the form of the differential equation. Note that if $y = y(x)$ denotes the solution of the initial-value problem (2.26), then $y'(u) = f(u, v)$; therefore, since $y'(u)$ is the slope of the line tangent to the graph of $y = y(x)$ at the point (u, v), it follows that $f(u, v)$ is also that slope. Thus, even if we cannot easily find the solutions of the differential equation, we can easily find all their tangent lines. The collection of all these tangent lines is called the *direction field* of the differential equation, and we can sometimes use the direction field to construct approximate graphs of the solutions. For example, if we select a large number of points (u, v) in the domain of f and if we draw a short line segment with slope $f(u, v)$ through each selected point (u, v), we may be able to draw curves through the selected points and tangent to the line segments. These curves should be reasonable approximations to the actual solution curves.

EXAMPLE 2.7 Use the direction field of the differential equation $y' = 1 - 2xy$ to draw some approximations to the actual solution curves.

We build up our picture by the following stages. First, we draw several curves along which the slope $f(x, y) = 1 - 2xy$ is constant; then, through several points on each curve, we draw short line segments with (constant) slope $f(x, y)$; and finally, we draw smooth curves tangent to the line segments at the selected points.

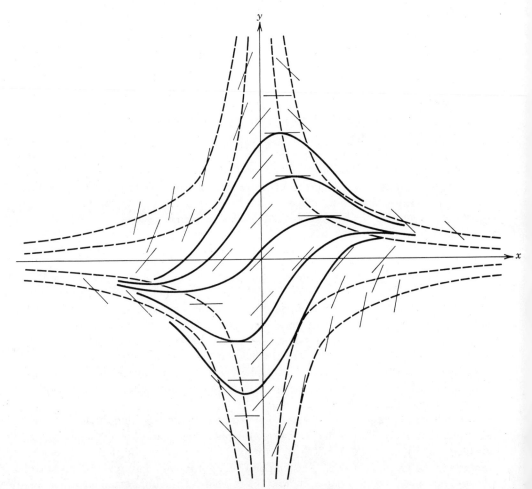

Figure 2.14 Direction field and solution curves of $y' = 1 - 2xy$

For the given differential equation $y' = 1 - 2xy$, we graph the curves for which the slope $f(x, y)$ has the values $-1, 0, 1, 2, 3$. These choices yield the respective curves $xy = 1$, $xy = \frac{1}{2}$, $xy = 0$, $xy = -\frac{1}{2}$, and $xy = -1$. For example, if $f(x, y) = -1$, then

$$-1 = 1 - 2xy$$

and so

$$xy = 1$$

All the curves except the third one are hyperbolas; they are indicated by dashes in Figure 2.14. The third curve, described by $xy = 0$, consists of the points on both axes. Then, though each of the five curves, several short line segments of slope $f(x, y)$ have been drawn. Finally, a few solutions have been graphed; they are the solid curves.

We now use these geometric insights to derive the *Euler* (or *Cauchy–Euler*) *method* for approximating the solution of an initial-value problem (2.25). First, we approximate the graph of the solution by its tangent line at (a, y_0). Since $f(a, y_0)$ is the slope of this tangent line, the equation of the line is

$$y = y_0 + f(a, y_0)(x - a)$$

In other words, we approximate the solution of the initial-value problem by its first Taylor polynomial at a. As x gets farther from x_0, however, the errors

$$y(x) - [y_0 + f(a, y_0)(x - a)]$$

are likely to grow. So, as in the preceding two sections, we partition the interval $[a, b]$ into n congruent subintervals and use a new approximation for each subinterval. As our initial value of y for each new subinterval, we use the approximate value of y at the right-hand endpoint of the preceding subinterval (see Figure 2.15). In symbols, we define the endpoints of the subintervals by

$$x_k = a + kh \quad \text{for } k = 0, \ldots, n \quad \text{where } h = \frac{b - a}{n} \tag{2.27}$$

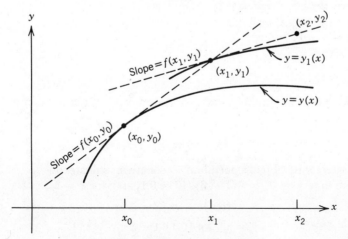

Figure 2.15 The Euler method

and the approximations to $y(x_0)$, $y(x_1), \ldots, (x_n)$ by

$$y_0 = \text{given value of } y \text{ at } x_0$$
$$y_k = y_{k-1} + hf(x_{k-1}, y_{k-1}) \quad \text{for } k = 1, \ldots, n \tag{2.28}$$

Note that $y_1, y_2, \ldots, y_n$ are actually approximations to different solution curves. The value y_1 arises from the tangent to the graph of $y = y(x)$, the solution which passes through the initial point (x_0, y_0). But y_2 is an approximation to a different solution of the same differential equation—the solution $y = y_1(x)$ which passes through (x_1, y_1). More generally, each y_k approximates a different solution $y = y_k(x)$. Nevertheless, we hope that all the solutions $y = y_k(x)$ are fairly close to the solution $y = y(x)$ of the original initial-value problem and, therefore, that each y_k is fairly close to $y(x_k)$.

A more algebraic way of deriving formulas (2.28) is to replace the differential equation $y' = f(x, y)$ by a difference equation. Note that since

$$y'(x_{k-1}) = \lim_{h \to 0} \frac{y(x_{k-1} + h) - y(x_{k-1})}{h}$$

we can say that, for small values of h,

$$hy'(x_{k-1}) = \text{(approximately)} \quad y(x_{k-1} + h) - y(x_{k-1})$$

So if we replace $x_{k-1} + h$ by x_k and use the fact that $y'(x_{k-1}) = f(x_{k-1}, y(x_{k-1}))$, we get

$$y(x_k) - y(x_{k-1}) = \text{(approximately)} \quad hf(x_{k-1}, y(x_{k-1}))$$

Thus, if we replace $y(x_{k-1})$ by y_{k-1} and $y(x_k)$ by y_k, we have the difference equation (2.28).

Algorithm 2.9   The Euler method

   1. **Input** a, b, y_0, n
   2. $h \Leftarrow (b - a)/n$
   3. $y \Leftarrow y_0$
   4. $x \Leftarrow a$
   5. **Repeat steps** 6–8 **with** $k \Leftarrow 1, \ldots, n$
       6. $y \Leftarrow y + hf(x, y)$
       7. $x \Leftarrow a + kh$
       8. **Print** k, x, y
   9. **End**

EXAMPLE 2.8   Use the Euler method to approximate the solution of the initial-value problem

$$\frac{dy}{dx} = 1 - 2xy, \qquad y(0) = 0$$

on the interval $[0, 1]$ partitioned into 10 congruent subintervals.

Note that $y_0 = 0$, $h = (1 - 0)/10 = 0.1$, and, for $k = 1, \ldots, 10$,

$$x_k = (0.1)k$$
$$y_k = y_{k-1} + (0.1)(1 - 2x_{k-1}y_{k-1})$$

Table 2.12 Approximate Solution of $y' = 1 - 2xy$, $y(0) = 0$

x_k	y_k	$y(x_k)$ (actual)	$y_k - y(x_k)$
0.0	0.0000	0.0000	0.0000
0.1	0.1000	0.0993	0.0007
0.2	0.1980	0.1948	0.0032
0.3	0.2901	0.2826	0.0075
0.4	0.3727	0.3599	0.0128
0.5	0.4429	0.4244	0.0185
0.6	0.4986	0.4748	0.0238
0.7	0.5387	0.5105	0.0282
0.8	0.5633	0.5321	0.0312
0.9	0.5732	0.5407	0.0325
1.0	0.5700	0.5381	0.0319

The resulting values of y_k are given in Table 2.12. The actual values (rounded to 4 digits) were found by other methods.

Several observations about this approximate solution are in order. First, although the actual solution is a function $y = y(x)$ defined over an interval, our approximate solution consists of approximations to y at a discrete set of values of x. We could, of course, fit a function to these points. For example, we could fit one or more Lagrange interpolation polynomials. In Figure 2.16, we did just that. We joined each pair of successive points (x_{k-1}, y_{k-1}) and (x_k, y_k) by a line segment; in other words, we interpolated between each pair of successive points by using a Lagrange polynomial of degree at most 1. The resulting function is called a *Cauchy polygon approximation*. We could get a smoother fit by using Lagrange polynomials of higher degree. In practice,

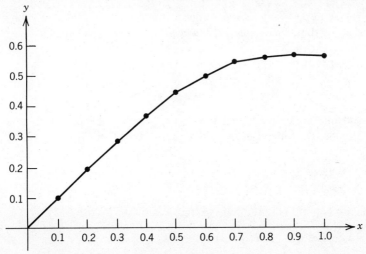

Figure 2.16 Approximate solution of $y' = 1 - 2xy$, $y(0) = 0$

however, the points (x_k, y_k) alone usually provide sufficient information. Next, note that the errors, listed in the fourth column of Table 2.12, increase with increasing k, except for the last one. As Figure 2.15 suggests, we should expect increasing errors as we move away from the initial point (x_0, y_0). It is the smaller error of 0.0319 at $x = 1.0$ that needs explaining. Figure 2.14 suggests a reason: the solution curves for $y' = 1 - 2xy$ get closer to one another for large positive values of x. Therefore, even though each stage of Euler's method seems to take us further away from the original solution curve, the approximations cannot diverge too badly from the actual values, since the solution curves converge toward one another for large positive values of x. The opposite behavior can also occur. In fact, as x approaches 0 in Figure 2.14, the solution curves do move rapidly away from one another. For some differential equations, the error $|y(x_k) - y_k|$ actually grows exponentially as a function of k. Theorem 2.3, however, shows that there is a reassuring bound for the errors.

Theorem 2.3 Suppose that the solution $y = y(x)$ of the initial-value problem

$$\frac{dy}{dx} = f(x, y), \qquad y(a) = y_0$$

exists and is twice differentiable on the interval $[a, b]$. Also, suppose there are constants M and L such that

$$|y''(x)| \le M$$
$$|f(x, u_1) - f(x, u_2)| \le L|u_1 - u_2|$$

whenever $a \le x \le b$ and $-\infty < u_1, u_2 < \infty$. If h, x_k, and y_k are defined by formulas (2.27) and (2.28) and if

$$E_k = |y(x_k) - y_k|, \qquad k = 0, 1, \ldots, n$$

denotes the kth error, then

$$E_k \le \frac{hM}{2L}\left(e^{L(x_k - a)} - 1\right), \qquad k = 0, 1, \ldots, n \tag{2.29}$$

Proof First we apply Taylor's remainder formula to the first Taylor polynomial of $y = y(x)$ at x_{k-1}; thus, there exists z_k between y_{k-1} and y_k such that

$$y(x_k) = y(x_{k-1}) + (x_k - x_{k-1})y'(x_{k-1}) + (x_k - x_{k-1})^2 \frac{y''(z_k)}{2}$$

$$= y(x_{k-1}) + hf(x_{k-1}, y(x_{k-1})) + h^2 \frac{y''(z_k)}{2}$$

In the second line we used the fact that $y'(x_{k-1}) = f(x_{k-1}, y(x_{k-1}))$, which is true since $y(x)$ satisfies the differential equation. Then, by Euler's formula (2.28),

$$E_k = \left| y(x_{k-1}) + hf(x_{k-1}, y(x_{k-1})) + h^2 \frac{y''(z_k)}{2} - y_{k-1} - hf(x_{k-1}, y_{k-1}) \right|$$

$$\le |y(x_{k-1}) - y_{k-1}| + h|f(x_{k-1}, y(x_{k-1})) - f(x_{k-1}, y_{k-1})| + h^2 \frac{|y''(z_k)|}{2}$$

$$\le |y(x_{k-1}) - y_{k-1}| + hL|y(x_{k-1}) - y_{k-1}| + \frac{h^2 M}{2}$$

$$= E_{k-1}(1 + hL) + \frac{h^2 M}{2}$$

We can convert the inequality $E_k \leq E_{k-1}(1 + hL) + h^2M/2$ into an explicit inequality for E_k by using the methods of difference equations in Section 1.3. We define the sequence $\{e_k\}$ by

$$e_k = e_{k-1}(1 + hL) + \frac{h^2M}{2} \quad \text{and} \quad e_0 = 0$$

Then (see Exercise 19)

$$e_k = \frac{hM}{2L}\left[(1 + hL)^k - 1\right]$$

Therefore, since $E_k \leq e_k$,

$$E_k \leq \frac{hM}{2L}\left[(1 + hL)^k - 1\right]$$

Finally, we apply Taylor's remainder formula to the first Taylor polynomial of e^x at 0:

$$e^x = 1 + x + \frac{e^z x^2}{2} \quad \text{for some } z$$

$$\geq 1 + x$$

Hence, $1 + hL \leq e^{hL}$ and so

$$E_k \leq \frac{hM}{2L}\left[e^{hLk} - 1\right]$$

$$= \frac{hM}{2L}\left[e^{L(x_k - a)} - 1\right]$$

The Euler method is sometimes referred to as a "method of order 1" since the errors are, at worst, proportional to the first power of h. In particular, inequality (2.29) implies that the errors E_k all converge to zero when h goes to zero. It can also be shown that the difference between the exact solution $y = y(x)$ and the Cauchy polygon approximations satisfies a similar inequality. Thus, the Cauchy polygon approximations converge to the solution $y = y(x)$ at every point of $[a, b]$; furthermore, the convergence is uniform on $[a, b]$ since the right side of inequality (2.29) is bounded by the constant

$$\frac{(b - a)M}{2nL}\left[e^{L(b-a)} - 1\right]$$

which has limit zero. Inequality (2.29) can also be interpreted as giving a warning: In those cases when the inequality is close to being an equality, the error grows exponentially with increasing k.

Unfortunately, inequality (2.29) is rarely helpful when we use the Euler method to approximate the solution of a specific differential equation, since only rarely can we find the values of the constants M and L. And when we can, the error bound provided by (2.29) is usually far too conservative; that is, the actual error is usually far less. Instead, we use an error estimate much like the one we used to estimate the error of an approximation to an integral. We double n (and thus halve h) and compare corresponding approximations; so we have the collection of error estimates

$$\left| y(x_k) - y_{2k}\left(\frac{h}{2}\right) \right| = \text{(approximately)} \left| y_k(h) - y_{2k}\left(\frac{h}{2}\right) \right| \qquad (2.30)$$

for $k = 1, 2, \ldots, n$. By using the notation $y_k(h)$, we emphasize that the approximations y_k depend on both k and h; thus, $y_k(h/2)$ denotes the kth approximation when the length of every subinterval is $h/2$. Note that we compare $y_k(h)$ with $y_{2k}(h/2)$, not $y_k(h/2)$, since $y_k(h)$ and $y_{2k}(h/2)$ are approximations of $y = y(x)$ for the same value of x: $a + hk = a + (h/2)(2k)$.

Programming note: Your program should ask the user how many of the n points $y_1, \ldots, y_n$ he or she wishes to print. Then, if the user asks for p points, the program should print y_k only when $k = n/p, 2n/p, \ldots, (p - 1)n/p, pn/p$. This feature has two advantages. First, it enables you to keep the number of printed values of y_k modest even when n is very large. Second, by requesting the same number p of points every time you double n, you can be certain that you are applying the error estimate (2.30) only to corresponding approximations of y. The number 10 is often a reasonable value for p. Incidentally, it is far easier to run your program repeatedly, printing p values of y_k each time and checking the error estimate (2.30) yourself, than it is to have the program make this check.

EXERCISES

In Exercises 1–4, for each differential equation, use its direction field to sketch some approximations of the solution curves.

1. $y' = x + y$ 2. $y' = x^2 - y^2$
3. $y' = x^2 + y^2$ 4. $y' = y - x^2$

In Exercises 5–10, (a) Use the Euler method to approximate the solution of the given initial-value problem on the given interval. (A calculator will be useful.) Choose n so that $h = 0.1$. (b) Solve the equation exactly and compute the value of the solution at the right endpoint of the indicated interval as a check on part (a).

5. $y' = 3x^2$, $y(0) = 0$ on $[0, 0.5]$
6. $y' = x + y$, $y(0) = 0$ on $[0, 0.5]$
7. $y' = y^2 + 4$, $y(0) = 0$ on $[0, 0.5]$
8. $y' = -y/3x$, $y(1) = 2$ on $[1, 1.5]$
9. $y' = y - 2x/y$, $y(0) = 1$ on $[0, 0.5]$

   Hint: To find the exact solution, convert the differential equation to one in which the unknown function is $u(x) = y(x)^2$ instead of y, and then solve this new differential equation.

10. $y' = 2xy/(x^2 - y^2)$, $y(1) = 2$ on $[1, 1.5]$.

    Hint: To find the exact solution, convert the differential equation to one in which the unknown function is $u(x) = y(x)/x$ instead of y, and then solve this new differential equation. Use your exact solution to explain why some of the terms of your numerical approximations are absurd.

For Exercises 11–18, first write a computer program that implements Algorithm 2.9 for an arbitrary initial-value problem. When you use the program to approximate a solution, apply the error estimates (2.30) with an error tolerance of 0.0001; that is, double n repeatedly until all the corresponding approximations of y for successive values of n differ by no more than 0.0001.

Print out only about 10 values of y_k, as suggested in the programming note at the end of the section. For each assigned function, describe your conclusion in words, following the recommendations at the end of Section 2.1.

11. $y' = 3x^2$, $y(0) = 0$ on $[0, 1]$

12. $y' = x + y$, $y(0) = 0$ on $[0, 1]$

13. $y' = y^2 + 4$, $y(0) = 0$ on $[0, 0.5]$

14. $y' = -y/3x$, $y(1) = 2$ on $[1, 8]$

15. $y' = y - 2x/y$, $y(0) = 1$ on $[0, 1.5]$

16. $y' = 3yx^2 \cos(x^3)$, $y(0) = 1$ on $[0, 3]$

17. $y' = 2xy/(x^2 - y^2)$, $y(1) = 2$ on $[1, 1.24]$

18. $y' = x^2 + y^2$, $y(0) = 1$ on $[0, 0.8]$

19. Solve the difference equation

$$e_k = e_{k-1}(1 + hL) + \frac{h^2 M}{2}, \qquad e_0 = 0$$

20. **(a)** For the initial-value problem $y' = y$, $y(0) = 1$, $0 \le x \le x_0$, derive the formula $y_k(h) = (1 + h)^k$ for the Euler method approximations.

(b) Find the exact solution $y = y(x)$ of the initial-value problem in (a).

(c) Use your answers to (a) and (b) to show that $\lim_{h \to 0} y_n(h) = y(x_0)$.

2.8 Runge–Kutta Methods

The Euler method is rarely used in practice. It converges so slowly that it usually takes an unacceptably long time to produce sufficiently accurate approximations. Furthermore, increasing n (the number of subintervals) to get greater accuracy also increases the number of arithmetic operations in the computation of the approximations y_k and hence increases the accumulated round-off error. Eventually, this round-off error may significantly affect the reliability of the approximations. But perhaps the best reason for not using the Euler method is that there are others that are not much more complicated to use but converge much more rapidly.

The methods we consider in this section are similar to the Euler method in the sense that they too involve a recursion formula of the form

$$y_k = y_{k-1} + hm_k'$$

where m_k' is an approximation to the slope m_k of the secant line joining $(x_{k-1}, y(x_{k-1}))$ and $(x_k, y(x_k))$ (see Figure 2.17). If we knew the exact value of the slope of that secant line, we would not need to use a numerical method—we could find the exact value of y. When we do not know the exact value of the slope, however, we can find much better approximations to it than $f(x_{k-1}, y_{k-1})$, the value used in the Euler method.

In the *improved Euler method*, we approximate the slope of the secant line in Figure 2.17 by the average of two values of f. As in the Euler method, we are approximating

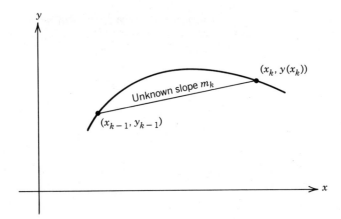

Figure 2.17 The secant line whose slope is to be approximated

the solution of the initial-value problem

$$\frac{dy}{dx} = f(x, y), \qquad y(a) = y_0 \tag{2.31}$$

on the interval $[a, b]$, and we begin by partitioning $[a, b]$ into n congruent subintervals with endpoints

$$x_k = a + kh \quad \text{for } k = 0, \ldots, n \quad \text{where } h = \frac{b - a}{n} \tag{2.32}$$

This time, however, we define y_k, for $k = 0, \ldots, n$, by

$$y_0 = \text{given value of } y \text{ at } x_0$$
$$z_k = y_{k-1} + hf(x_{k-1}, y_{k-1}) \tag{2.33}$$
$$y_k = y_{k-1} + \frac{h}{2} \left[f(x_{k-1}, y_{k-1}) + f(x_k, z_k) \right]$$

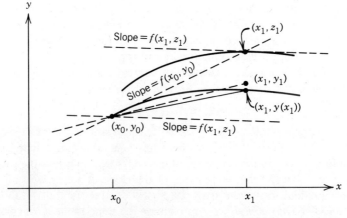

Figure 2.18 The improved Euler method

Figure 2.18 can help us see why the improved Euler method produces more accurate approximations than the original Euler method. The first of our two approximations to the slope m_1 of the secant line joining (x_0, y_0) and $(x_1, y(x_1))$ is $f(x_0, y_0)$. That is, z_1 is the Euler method approximation to $y(x_1)$, since $f(x_0, y_0)$ is the slope used in the Euler method. But here we use z_1 only as a means to producing a second approximation to m_1: the slope $f(x_1, z_1)$. Since it is often the case that $f(x_0, y_0)$ and $f(x_1, z_1)$ cause errors of opposite sign in approximating m_1 (see Figure 2.18), their average might well cause less error than either one of them alone.

Algorithm 2.10   *The improved Euler method*

    1. Input a, b, y_0, n
    2. $h \Leftarrow (b - a)/n$
    3. $y \Leftarrow y_0$
    4. $x \Leftarrow a$
    5. Repeat steps 6–10 *with* $k \Leftarrow 1, \ldots, n$
        6. $oldf \Leftarrow f(x, y)$
        7. $z \Leftarrow y + h\,oldf$
        8. $x \Leftarrow a + kh$
        9. $y \Leftarrow y + h[oldf + f(x, z)]/2$
        10. Print k, x, y
    11. End

EXAMPLE 2.9   Use the improved Euler method to approximate the solution of the initial-value problem

$$y' = 1 - 2xy, \qquad y(0) = 0$$

on the interval $[0, 1]$ partitioned into 10 congruent subintervals.

    Thus, $y_0 = 0$, $h = (1 - 0)/10 = 0.1$, and, for $k = 1, \ldots, n$,

$$x_k = (0.1)k$$

$$z_k = y_{k-1} + (0.1)(1 - 2x_{k-1}y_{k-1})$$

$$y_k = y_{k-1} + \frac{(0.1)}{2}(1 - 2x_{k-1}y_{k-1} + 1 - 2x_k z_k)$$

$$= y_{k-1} + (0.1)(1 - x_{k-1}y_{k-1} - x_k z_k)$$

The resulting values of y_k are given in the third column of Table 2.13. The second column contains the Euler method approximations reported previously in Table 2.12, and the fourth column contains approximations to be computed in the next example.

The most widely used of the class of related methods we are discussing is the so-called *classical Runge–Kutta method*. In this method, we approximate the slope of the secant line by a weighted average of four values of f. The numbers x_k are defined as

Table 2.13 Approximate Solutions of $y' = 1 - 2xy$, $y(0) = 0$

x_k	y_k (Euler)	y_k (Improved Euler)	y_k (Runge–Kutta)	$y(x_k)$ (actual)
0.0	0.0000	0.0000	0.0000	0.0000
0.1	0.1000	0.0990	0.0993	0.0993
0.2	0.1980	0.1941	0.1948	0.1948
0.3	0.2901	0.2816	0.2826	0.2826
0.4	0.3727	0.3586	0.3599	0.3599
0.5	0.4429	0.4227	0.4244	0.4244
0.6	0.4984	0.4728	0.4748	0.4748
0.7	0.5387	0.5083	0.5105	0.5105
0.8	0.5633	0.5297	0.5321	0.5321
0.9	0.5732	0.5383	0.5407	0.5407
1.0	0.5700	0.5387	0.5381	0.5381

before, but the approximations y_k are defined, for $k = 0, \ldots, n$, by

$$y_0 = \text{given value of } y \text{ at } x_0$$

$$r_k = f(x_{k-1}, y_{k-1})$$

$$s_k = f(x_{k-1} + 0.5h, y_{k-1} + 0.5hr_k)$$

$$t_k = f(x_{k-1} + 0.5h, y_{k-1} + 0.5hs_k) \qquad (2.34)$$

$$u_k = f(x_{k-1} + h, y_{k-1} + ht_k)$$

$$y_k = y_{k-1} + \frac{h}{6}\left[r_k + 2s_k + 2t_k + u_k\right]$$

Figure 2.19 illustrates the classical Runge–Kutta method. The four dashed line segments represent tangent lines to various solution curves (each of which we presume

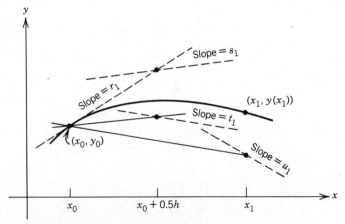

Figure 2.19   The classical Runge–Kutta method

is close to the actual solution curve, represented as a solid curve in the figure); these line segments have slopes r_1, s_1, t_1, u_1. The slope r_1 is the one used in the Euler method. The next slope, s_1, is a correction to r_1 derived in the same way $f(x_1, z_1)$ was in the improved Euler method, but where f is evaluated at a point whose x coordinate is midway between x_0 and x_1, rather than at x_1. Next, t_1 is a midpoint correction to s_1, just as s_1 was a midpoint correction to t_1. Finally, u_1 is a correction to t_1 but where f is evaluated at a point whose x coordinate is x_1. We assume that all four of these slopes are reasonable approximations to the slope m_1 of the secant line, but that some may overestimate m_1 while others underestimate it. Thus, the weighted average $[r_1 + 2s_1 + 2t_1 + u_1]/6$ may be a better estimate of m_1 than any of the four individually. Note that the weights (i.e., coefficients) $\frac{1}{6}, \frac{2}{6}, \frac{2}{6}, \frac{1}{6}$ add up to 1, as they must, and that the two midpoint slopes are given twice the weight of the other two slopes.

Algorithm 2.11 *The classical Runge–Kutta method*

    1. Input a, b, y_0, n
    2. $h \Leftarrow (b - a)/n$
    3. $y \Leftarrow y_0$
    4. $x \Leftarrow a$
    5. Repeat steps 6–12 **with** $k \Leftarrow 1, \ldots, n$
        6. $r \Leftarrow f(x, y)$
        7. $s \Leftarrow f(x + 0.5h, y + 0.5hr)$
        8. $t \Leftarrow f(x + 0.5h, y + 0.5hs)$
        9. $x \Leftarrow a + kh$
     10. $u \Leftarrow f(x, y + ht)$
     11. $y \Leftarrow y + h[r + 2s + 2t + u]/6$
     12. Print k, x, y
  13. End

EXAMPLE 2.10 Use the classical Runge–Kutta method to approximate the solution of the initial-value problem

$$y' = 1 - 2xy, \qquad y(0) = 0$$

on the interval $[0, 1]$ partitioned into 10 congruent subintervals.

Thus, $y_0 = 0$, $h = (1 - 0)/10 = 0.1$, and, for $k = 1, \ldots, n$,

$$x_k = (0.1)k$$
$$r_k = 1 - 2x_{k-1}y_{k-1}$$
$$s_k = 1 - 2(x_{k-1} + 0.05)(y_{k-1} + 0.05r_k)$$
$$t_k = 1 - 2(x_{k-1} + 0.05)(y_{k-1} + 0.05s_k)$$
$$u_k = 1 - 2x_k(y_{k-1} + 0.1t_k)$$
$$y_k = y_{k-1} + \frac{(0.1)}{6}(r_k + 2s_k + 2t_k + u_k)$$

The resulting values of y_k are given in the fourth column of Table 2.13.

Table 2.13 summarizes the results of Examples 2.8, 2.9, and 2.10. For the initial-value problem $y' = 1 - 2xy$, $y(0) = 0$, the Euler method produced the least accurate results, the improved Euler method produced more accurate results, and the classical Runge–Kutta method produced such excellent results that the approximations equal the exact values to at least four places. The more accurate results produced by the latter two methods did, however, take somewhat longer to compute even though we used the same value of h for all three. Note that the improved Euler method requires that we evaluate f twice for each computation of an approximation y_k, and the classical Runge–Kutta method requires that we evaluate f four times for each y_k. If f is a function that is time-consuming to evaluate, these methods might be undesirable. In Exercise 1, we outline a method that converges about as rapidly as the improved Euler method but requires only one evaluation of f for each y_k.

Our claim that the improved Euler and classical Runge–Kutta methods converge more rapidly than the Euler method can be supported by stronger evidence than numerical examples. It can also be supported by theorems and error bounds analogous to Theorem 2.3 and error bound (2.29). Specifically, the Euler, improved Euler, and classical Runge–Kutta methods satisfy error bounds of the following forms, respectively [see Birkhoff and Rota (1978)]:

$$|y_k - y(x_k)| \le \frac{hM_1}{L}\left(e^{L(x_k-a)} - 1\right)$$

$$|y_k - y(x_k)| \le \frac{h^2M_2}{L}\left(e^{L(x_k-a)} - 1\right)$$

$$|y_k - y(x_k)| \le \frac{h^4M_3}{L}\left(e^{L(x_k-a)} - 1\right)$$

Thus, we say that the three methods are, respectively, methods of order 1, 2, and 4, where these numbers refer to the exponents of h. Note that h^2 converges to zero more rapidly than h, and h^4 converges to zero more rapidly yet.

EXERCISES

1. The midpoint method for approximating the solution of the initial-value problem (2.31) proceeds as follows. As usual, n denotes a positive integer, and h and k are defined as in (2.32). Suppose y_{k-1} and y_{k-2}, the approximations to $y(x_{k-1})$ and $y(x_{k-2})$, have been found. Define y_k so that (x_k, y_k) is on the line through (x_{k-2}, y_{k-2}) with slope $f(x_{k-1}, y_{k-1})$ (see Figure 2.20). This defines y_k for $k \ge 2$; define y_1 by the improved Euler method. Your problem is to derive formulas analogous to (2.33) for the midpoint method approximations $y_1, \ldots, y_n$.

In Exercises 2–7, use the improved Euler method to approximate the solution of the given initial-value problem on the given interval. (A calculator will be useful.) Choose n so that $h = 0.1$. Compare your approximate y values with the exact y values.

2. $y' = 3x^2$, $y(0) = 0$ on $[0, 0.5]$
3. $y' = x + y$, $y(0) = 0$ on $[0, 0.5]$
4. $y' = y^2 + 4$, $y(0) = 0$ on $[0, 0.5]$

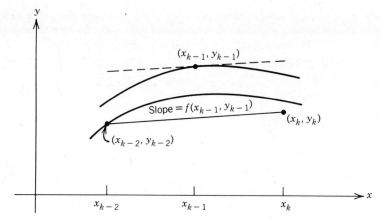

Figure 2.20   The midpoint method

5. $y' = -y/3x,$ $y(1) = 2$ on $[1, 1.5]$
6. $y' = y - 2x/y,$ $y(0) = 1$ on $[0, 0.5]$
7. $y' = 2xy/(x^2 - y^2),$ $y(1) = 2$ on $[1, 1.5]$

In Exercises 8 and 9, use the Runge–Kutta method to approximate the solution of the given initial-value problem on the given interval. (A calculator will be useful.) Choose n so that $h = 0.1$. Compare your approximate y values with the exact y values.

8. $y' = 3x^2,$ $y(0) = 0$ on $[0, 0.2]$
9. $y' = x + y,$ $y(0) = 0$ on $[0, 0.2]$

For Exercises 10–17, first write a computer program that implements Algorithm 2.10 for an arbitrary initial-value problem. When you use the program to approximate a solution, apply the error estimates (2.30) with an error tolerance of 0.0001; that is, double n repeatedly until all the corresponding approximations of y for successive values of n differ by no more than 0.0001. Print out only about 10 values of y_k, as suggested in the programming note at the end of Section 2.7. For each assigned function, describe your conclusion in words, following the recommendations at the end of Section 2.1.

10. $y' = 3x^2,$ $y(0) = 0$ on $[0, 1]$
11. $y' = x + y,$ $y(0) = 0$ on $[0, 1]$
12. $y' = y^2 + 4,$ $y(0) = 0$ on $[0, 0.5]$
13. $y' = x^2 - y^2,$ $y(1) = 0$ on $[1, 2]$
14. $y' = y - 2x/y,$ $y(0) = 1$ on $[0, 1.5]$
15. $y' = 3yx^2 \cos(x^3),$ $y(0) = 1$ on $[0, 3]$
16. $y' = 2xy/(x^2 - y^2),$ $y(1) = 2$ on $[1, 1.25]$

(You might have to settle for less accuracy at $x = 1.25$.)

17. $y' = x^2 + y^2,$ $y(0) = 1$ on $[0, 0.9]$

For Exercises 18–24, first write a computer program that implements Algorithm 2.11 for an arbitrary initial-value problem. When you use the program to approximate a solution, apply the error estimates (2.30) with an error tolerance of 0.0001; that is, double n repeatedly until all the corresponding approximations of y for successive values of n differ by no more than 0.0001. Print out only about 10 values of y_k, as suggested in the programming note at the end of Section 2.7. For each assigned function, describe your conclusion in words, following the recommendations at the end of Section 2.1.

18. $y' = x + y$, $y(0) = 0$ on $[0, 1]$

19. $y' = y^2 + 4$, $y(0) = 0$ on $[0, 0.5]$

20. $y' = x^2 - y^2$, $y(1) = 0$ on $[1, 2]$

21. $y' = y - 2x/y$, $y(0) = 1$ on $[0, 1.5]$

22. $y' = 3yx^2 \cos(x^3)$, $y(0) = 1$ on $[0, 3]$

23. $y' = 2xy/(x^2 - y^2)$, $y(1) = 2$ on $[1, 1.25]$
     (You might have to settle for less accuracy at $x = 1.25$.)

24. $y' = x^2 + y^2$, $y(0) = 1$ on $[0, 0.9]$

25. Under certain favorable conditions, the growth of a fish population is described fairly accurately by the differential equation

$$\frac{dy}{dt} = k_1 y^{3/2} - k_2 y$$

where t is time in months, y is the total weight of the fish population, and k_1, k_2 are positive constants. Assume that, for a particular fish population, $k_1 = 0.11$ and $k_2 = 1.00$. Make a table of values of the weight as a function of time t, where $0 \le t \le 4.5$ and where the initial weight is 100 pounds. What trends do you see? (The given differential equation can be solved by hand, but not easily; instead, use a computer program that you have written.)

26. Epidemics of many communicable diseases are described fairly accurately by the differential equation

$$I'(t) = kI(t)S(t)$$

where t is time in days, I is the number of people who have been infected by the disease, and S is the number who are are still susceptible. We make the following additional assumptions: Once an epidemic breaks out, the medical community begins to innoculate the population at a constant rate r; any person who becomes infected is never susceptible again; and anyone who is not infected or innoculated is susceptible. Then

$$S(t) = N - rt - I(t)$$

where N is the total population under consideration. Assume that for a particular epidemic, $k = 10^{-5}$, $r = 3500$, $N = 100,000$, and $I(0) = 1$, which means that the epidemic is triggered by a single infected individual. Make a table of the number of infected people as a function of time, and locate the approximate time it takes for the epidemic to halt, which is when I reaches its maximum value and S becomes zero. (The given differential equation can be solved by hand, but not easily; instead, use a computer program that you have written.)

CHAPTER 3

Series

We often find it useful to represent a function f as a sum of an infinite series of simpler functions:

$$f(x) = f_1(x) + f_2(x) + f_3(x) + \cdots + f_n(x) + \cdots$$

For example, if we can write f as a sum of infinitely many monomials, as in

$$f(x) = c_0 + c_1 x + c_2 x^2 + \cdots + c_n x^n + \cdots$$

then f can be thought of as a "polynomial of infinite degree" and so, like a polynomial, should be easy to differentiate and integrate. Such a representation can give us new insights into familiar functions or can provide a means for defining new functions. In Chapter 4 we use this kind of representation to solve differential equations.

In Chapter 5, on the other hand, we represent functions in the form

$$f(x) = a_0 + a_1 \cos x + a_2 \cos 2x + \cdots + a_n \cos(nx) + \cdots$$
$$+ b_1 \sin x + b_2 \sin 2x + \cdots + b_n \sin(nx) + \cdots$$

This kind of representation is especially helpful in studying periodic behavior, such as rotation and vibration.

We can also use infinite series to approximate values of $f(x)$; in practice, the sum of just the first few terms in an infinite sum is often quite close to the entire sum, $f(x)$. In Sections 3.6 and 3.7, we explore connections between the numerical approximations in Chapter 2 and infinite series representations.

These are only some of the uses of infinite series of functions or numbers; others appear in the exercises. First, however, we discuss what is meant by the sum of an infinite series.

3.1 The Sum of a Series

Let us try to calculate the specific sum

$$1 + \frac{1}{2} + \frac{1}{4} + \cdots + \frac{1}{2^n} + \cdots$$

We might look for a pattern by adding together the first two terms of the sum, then the first three terms, then the first four, and so on. A geometric argument will help us see a pattern in this example. Suppose we are running a 2-mile race and we keep track of how much distance we have left to run as follows:

When we have run half the distance, 1 mile is left;
when we have run half the remaining mile, $\frac{1}{2}$ mile is left;
when we have run half the remaining $\frac{1}{2}$ mile, $\frac{1}{4}$ mile is left;

and so on. This suggests the following pattern for the total distance we have run:

$$1 = 2 - 1$$
$$1 + \frac{1}{2} = 2 - \frac{1}{2}$$
$$1 + \frac{1}{2} + \frac{1}{4} = 2 - \frac{1}{4}$$

and so on. More generally,

$$1 + \frac{1}{2} + \frac{1}{4} + \cdots + \frac{1}{2^n} = 2 - \frac{1}{2^n}$$

which converges to 2 as $n \to \infty$. Therefore, we assert that

$$1 + \frac{1}{2} + \frac{1}{4} + \cdots + \frac{1}{2^n} + \cdots = 2$$

by which we simply mean that the sum of the first n terms gets arbitrarily close to 2 for n sufficiently large. We imagine, however, that we have subdivided the distance 2 into infinitely many pieces whose lengths add up to 2 (see Figure 3.1). We now make the above technique into a formal definition.

Definition 3.1   To any sequence of numbers $\{a_n\}$ we associate another sequence $\{s_n\}$ by the formula

$$s_n = \sum_{k=1}^{n} a_k = a_1 + a_2 + \cdots + a_n$$

Our notation for the sequence $\{s_n\}$ will be $a_1 + a_2 + \cdots + a_k + \cdots$ or, more compactly,

$$\sum_{k=1}^{\infty} a_k \qquad (3.1)$$

or even

$$\sum a_k$$

We refer to the sequence (3.1) as an *infinite series* (or just *series*) and to the number s_n as the *nth partial sum* of the series. If the sequence of partial sums $\{s_n\}$ converges to a limit s, we say that the series (3.1) *converges to the sum s*, and we write

$$\sum_{k=1}^{\infty} a_k = s$$

If the sequence $\{s_n\}$ diverges, we say that the series *diverges*.

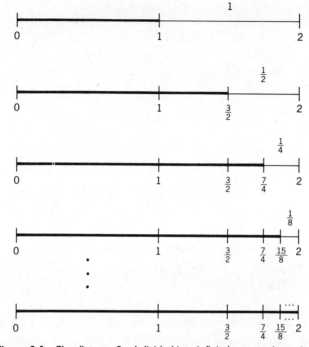

Figure 3.1 The distance 2 subdivided into infinitely many pieces.

The initial value of the index k need not be 1, as in

$$\sum_{k=4}^{\infty} a_k \qquad \sum_{k=-7}^{\infty} a_k \qquad \sum_{k=0}^{\infty} a_k$$

EXAMPLE 3.1   Find the sum of the series $\sum_{k=1}^{\infty} 1/k(k+1)$.

The nth partial sum of this series is

$$s_n = \frac{1}{1(2)} + \frac{1}{2(3)} + \cdots + \frac{1}{n(n+1)}$$

$$= \left(\frac{1}{1} - \frac{1}{2}\right) + \left(\frac{1}{2} - \frac{1}{3}\right) + \cdots + \left(\frac{1}{n} - \frac{1}{n+1}\right)$$

$$= 1 - \frac{1}{n+1}$$

Therefore, $s_n \to 1$, which means that $\sum_{k=1}^{\infty} 1/k(k+1) = 1$.

Any series of the form $\Sigma(b_k - b_{k+1})$ is called a *telescoping series*. We can always use the trick in Example 3.1 to find the sum of such a series, provided the sequence $\{b_k\}$ converges (Exercise 17). In practice, however, very few series are telescoping.

Most of our theorems in Chapter 1 on sequences have immediate consequences for series. First we apply Theorems 1.4b–d.

Theorem 3.1   Suppose $\sum_{k=1}^{\infty} a_k = s$ and $\sum_{k=1}^{\infty} b_k = t$. Then

(a) $\displaystyle\sum_{k=1}^{\infty} (a_k + b_k) = s + t$

(b) $\displaystyle\sum_{k=1}^{\infty} (a_k - b_k) = s - t$

(c) $\displaystyle\sum_{k=1}^{\infty} ca_k = cs$ for every number c

Proof

(a) The nth partial sum of $\sum_{k=1}^{\infty}(a_k + b_k)$ is

$$s_n = \sum_{k=1}^{n} (a_k + b_k) = \sum_{k=1}^{n} a_k + \sum_{k=1}^{n} b_k$$

But since $\sum_{k=1}^{n} a_k \to s$ and $\sum_{k=1}^{n} b_k \to t$, it follows that $s_n \to s + t$.

(b) This theorem is proved similarly.

(c) The nth partial sum of $\sum_{k=1}^{\infty} ca_k$ is $s_n = \sum_{k=1}^{n} ca_k = c\sum_{k=1}^{n} a_k$. But since $\sum_{k=1}^{n} a_k \to s$, it follows that $s_n \to cs$.

Theorem 3.2   If the series $\sum a_k$ converges, then the sequence $\{a_k\}$ has limit zero. (Thus, if the sequence $\{a_k\}$ does not have limit zero, the series $\sum a_k$ diverges.)

Proof $s_n = a_1 + a_2 + \cdots + a_{n-1} + a_n = s_{n-1} + a_n$. So $a_n = s_n - s_{n-1} \to s - s = 0$, where s denotes the sum of the series $\sum a_k$.

EXAMPLE 3.2   The series $\sum k/(k + 1)$ diverges since the sequence $\{k/(k + 1)\}$ has limit 1, not 0.

Theorem 3.3   The *geometric series*, $\sum_{k=0}^{\infty} x^k$, converges iff $|x| < 1$; Furthermore,

$$\sum_{k=0}^{\infty} x^k = \frac{1}{1 - x} \quad \text{if } |x| < 1 \tag{3.2}$$

Proof   If $|x| \geq 1$, then the sequence $\{x^k\}$ does not have limit zero; so, by Theorem 3.2, the geometric series diverges. Now assume that $|x| < 1$. Then the nth partial sum of the series is

$$s_n = 1 + x + x^2 + \cdots + x^n$$

and so

$$(1 - x)s_n = (1 + x + x^2 + \cdots + x^n) - (x + x^2 + \cdots + x^n + x^{n+1})$$

$$= 1 - x^{n+1}$$

$$s_n = \frac{1 - x^{n+1}}{1 - x} \to \frac{1}{1 - x}$$

EXAMPLE 3.3 Find the sum of the series $\sum_{k=0}^{\infty}(3/2^k - 5(2/3)^k)$. We can find the sum of the two geometric series

$$\sum_{k=0}^{\infty} \frac{1}{2^k} = \sum_{k=0}^{\infty} \left(\frac{1}{2}\right)^k = \frac{1}{1-1/2} = 2$$

$$\sum_{k=0}^{\infty} \left(\frac{2}{3}\right)^k = \frac{1}{1-2/3} = 3$$

and so, by Theorem 3.1,

$$\sum_{k=0}^{\infty} \left(\frac{3}{2^k} - 5\left(\frac{2}{3}\right)^k\right) = 3(2) - 5(3) = -9$$

EXAMPLE 3.4 An infinite decimal, $0.a_1a_2 \ldots a_k \ldots$, where each a_k is a digit, can be thought of as an infinite series:

$$\frac{a_1}{10^1} + \frac{a_2}{10^2} + \cdots + \frac{a_k}{10^k} + \cdots$$

Express the repeating decimal $6.\overline{37} = 6.373737\ldots$ as a rational number.

$$6.\overline{37} = 6 + \frac{37}{100^1} + \frac{37}{100^2} + \cdots$$

$$= 6 + \sum_{k=0}^{\infty} \frac{37}{100^k} - 37 \quad \sim \frac{37}{100}^0$$

$$= -31 + 37\left(\frac{1}{1-\frac{1}{100}}\right) = -31 + \frac{3700}{99}$$

$$= \frac{3700 - 3069}{99} = \frac{631}{99}$$

In property (1.7), we saw that changing the starting index of a sequence affects neither its convergence nor its limit. The next theorem, however, says that changing the starting index of a series does affect its sum. This is not surprising, since the sum of all the terms should be the sum of the first few plus the sum of all that remain.

Theorem 3.4 For every positive integer n, $\sum_{k=1}^{\infty} a_k$ converges iff $\sum_{k=n+1}^{\infty} a_k$ converges. And, if these series converge, then

$$\sum_{k=1}^{\infty} a_k = \sum_{k=1}^{n} a_k + \sum_{k=n+1}^{\infty} a_k \qquad (3.3)$$

Proof   Exercise 21.

Unfortunately, it is often quite difficult or even impossible to calculate the sum of an infinite series exactly. As the next example shows, it can even be difficult to guess whether or not a series converges.

Table 3.1 Partial Sums of $\sum_{k=1}^{\infty} 1/k^2$ and $\sum_{k=1}^{\infty} 1/k$
(Denoted s_n and t_n, Respectively)

n	s_n	t_n
1	1	1
2	1.25	1.5
10	1.54977	2.92897
100	1.63498	5.18738
1000	1.64393	7.48547
10000	1.64483	9.78761

EXAMPLE 3.5   Decide whether the series $\sum_{k=1}^{\infty} 1/k^2$ and $\sum_{k=1}^{\infty} 1/k$ converge; if they do, guess at their sums. As evidence, use Table 3.1, which shows the values of their partial sums (found with the aid of a computer).

The evidence in Table 3.1 is not completely compelling, but it appears that the sum of $\sum_{k=1}^{\infty} 1/k^2$ is between 1.64 and 1.65 whereas $\sum_{k=1}^{\infty} 1/k$ appears to have no finite sum since its partial sums increase by about 2.3 whenever n increases by a factor of 10. In fact, later we will see that $\sum_{k=1}^{\infty} 1/k^2 = \pi^2/6$ and that the partial sums $\sum_{k=1}^{n} 1/k$ grow at about the same rate as $\log n$.

Definition 3.1 and Theorems 3.1–3.4 carry over without change to infinite series of complex numbers, provided the absolute value signs in Theorem 3.3 are interpreted as indicating the absolute value of a complex number.

EXERCISES

For each of the infinite series in Exercises 1–10, find the sum if the series converges.

1. $\displaystyle\sum_{k=1}^{\infty} (0.3)^k$

2. $\displaystyle\sum_{k=0}^{\infty} 5(-3/4)^k$

3. $\displaystyle\sum_{k=0}^{\infty} (2(0.1)^k - 7(-0.2)^k)$

4. $\displaystyle\sum_{k=0}^{\infty} 2^k/(2^k + k)$

5. $\displaystyle\sum_{k=1}^{\infty} 1/(2k - 1)(2k + 1)$

6. $\displaystyle\sum_{k=1}^{\infty} (2k + 1)/k^2(k + 1)^2$

7. $\displaystyle\sum_{k=2}^{\infty} \log(1 - 1/k)$

8. $\displaystyle\sum_{k=0}^{\infty} (\sin x)^k$ (For which values of x does this series converge?)

9. $\displaystyle\sum_{k=0}^{\infty} (\log x)^k$ (For which values of x does this series converge?)

10. $\displaystyle\sum_{k=0}^{\infty} 1/(1 - x)^k$ (For which values of x does this series converge?)

In Exercises 11–13, use Theorem 3.3 to express the infinite repeating decimal as a rational number.

11. $0.\overline{3}$ **12.** $5.\overline{26}$ **13.** $4.\overline{361}$

14.   A rubber ball rebounds to two-thirds the height from which it falls. If it is dropped from a height of 4 ft and is allowed to continue bouncing indefinitely, what is the total distance it travels?

15.   Two cyclists, 60 miles apart, approach each other, each pedaling at 10 miles/hr. A fly starts at one cyclist and flies back and forth between the cyclists at 20 miles/hr. When the cyclists come together, how far has the fly flown? (Solve the problem two ways: one using geometric series and one much easier way.)

16.   Achilles runs a 2-mile race with a tortoise who has a 1-mile head start. Even though Achilles runs 10 times as fast as the tortoise, he seems unable to catch up with it for the following reason. When Achilles gets to the spot where the tortoise started, the tortoise is one-tenth of a mile further on; when Achilles gets to the one-and-one-tenth-mile point, the tortoise is further on yet; and so on. Use geometric series to explain this seeming paradox.

17.   If the sequence $\{b_k\}$ converges, prove that the telescoping series $\sum_{k=1}^{\infty}(b_k - b_{k+1})$ converges.

18.   If $\sum_{k=1}^{\infty} a_k = s$ and $\sum_{k=1}^{\infty} b_k = t$, prove that $\sum_{k=1}^{\infty}(a_k - b_k) = s - t$.

19.   Show that every sequence is a sequence of partial sums.

20.   Use the techniques of Section 1.3 to find a compact formula for the nth partial sum of the geometric series.

21.   Prove Theorem 3.4.

22.   A set A of points on the real line is said to have "length" equal to zero if for every $\varepsilon > 0$, there exists a sequence of bounded intervals $[a_1, b_1], [a_2, b_2], \ldots, [a_k, b_k], \ldots$ whose union contains A and such that $\sum_{k=1}^{\infty}(b_k - a_k) < \varepsilon$. Prove that the set of terms of any sequence of real numbers has length equal to zero, in this sense.

For Exercises 23–26, first write a computer program that computes partial sums $\sum_{k=p}^{n} a_k$ of an infinite series $\sum_{k=p}^{\infty} a_k$. For each of the assigned infinite series, use your program to help you guess whether the series converges and, if so, what the sum is. *Programming hints:* Input p and n, and define the kth term a_k as a function $a(k)$ so that your program is quite general. Add the smallest terms first and largest terms last to minimize round-off error. If the terms are alternately positive and negative, apply your program to the series in which two of these terms are added at a time, and simplify the resulting formula for the terms so that cancellation errors are avoided.

23. $\displaystyle\sum_{k=2}^{\infty} 1/\log k$ **24.** $\displaystyle\sum_{k=1}^{\infty} 1/k^3$

25. $\displaystyle\sum_{k=1}^{\infty} 1/k^{1.5}$ **26.** $\displaystyle\sum_{k=1}^{\infty} (-1)^{k+1}/k$

3.2 Taylor Series

Given a function f, we would like to represent f as a "polynomial of infinite degree":

$$f(x) = c_0 + c_1 x + c_2 x^2 + \cdots + c_n x^n + \cdots$$

or, more generally,

$$f(x) = c_0 + c_1(x - x_0) + \cdots + c_n(x - x_0)^n + \cdots \tag{3.4}$$

Series of this form are called *power series*. They are not polynomials (except those for which all but a finite number of the coefficients are zero), but they do have many properties analogous to those possessed by polynomials. In particular, it is easy to differentiate, integrate, add, and multiply constants by power series, and these operations are performed in the same way as they are on polynomials. These four operations are especially useful since they are exactly the ones needed to solve linear differential equations, which is one of the applications we are aiming toward. However, we do not take up the general theory of power series until Section 3.9.

For now we learn only how to represent a function as a sum of a power series. That is, if f is given, how might we choose the coefficients c_k to make (3.4) true? The theory of Taylor polynomials in Section 1.4 suggests how: Treat the Taylor polynomials P_n as the nth partial sums of a series.

Definition 3.2   If f has derivatives of all orders at x_0, then the series

$$\sum_{k=0}^{\infty} \frac{f^{(k)}(x_0)}{k!}(x - x_0)^k = f(x_0) + f'(x_0)(x - x_0) + \frac{f''(x_0)}{2}(x - x_0)^2$$

$$+ \cdots + \frac{f^{(k)}(x_0)}{k!}(x - x_0)^k + \cdots$$

is called the *Taylor series of f at x_0*.

EXAMPLE 3.6   Find the Taylor series of $f(x) = e^x$ at 0.

$$f^{(k)}(x) = e^x$$

and so

$$f^{(k)}(0) = 1 \quad \text{for all } k \geq 0$$

Therefore, by Definition 3.2, the Taylor series of e^x at 0 is

$$\sum_{k=0}^{\infty} \frac{x^k}{k!} = 1 + x + \frac{x^2}{2} + \cdots + \frac{x^n}{n!} + \cdots$$

Furthermore, by Taylor's remainder formula in Section 1.4, for each n there exists z between 0 and x such that

$$e^x - \sum_{k=0}^{n} \frac{x^k}{k!} = \frac{e^z x^{n+1}}{(n + 1)!}$$

Since $x^{n+1}/(n + 1)! \to 0$, the Taylor polynomials of e^x at 0 converge to e^x; in other words, $e^x = \sum_{k=0}^{\infty} x^k/k!$, for every value of x.

The Taylor series we most often use are the Taylor series at 0, as in Example 3.6. Such a Taylor series is called a *Maclaurin series*.

EXAMPLE 3.7   Find the Maclaurin series of $f(x) = \sin x$.

$$f(x) = \sin x \qquad f(0) = 0$$

$$f'(x) = \cos x \qquad f'(0) = 1$$

$$f^{(2)}(x) = -\sin x \qquad f^{(2)}(0) = 0$$

$$f^{(3)}(x) = -\cos x \qquad f^{(3)}(0) = -1$$

$$f^{(4)}(x) = \sin x \qquad f^{(4)}(0) = 0$$

$$\vdots \qquad\qquad \vdots \quad \vdots$$

Therefore

$$\sum_{k=0}^{\infty} \frac{f^{(k)}(0)}{k!} x^k = f(0) + f'(0)x + \frac{f^{(2)}(0)}{2!} x^2 + \frac{f^{(3)}(0)}{3!} x^3 + \cdots$$

$$= x - \frac{x^3}{3!} + \frac{x^5}{5!} - \frac{x^7}{7!} + \cdots$$

$$= \sum_{k=0}^{\infty} \frac{(-1)^k x^{2k+1}}{(2k+1)!}$$

Again, Taylor's remainder formula can be used to show that this series does converge to $\sin x$ for every value of x. However, we cannot take it for granted that every function is the sum of its Taylor series. For example, if f is defined by $f(x) = \exp(-1/x^2)$ when $x \neq 0$ and $f(0) = 0$, it can be shown (Exercise 23) that f has derivatives of all orders at $x = 0$ and that $f^{(k)}(0) = 0$ for all $k \geq 0$. But then the Maclaurin series of f has the sum 0 for all x, and so $f(x)$ is not the sum of its Maclaurin series when $x \neq 0$. Despite this example, the elementary functions of calculus are, as we will see, the sums of their Taylor series. In Table 3.2, we list some of these Taylor series for future reference.

The last of the series in Table 3.2 (in which p is any real number) is called a *binomial series* because of its close resemblance to the finite sum in the binomial theorem of Appendix B (in which p is a nonnegative integer). Its interval of convergence is $[-1, 1]$ if $-1 < p < 0$, $[-1, 1]$ if $p > 0$ but is not an integer, and $(-\infty, \infty)$ if p is a nonnegative integer. [For a proof see Knopp (1956).]

The facts displayed in Table 3.2 are not easy to establish. The first three series, as we have noted, can be shown to converge to the given function on the indicated domain by using Taylor's remainder formula. That formula, however, is not as effective when applied to the last four Taylor series in Table 3.2; knowing more about the theory of power series would be useful. For now, we will simply assume that we may perform the usual operations of calculus and algebra on power series just as we would on polynomials, and we will not worry too much about whether the methods are valid. In Section 3.9, we will verify both our methods and these particular results.

Table 3.2 Some Taylor Series

$f(x)$	Maclaurin series	Sum is $f(x)$ on
e^x	$\sum_{k=0}^{\infty} x^k/k!$	$(-\infty, \infty)$
$\cos x$	$\sum_{k=0}^{\infty} (-1)^k x^{2k}/(2k)!$	$(-\infty, \infty)$
$\sin x$	$\sum_{k=0}^{\infty} (-1)^k x^{2k+1}/(2k+1)!$	$(-\infty, \infty)$
$1/(1-x)$	$\sum_{k=0}^{\infty} x^k$	$(-1, 1)$
$\log(1+x)$	$\sum_{k=1}^{\infty} (-1)^{k+1} x^k/k$	$(-1, 1)$
$\arctan x$	$\sum_{k=0}^{\infty} (-1)^k x^{2k+1}/(2k+1)$	$[-1, 1]$
$(1+x)^p$	$1 + \sum_{k=1}^{\infty} p(p-1)\cdots(p-k+1)x^k/k!$	$(-1, 1)$ (at least)

binomial series

EXAMPLE 3.8 Derive the sixth line of Table 3.2 from the fourth line; use the operations of calculus and algebra freely.

$$\frac{1}{1-x} = \sum_{k=0}^{\infty} x^k$$

Replace x by $-x^2$:

$$\frac{1}{1+x^2} = \sum_{k=0}^{\infty} (-x^2)^k = \sum_{k=0}^{\infty} (-1)^k x^{2k}$$

Antidifferentiate:

$$\arctan x = \sum_{k=0}^{\infty} \frac{(-1)^k x^{2k+1}}{2k+1} + c$$

The term c must equal zero, since the above equation becomes $0 = 0 + c$ when we replace x by zero.

We know that the first line of this derivation is valid since $\sum_{k=0}^{\infty} x^k$ is the geometric series and its sum is $1/(1-x)$ when $|x| < 1$ (Theorem 3.3). Furthermore, the substitution of $-x^2$ for x in the second line is certainly valid and the set of values of x for which that equation holds is still $(-1, 1)$ since $|-x^2| < 1$ iff $|x| < 1$. However, it does not follow that $\sum_{k=0}^{\infty}(-1)^k x^{2k}$ has to be the Taylor series of $1/(1+x^2)$ at 0, although it is implied by the general theory in Section 3.9. We also do not yet know that the third line of the derivation is valid; conceivably, the integral of the *infinite* sum of functions $(-1)^k x^{2k}$ might not be the sum of the integrals of these functions. In fact, examples we discussed in Section 1.6 suggest that this could be a significant difficulty. Furthermore, it is not yet clear why the domain of the Taylor series of arctan x should include the extra points $x = \pm 1$.

We may also use Taylor series to approximate functions. In fact, we did just this in Section 1.4; in our new vocabulary we could describe that method as follows: Approximate f by the nth partial sum of the Taylor series of f at x_0, and find a bound for the error of that approximation by using Taylor's remainder formula. But this is not the only way to estimate the error in a series approximation; other methods will be taken up in Sections 3.6 and 3.7.

EXAMPLE 3.9 If z is a complex variable, e^z is defined (by analogy with the result in Example 3.6) as follows:

$$e^z = \sum_{k=0}^{\infty} \frac{z^k}{k!}$$

Show that $e^{it} = \cos t + i \sin t$ for all real numbers t, where $i^2 = -1$.

$$e^{it} = \sum_{k=0}^{\infty} \frac{(it)^k}{k!} = \sum_{k=0}^{\infty} \frac{i^k t^k}{k!}$$

$$= 1 + it - \frac{t^2}{2!} - \frac{it^3}{3!} + \frac{t^4}{4!} + \frac{it^5}{5!} - \frac{t^6}{6!} - \frac{it^7}{7!} + \cdots$$

$$= \left(1 - \frac{t^2}{2!} + \frac{t^4}{4!} - \frac{t^6}{6!} + \cdots\right) + i\left(t - \frac{t^3}{3!} + \frac{t^5}{5!} - \frac{t^7}{7!} + \cdots\right)$$

$$= \cos t + i \sin t \quad \text{(by second and third entries in Table 3.2)} .$$

EXERCISES

In Exercises 1–6, find, directly from Definition 3.2, the Taylor series of the given function at x_0.

1. $\cos x$, $x_0 = 0$

2. $\sqrt{x}$, $x_0 = 1$

3. $\log x$, $x_0 = 1$

4. $1/x$, $x_0 = 1$

5. $\sin x$, $x_0 = \pi/2$

6. $(1 + x)^p$, $x_0 = 0$

In Exercises 7–13, derive the Maclaurin series of the function from one of the series in Table 3.2; use operations of calculus and algebra freely.

7. $\log(1 - x)$

8. $\sin(3x)$

9. $\exp(x^2)$

10. $\cos(\sqrt{x})$ (where $x \geq 0$)

11. $1/\sqrt{1 - x^2}$

12. $\arcsin x$

13. $\int_0^x \exp(-t^2)\, dt$

14. $\int_0^x \cos\sqrt{t}\, dt$ (where $x \geq 0$)

In Exercises 15–21, find the sum of the series by performing operations of algebra or calculus on one of the series in Table 3.2.

15. $x + x^2 + x^3/2! + x^4/3! + \cdots$

16. $1 - x + x^2/2! - x^3/3! + \cdots$

17. $1 + x/2! + x^2/3! + x^3/4! + \cdots$

18. $1 + x/2 + x^2/2^2 2! + x^3/2^3 3! + x^4/2^4 4! + \cdots$

19. $1 + x^2/2! + x^4/4! + x^6/6! + \cdots$

20. $1 + \sum_{k=1}^{\infty} (-1/2)(-3/2)(-5/2) \cdots (1/2 - k) x^k / 2^k k!$

21. $\sum_{k=0}^{\infty} (-1)^k (2k + 1) x^{2k} / (2k)!$

22. If p is a positive integer, prove that the binomial series for $(1 + x)^p$ has exactly the same nonzero terms as the finite sum in the binomial theorem (Appendix B) applied to $(1 + x)^p$.

23. If $f(x) = \exp(-1/x^2)$ when $x \neq 0$ and if $f(0) = 0$, show that $f^{(k)}(0) = 0$ for all $k \geq 0$. *Hint:* Use induction to show that, for each $k \geq 0$, there exists a polynomial q_k such that $f^{(k)}(x) = \exp(-1/x^2) q_k(x^{-1})$ for all $x \neq 0$.

24. **(a)** From Table 3.2, deduce that $\sum_{k=1}^{\infty} (-1)^{k+1}/k = \log 2$. **(b)** Use the computer program you wrote for Section 3.1 to approximate $\log 2$ by computing the nth partial sum of this series for a large value of n.

In Exercises 25–27, find the sum of the given series by performing operations of algebra or calculus on one of the series in Table 3.2 and then replacing x by an appropriate constant.

25. $\sum_{k=1}^{\infty} k/2^k$ 26. $\sum_{k=1}^{\infty} 1/k2^k$ 27. $\sum_{k=0}^{\infty} 2^k/(k + 1)!$

28. The total relativistic energy of a particle is

$$E = mc^2 \left(1 - \frac{v^2}{c^2} \right)^{-1/2}$$

where m = mass, v = velocity, c = speed of light. Expand this expression for E as a Maclaurin series in the variable v^2/c^2 and show that the second term in the expansion is the classical formula for kinetic energy, $mv^2/2$. *Hint:* The expansion should have the form of a binomial series times the constant mc^2.

29. The *Poisson distribution* is the sequence of probabilities

$$P_k(x) = \frac{x^k e^{-x}}{k!} \quad \text{where } x > 0 \text{ and } k = 0, 1, 2, \ldots$$

(a) Prove that $\sum_{k=0}^{\infty} P_k(x) = 1$.

(b) Prove that $\sum_{k=0}^{\infty} k P_k(x) = x$. (This sum is called the *average* of the distribution.)

(c) In n independent trials of an event with probability p of success, the probability of exactly k successes is given by the *binomial distribution*

$$B_k(n, p) = \binom{n}{k} p^k (1 - p)^{n-k}$$

If $p_n = x/n$, prove that $\lim_{n \to \infty} B_k(n, p_n) = P_k(x)$.

(d) If, on the average, x calls arrive at a telephone switchboard every hour, give an argument to explain why $P_k(x)$ should be the probability of exactly k calls arriving at the switchboard every hour. *Hint:* Divide the hour into n equal time periods and use part (c).

3.3 Two Fundamental Convergence Criteria

In practice, it is not often possible to find an exact numerical value or a simple expression for the sum of a series. It can even be difficult to tell whether or not a series converges, as we saw in Example 3.5 and in parts of Table 3.2. In this section we prove two criteria (Corollaries 3.7 and 3.9) that help us decide whether or not a series converges. These convergence criteria can only be proved if one assumes the Axiom of Completeness (stated below) or some property of the real numbers equivalent to it.

Definition 3.3   A number M is an *upper bound* of a set A of real numbers if $x \le M$ for all x in A; a number m is a *lower bound* of A if $m \le x$ for all x in A. The *least upper bound* or *supremum* of A (denoted by lub A or sup A) is the upper bound of A which is less than all other upper bounds of A; the *greatest lower bound* or *infimum* of A (denoted by glb A or inf A) is the lower bound of A which is greater than all other lower bounds of A.

EXAMPLE 3.10   The interval $[1.4, 3)$ has many upper bounds: 3, 3.2, 17, among others. But 3 is less than all other upper bounds and so $3 = \text{lub}[1.4, 3)$. Similarly, $1.4 = \text{glb}[1.4, 3)$. On the other hand, the set of all integers has no upper bounds and no lower bounds.

Axiom of Completeness   If A is a nonempty set of real numbers with an upper bound, then A has a least upper bound.

Roughly speaking, therefore, unless A contains arbitrarily large numbers, it has an upper limit. This upper limit, of course, might or might not be in A, as we saw in Example 3.10.

Our first convergence criterion concerns series whose partial sums form bounded monotonic sequences.

Definition 3.4   A sequence is *bounded* if the set of its terms has a lower bound and an upper bound.

Definition 3.5   If $x_n \le x_{n+1}$ for all n, the sequence $\{x_n\}$ is *monotonically increasing*. If $x_n \ge x_{n+1}$ for all n, the sequence $\{x_n\}$ is *monotonically decreasing*. In either case, $\{x_n\}$ is said to be *monotonic*.

EXAMPLE 3.11   The sequence $\{1/n\}$ is bounded and monotonically decreasing; it converges to 0. The sequence $\{n\}$ is not bounded but is monotonically increasing; it does not converge. The sequence $\{(-1)^n/n\}$ is bounded but is not monotonic; it converges to 0. The sequence $\{(-1)^n\}$ is bounded but is not monotonic; it does not converge.

The next two theorems show some relationships among the concepts of bounded, monotonic, and convergent sequences.

Theorem 3.5   Every convergent sequence is bounded.

Proof   Exercise 17.

Theorem 3.6   Every bounded monotonic sequence has a limit.

Proof   Suppose $\{x_n\}$ is a monotonically increasing bounded sequence; let A be the set of its terms. Then, by the Axiom of Completeness, A has a least upper bound x. We now show that x is the limit of $\{x_n\}$. For every $\varepsilon > 0$ we must find N such that $|x_n - x| < \varepsilon$ for all $n \geq N$. We choose N such that $x - \varepsilon < x_N$. Such an x_N must exist, for otherwise $x - \varepsilon$ would be a smaller upper bound of A than x. Furthermore, for all $n \geq N$, $x_N \leq x_n \leq x$ since $\{x_n\}$ is monotonically increasing and since x is an upper bound for A. Therefore, for all $n \geq N$, $|x - x_n| = x - x_n \leq x - x_N < \varepsilon$. We leave it as an exercise to show that monotonically decreasing bounded sequences converge (Exercise 12).

Theorem 3.6 has an immediate corollary for series.

Corollary 3.7   Suppose $a_k \geq 0$ for all k. Then the series $\sum_{k=1}^{\infty} a_k$ converges iff its sequence $\{s_n\}$ of partial sums is bounded.

Proof $s_n = \sum_{k=1}^{n} a_k = s_{n-1} + a_n \geq s_{n-1}$ since $a_n \geq 0$, which proves that the sequence $\{s_n\}$ is monotonically increasing. If $\{s_n\}$ is also bounded, it follows from Theorem 3.6 that this sequence of partial sums converges; that is, the series $\sum a_k$ converges. If, on the other hand, the series $\sum a_k$ converges, then $\{s_n\}$ is bounded, according to Theorem 3.5.

EXAMPLE 3.12   Use Corollary 3.7 to prove that $\sum_{k=1}^{\infty} 1/k^2$ converges.

$$\frac{1}{k^2} \leq \frac{1}{k(k-1)} = \frac{1}{k-1} - \frac{1}{k} \quad \text{for } k \geq 2$$

and so

$$\sum_{k=1}^{n} \frac{1}{k^2} \leq 1 + \sum_{k=2}^{n} \left(\frac{1}{k-1} - \frac{1}{k} \right)$$

$$= 1 + \left(1 - \frac{1}{2}\right) + \left(\frac{1}{2} - \frac{1}{3}\right) + \cdots + \left(\frac{1}{n-1} - \frac{1}{n}\right)$$

$$= 2 - \frac{1}{n} < 2$$

Therefore $\sum 1/k^2$ converges, since its partial sums are bounded and $1/k^2 \geq 0$.

Our second convergence criterion says, roughly, that if the terms of a sequence are getting close to one another, they are approaching a limit. Note its resemblance to the error estimate $|x_n - x_{2n}|$ which we used often in Chapter 2.

Theorem 3.8   The sequence $\{x_n\}$ converges iff it satisfies the following *Cauchy criterion*: For every $\varepsilon > 0$ there exists N such that

$$|x_n - x_m| < \varepsilon \quad \text{for all } n \geq N \text{ and } m \geq N$$

As $x_n\uparrow$, subsequent increments get smaller.

Proof   We assume that $\{x_n\}$ satisfies the Cauchy criterion and prove that it converges. First we show that $\{x_n\}$ is a bounded sequence. By the Cauchy criterion, there exists N such that $|x_N - x_m| < 1$ for all $m \geq N$. (We used $\varepsilon = 1$ and then $n = N$.) It therefore follows that, for $m \geq N$,

$$|x_m| = |(x_m - x_N) + x_N| \leq |x_m - x_N| + |x_N| < 1 + |x_N|$$

Thus, the terms $|x_m|$, for $m \geq N$, have an upper bound. But, since there are only finitely many other terms, $|x_1|, \dots, |x_{N-1}|$, of the sequence $\{|x_n|\}$, the entire sequence $\{|x_n|\}$ has an upper bound. If M denotes this upper bound, then $-M \leq x_n \leq M$ for all n.

Next, we let $y_n = \text{lub}\{x_n, x_{n+1}, x_{n+2}, \dots\}$, which we know exists by the Axiom of Completeness. But, by Exercise 8, $y_{n+1} \leq y_n$. Furthermore, the terms y_n are bounded by the same numbers $-M$ and M that bound the terms x_n. Therefore, by Theorem 3.6, $\{y_n\}$ converges to some limit x.

Finally, we show that $x_n \to x$. For every $\varepsilon > 0$ there exists, by the Cauchy criterion, an integer N such that

$$|x_n - x_m| < \varepsilon/3 \quad \text{for all } n \geq N \text{ and } m \geq N$$

For each fixed value of n, where $n \geq N$, we can select $p \geq n$ such that $|y_p - x| < \varepsilon/3$, since $y_p \to x$. And, since $y_p = \text{lub}\{x_p, x_{p+1}, \dots\}$, we can select $m \geq p$ such that $|y_p - x_m| < \varepsilon/3$ (see Exercise 9). Since $m \geq p \geq n \geq N$,

$$|x_n - x| = |(x_n - x_m) + (x_m - y_p) + (y_p + x)|$$
$$\leq |x_n - x_m| + |x_m - y_p| + |y_p - x|$$
$$< \frac{\varepsilon}{3} + \frac{\varepsilon}{3} + \frac{\varepsilon}{3} = \varepsilon$$

for all $n \geq N$. Therefore, $x_n \to x$.

The other direction of the proof, in which a convergent sequence is shown to satisfy the Cauchy criterion, is left to Exercise 14.

Theorem 3.8 has an immediate corollary for series.

Corollary 3.9   The series Σa_k converges iff it satisfies the following Cauchy criterion: For every $\varepsilon > 0$ there exists N such that

$$\left| \sum_{k=n+1}^{m} a_k \right| < \varepsilon \quad \text{for all } n \geq N \text{ and all } m > n$$

Proof   If s_n denotes the nth partial sum of Σa_k, then for $m > n$

$$|s_m - s_n| = \left| \sum_{k=1}^{m} a_k - \sum_{k=1}^{n} a_k \right| = \left| \sum_{k=n+1}^{m} a_k \right|$$

Therefore, the sequence $\{s_n\}$ satisfies the Cauchy criterion of Theorem 3.8 iff the series Σa_k satisfies the Cauchy criterion of this theorem. (We may assume that $m > n$ since $|s_n - s_m| = |s_m - s_n|$.)

EXAMPLE 3.13   Use Corollary 3.9 to prove that $\Sigma 1/k$ diverges.

$$\sum_{k=2^n+1}^{2^{n+1}} \frac{1}{k} = \frac{1}{2^n + 1} + \frac{1}{2^n + 2} + \cdots + \frac{1}{2^{n+1}}$$

$$\geq \frac{1}{2^{n+1}} + \frac{1}{2^{n+1}} + \cdots + \frac{1}{2^{n+1}} \quad (2^n \text{ terms})$$

$$= \frac{2^n}{2^{n+1}} = \frac{1}{2}$$

Therefore, the Cauchy criterion fails for every $\varepsilon \leq \frac{1}{2}$ and so $\Sigma 1/k$ diverges.

EXERCISES

Which of the sequences in Exercises 1–4 (a) are bounded, (b) are monotonic, (c) converge? (Assume that $n \geq 1$.)

1. $e^{1/n}$

2. $n^2 - 2n$

3. $\sin(n)$

4. $3 - (-1)^n/n$

Prove that the recursively defined sequences in Exercises 5–7 converge; use Theorem 3.6.

5. $x_{n+1} = \sqrt{x_n}, \qquad x_1 = 4$

6. $x_{n+1} = \sqrt{6 + x_n}, \qquad x_1 = 0$

7. $x_{n+1} = (x_n + 9/x_n)/2, \qquad x_1 = 5$

8. If A is a nonempty subset of a set B of real numbers and if B has an upper bound, prove that lub $A \leq$ lub B.

9. If A is a nonempty set of real numbers with the least upper bound M and if $\varepsilon > 0$, prove there exists a number x in A such that $|x - M| < \varepsilon$.

10. If A is a nonempty set of real numbers with the least upper bound M, prove there exists a sequence $\{x_n\}$ of numbers in A such that $x_n \to M$.

11. If A is a nonempty set of real numbers with a lower bound m, prove that A has a greatest lower bound. *Hint:* Let $B = \{-x | x$ is in $A\}$, show that $-m$ is an upper bound of B, and apply the Axiom of Completeness to B.

12. If $\{x_n\}$ is a monotonically decreasing and bounded sequence, prove that $\{x_n\}$ converges. *Hint:* Apply Theorem 3.6 to the sequence $\{-x_n\}$. (This completes the proof of Theorem 3.6.)

13. Suppose that $\sum_{k=1}^{n} a_k \leq y_n$ for all n and that $a_k \geq 0$. If the sequence $\{y_n\}$ has a limit y, prove that the series $\sum_{k=1}^{\infty} a_k$ converges and that its sum is no larger than y.

14. If the sequence $\{x_n\}$ converges, prove that it satisfies the Cauchy criterion of Theorem 3.8. (This proves the "only if" part of Theorem 3.8.)

15. Suppose that $|x_n - x_m| \leq y_n$ for all n and for all $m > n$. If the sequence $\{y_n\}$ has the limit 0, prove that the sequence $\{x_n\}$ converges.

16. Suppose that $|\sum_{k=n+1}^{m} a_k| \leq y_n$ for all n and for all $m > n$. If the sequence $\{y_n\}$ has the limit 0, prove that the series $\sum a_k$ converges.

17. Prove Theorem 3.5. *Hint:* Use Theorem 1.2.

18. The object of this several part exercise is to derive *Wallis' formula*. Let $I_n = \int_0^{\pi/2} \sin^n x \, dx$, for $n = 0, 1, 2, \ldots$.

    (a) Prove that $I_{n+1} \leq I_n$. *Hint:* Show that $\sin^{n+1} x \leq \sin^n x$ when $0 \leq x \leq \pi/2$.

    (b) Use integration by parts to prove that $I_n = I_{n-2}(n-1)/n$ for $n \geq 2$.

    (c) Use part (b) and induction to prove that, for $n \geq 1$

    $$I_{2n} = \frac{\pi}{2} \frac{1}{2} \frac{3}{4} \cdots \frac{2n-3}{2n-2} \frac{2n-1}{2n}$$

    $$I_{2n+1} = \frac{2}{3} \frac{4}{5} \cdots \frac{2n-2}{2n-1} \frac{2n}{2n+1}$$

    (d) Use parts (a) and (c) to prove that, for $n \geq 1$

    $$\frac{2n}{2n+1} = \frac{I_{2n+1}}{I_{2n-1}} \leq \frac{I_{2n+1}}{I_{2n}} \leq 1$$

    and, therefore, that $\lim_{n \to \infty} I_{2n+1}/I_{2n} = 1$ or (Wallis' formula)

    $$\lim_{n \to \infty} \frac{2 \cdot 2 \cdot 4 \cdot 4 \cdot 6 \cdot 6 \cdots (2n)(2n)}{1 \cdot 3 \cdot 3 \cdot 5 \cdot 5 \cdot 7 \cdots (2n-1)(2n+1)} = \frac{\pi}{2}$$

19. Let f be a function satisfying $|f(x) - f(z)| \leq L|x - z|$ for all x and z in the domain of f, where L is a fixed number less than 1. Define a sequence $\{x_n\}$ recursively by $x_{n+1} = f(x_n)$, where the first term x_0 is any point in the domain of f. (Assume that every x_n is also in the domain of f.) Prove

    (a) $|x_{n+1} - x_n| \leq L^n|x_1 - x_0|$ for all $n \geq 0$

    (b) $|x_m - x_n| \leq \left(\sum_{k=n}^{m-1} L^k \right) |x_1 - x_0|$ for all $m > n \geq 0$

    (c) $\{x_n\}$ converges

    (d) $f(x) = x$ if $x_n \to x$

    (e) $|x_n - x| \leq L^n|x_0 - x|$ if $x_n \to x$

3.4 Comparison Tests

In this section and the next, we prove several useful "convergence tests." These are theorems which help us decide whether or not a series converges. The tests in this section are based on Corollary 3.7.

Theorem 3.10 (Comparison test)   Suppose $0 \le a_k \le b_k$ for all k. If the series Σb_k converges, then Σa_k converges. (Thus, if the series Σa_k diverges, then Σb_k diverges.)

Proof   Let s_n and t_n denote the nth partial sums of $\Sigma_{k=1}^{\infty} a_k$ and $\Sigma_{k=1}^{\infty} b_k$, respectively. Then

$$s_n = a_1 + a_2 + \cdots + a_n \le b_1 + b_2 + \cdots + b_n = t_n$$

If Σb_k converges then $\{t_n\}$, its sequence of partial sums, is bounded, by Corollary 3.7. But since $0 \le s_n \le t_n$, $\{s_n\}$ is also bounded. Therefore, again by Corollary 3.7, Σa_k converges.

EXAMPLE 3.14   Show that $\Sigma 1/(k^2 + 1)$ converges.

The typical method for applying the comparison test is to find another series whose terms are very close to those of the given series and which we know either converges or diverges. In this case, we know from Example 3.12 that the series $\Sigma 1/k^2$ converges. Therefore, since $0 \le 1/(k^2 + 1) \le 1/k^2$ for all $k \ge 1$, the series $\Sigma 1/(k^2 + 1)$ also converges, by Theorem 3.10.

Note that, by Theorem 3.4, we can also apply the comparison test when we only know that $0 \le a_k \le b_k$ for all $k \ge n$, where n can be any integer. For example, even though $0 \le 5/(6k^2 - 5) \le 1/k^2$ is only true when $k \ge 3$, we may still conclude that $\Sigma 5/(6k^2 - 5)$ converges since $\Sigma 1/k^2$ converges.

Often, we can apply the comparison test without finding a specific inequality. Suppose we find that

$$a_k/b_k \to L \quad \text{where } a_k, b_k > 0 \text{ and } 0 < L < \infty$$

Then it can be shown (Exercise 20) that

$$\Sigma a_k \text{ converges iff } \Sigma b_k \text{ converges}$$

This result is known as the "limit comparison test." The reasoning behind it is roughly as follows. a_k is approximately equal to Lb_k for large k, and so the series Σa_k is approximately the same as the series $\Sigma L b_k$ (if we ignore sufficiently many initial terms of both series). But Σb_k converges iff $\Sigma L b_k$ does.

EXAMPLE 3.15   Find out whether $\Sigma 1/(3k^2 - k)$ converges or diverges.

Since $3k^2$ is the dominant term in the denominator, we suspect that the given series behaves much like $\Sigma 1/3k^2$ or even like $\Sigma 1/k^2$. In fact,

$$\frac{1/(3k^2 - k)}{1/k^2} = \frac{k^2}{3k^2 - k} = \frac{1}{3 - 1/k} \to \frac{1}{3}$$

Since $0 < \frac{1}{3} < \infty$, the limit comparison test assures us that the series $1/(3k^2 - k)$ converges iff $\Sigma 1/k^2$ converges. But, by Example 3.12, $\Sigma 1/k^2$ converges; therefore $\Sigma 1/(3k^2 - k)$ converges.

EXAMPLE 3.16   Show that $\Sigma k^p x^k$ converges whenever $0 \le x < 1$.

Let y be a number such that $x < y < 1$. Then $0 \le x/y < 1$, and so $k^p(x/y)^k \to 0$ by Exercise 12 in Section 1.1. But, since any convergent sequence is bounded, there exists a number M such that

$$k^p \left(\frac{x}{y} \right)^k \le M$$

or

$$k^p x^k \le M y^k$$

Therefore $\Sigma k^p x^k$ converges since Σy^k is a convergent geometric series.

We can also test a series for convergence by comparing the series with an improper integral.

Theorem 3.11 (Integral test)   Suppose f is a continuous, nonnegative, monotonically decreasing function on the interval $[1, \infty)$. (This last condition means that $f(u) \ge f(v)$ whenever $1 \le u \le v$.) Let $a_k = f(k)$ for all $k \ge 1$. Then the series Σa_k converges iff the sequence $\{ \int_1^n f(x) \, dx \}$ converges.

Proof   Let s_n denote the nth partial sum of Σa_k and let $t_n = \int_1^n f(x) \, dx$. Since $f(x) \le f(k) = a_k$ for all $x \ge k$,

$$\int_k^{k+1} f(x) \, dx \le \int_k^{k+1} a_k \, dx = a_k \quad \text{(Theorem A.24d)}$$

So, for $n \ge p$,

$$\int_p^{n+1} f(x) \, dx = \sum_{k=p}^{n} \int_k^{k+1} f(x) \, dx \le \sum_{k=p}^{n} a_k \tag{3.6}$$

Similarly (see Figure 3.2)

$$\sum_{k=p}^{n} a_k \le \int_{p-1}^{n} f(x) \, dx \tag{3.7}$$

If the sequence $\{ t_n \}$ converges, its terms have an upper bound M (Theorem 3.5). Hence, by (3.7) with $p = 2$,

$$s_n = \sum_{k=1}^{n} a_k = a_1 + \sum_{k=2}^{n} a_k \le a_1 + \int_1^n f(x) \, dx \le a_1 + M$$

and therefore Σa_k converges (Corollary 3.7). Conversely, if the series Σa_k converges then its partial sums s_n have an upper bound M. Hence, by (3.6) with $p = 1$,

$$t_n = \int_1^n f(x) \, dx \le \sum_{k=1}^{n-1} a_k = s_{n-1} \le M$$

Furthermore, $t_{n+1} - t_n = \int_n^{n+1} f(x) \, dx \ge 0$, and therefore $\{ t_n \}$ is a bounded monotonic sequence. By Theorem 3.6, $\{ t_n \}$ converges.

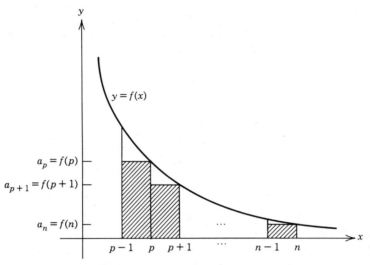

Figure 3.2 a_k is the area of the shaded rectangle with height a_k and base 1; hence $\sum_{k=p}^{n} a_k$ is the total shaded area, which is less than the area under the curve, $\int_{p-1}^{n} f(x)\, dx$

EXAMPLE 3.17   For what real numbers p does the *p-series* $\sum 1/k^p$ converge?

If $p \le 0$, the terms $1/k^p$ do not converge to 0 and so the series diverges (Theorem 3.2). Now assume that $p > 0$ and define

$$f(x) = \frac{1}{x^p} \quad \text{for } x \ge 1$$

Then f is continuous, nonnegative and monotonically decreasing. For $p \ne 1$

$$\int_1^n \frac{1}{x^p}\, dx = \frac{n^{1-p}}{1-p} - \frac{1}{1-p}$$

which converges iff $p > 1$. For $p = 1$

$$\int_1^n \frac{1}{x^p}\, dx = \int_1^n \frac{1}{x}\, dx = \log n$$

which diverges. Therefore

$$\sum \frac{1}{k^p} \text{ converges iff } p > 1 \tag{3.8}$$

In particular, the *harmonic series*

$$\sum \frac{1}{k} \text{ diverges.} \tag{3.9}$$

This last example points up a limitation of Theorem 3.2. Although the series $\sum a_k$ diverges when the sequence $\{a_k\}$ does not converge to 0, when $\{a_k\}$ does converge to 0 the series $\sum a_k$ might or might not converge. For example,

$$\frac{1}{k} \to 0 \quad \text{and} \quad \frac{1}{k^2} \to 0$$

but

$$\sum \frac{1}{k} \text{ diverges and } \sum \frac{1}{k^2} \text{ converges.}$$

EXAMPLE 3.18   Find out whether $\sum 1/\sqrt{4k+1}$ converges or diverges.

Since $4k$ is the dominant term inside the square root, we suspect that the given series behaves much like $\sum 1/2\sqrt{k}$ or even like $\sum 1/\sqrt{k}$. In fact

$$\frac{1/\sqrt{4k+1}}{1/\sqrt{k}} = \frac{\sqrt{k}}{\sqrt{4k+1}} = \frac{1}{\sqrt{4+1/k}} \rightarrow \frac{1}{\sqrt{4}} = \frac{1}{2}$$

Since $0 < \frac{1}{2} < \infty$, the limit comparison test assures us that the series $\sum 1/\sqrt{4k+1}$ converges iff the series $\sum 1/\sqrt{k}$ converges. But, by Example 3.17, $\sum 1/\sqrt{k} = \sum 1/k^{1/2}$ diverges; therefore, the given series diverges.

EXAMPLE 3.19   Show that $\sum (\log k)/k^2$ converges.

In Example 1.27 we showed that

$$\log x < \frac{x^p}{p} \quad \text{if } x > 1 \text{ and } p > 0$$

So if we replace x by k and p by $\frac{1}{2}$ and then divide by k^2, we get

$$\frac{\log k}{k^2} \le \frac{2k^{1/2}}{k^2} = \frac{2}{k^{3/2}}$$

Since the p-series $\sum 1/k^{3/2}$ (where $p = \frac{3}{2} > 1$) converges, Theorem 3.10 assures us that $\sum (\log k)/k^2$ converges.

EXERCISES

In Exercises 1–13, find out whether the series converges or diverges.

1. $\sum 3k^{-4/3}$
2. $\sum 5k^{-2/3}$
3. $\sum (2k+1)/(k^2+4)$
4. $\sum \sqrt{k}/(k^2+7)$
5. $\sum (\sqrt{k+1} - \sqrt{k})$
6. $\sum (\sqrt{k+1} - \sqrt{k})/k$
7. $\sum 1/k^k$
8. $\sum 1/(1+2^k)$
9. $\sum k/(1+e^k)$
10. $\sum (\log k)/k$
11. $\sum 1/(k \log k)$
12. $\sum 1/k(\log k)^2$
13. $\sum (\log k)/(1+k^{3/2})$
14. For what values of p and q does $\sum (\log k)^p/k^q$ converge?
15. Find two divergent series $\sum a_k$ and $\sum b_k$ such that $\sum (a_k + b_k)$ converges.
16. (Condensation test) Suppose $\{a_k\}$ is a monotonically decreasing sequence of nonnegative numbers. Prove that $\sum a_k$ converges iff $\sum 2^k a_{2^k}$ converges. *Hint:* Show that

$$\sum_{k=1}^{2^{n+1}-1} a_k \le a_1 + \sum_{k=1}^{n} 2^k a_{2^k} \quad \text{and} \quad \sum_{k=1}^{n} 2^k a_{2^k} \le 2 \sum_{k=1}^{2^n} a_k$$

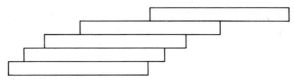

Figure 3.3   Stack of books

17. Use the condensation test to prove $\Sigma 1/k^2$ converges.

18. Use the condensation test to find out which of the p-series converge and which diverge.

19. Let H_n be the nth partial sum of the harmonic series:

$$H_n = 1 + \frac{1}{2} + \cdots + \frac{1}{n}$$

and let $\gamma_n = H_n - \log n$. Prove that the sequence $\{\gamma_n\}$ converges to a limit and that the limit is between 0 and 1. (The limit is called *Euler's constant*.) *Hint:* Show that $\gamma_n - \gamma_{n+1} \geq 0$ by picturing $\log(n + 1) - \log n$ as an area under a portion of the graph of $y = 1/x$.

20. (Limit comparison test) Suppose $a_k/b_k \to L$ where $a_k > 0$, $b_k > 0$, and $0 < L < \infty$.

    (a) If Σb_k converges, prove that Σa_k converges. *Hint:* Use Theorem 3.5 to deduce that there exists a constant M such that $a_k \leq Mb_k$ for all k.

    (b) If Σa_k converges, prove that Σb_k converges. *Hint:* Since $L > 0$, $b_k/a_k \to 1/L$. Then use part (a).

21. If $a_k \geq 0$ for all k and Σa_k converges, prove that Σa_k^2 converges.

22. Prove that a stack of identical books can be arranged so that the top book is as far as you like to the right of the bottom book (see Figure 3.3). *Hint:* Proceed recursively from the top of the pile downward. Find the center of mass, c_n, of the top n books and place the right edge of the $(n + 1)$st book at this point. Prove that c_n is the nth partial sum of a divergent series.

3.5 Absolute and Conditional Convergence

The comparison tests in the preceding section apply only to infinite series whose terms are non-negative, while the theorems in this section apply to series whose terms may change sign. Our principal tool will be Corollary 3.9.

Theorem 3.12   If the series $\Sigma |a_k|$ converges then Σa_k converges.

Proof   By the triangle inequality, $|\Sigma_{k=n+1}^{m} a_k| \leq \Sigma_{k=n+1}^{m} |a_k|$ for all $m > n$. Therefore, since $\Sigma |a_k|$ satisfies the Cauchy criterion of Corollary 3.9, so does Σa_k. Hence Σa_k converges.

   EXAMPLE 3.20 $\Sigma (-1)^k/k^2$ converges since $\Sigma 1/k^2$ converges.

   However, it can happen that Σa_k converges even though $\Sigma |a_k|$ diverges, as we will soon see. The following definition helps us distinguish such series.

Definition 3.6   If $\Sigma |a_k|$ converges, Σa_k is said to be *absolutely convergent*. If $\Sigma |a_k|$ diverges but Σa_k converges, Σa_k is said to be *conditionally convergent*.

The next theorem provides a crude but quick test for absolute convergence. Roughly speaking, it tests a series by comparing it with the geometric series of Theorem 3.3.

Theorem 3.13 (Ratio test)   Suppose the sequence $\{|a_{k+1}/a_k|\}$ converges to a limit r.

If $r < 1$, the series Σa_k converges absolutely.

If $r > 1$, the series Σa_k diverges.

If $r = 1$, the test gives no information.

Proof   First we assume $r < 1$. If we choose x so that $r < x < 1$, then, by Theorem 1.2a, there exists N such that

$$\left|\frac{a_{k+1}}{a_k}\right| < x \quad \text{for all } k \geq N$$

Thus

$$|a_{N+1}| < x|a_N|$$

$$|a_{N+2}| < x|a_{N+1}| < x^2|a_N|$$

$$\vdots$$

$$|a_k| < x^{k-N}|a_N| = \frac{x^k|a_N|}{x^N} \quad \text{for every } k > N$$

But the geometric series Σx^k converges, and therefore Σa_k converges absolutely by the comparison test.

If, on the other hand, $r > 1$, then (Theorem 1.2b) there exists N such that

$$\left|\frac{a_{k+1}}{a_k}\right| > 1 \quad \text{for all } k \geq N$$

But then $|a_{k+1}| > |a_k|$ and so the sequence $\{a_k\}$ cannot converge to zero. Therefore, by Theorem 3.2, the series Σa_k diverges.

EXAMPLE 3.21   Apply the ratio test to the series $\Sigma k^2/2^k$.

$$\frac{|a_{k+1}|}{|a_k|} = \frac{(k+1)^2/2^{k+1}}{k^2/2^k} = \frac{(k+1)^2}{2k^2} = \frac{(1+1/k)^2}{2} \to \frac{1}{2}$$

Since this limit is less than 1, the series converges.

EXAMPLE 3.22   Apply the ratio test to the series $\Sigma 1/k$.

$$\left|\frac{a_{k+1}}{a_k}\right| = \frac{k}{(k+1)} \to 1$$

and so the ratio test does not tell us whether the harmonic series converges.

Theorem 3.14 (Leibnitz' alternating series test)   If the terms a_k are alternately positive and negative and if the sequence $\{|a_k|\}$ decreases monotonically to zero, then the series Σa_k converges.

Proof Let $b_k = |a_k|$. Then $b_k - b_{k+1} \geq 0$ for all k, since $\{|a_k|\}$ is monotonically decreasing. Furthermore, either $a_k = (-1)^k b_k$ for all k or $a_k = (-1)^{k+1} b_k$ for all k; we will assume the former. Then

$$\left| \sum_{k=n+1}^{m} a_k \right| = |a_{n+1} + a_{n+2} + a_{n+3} + \cdots + a_m|$$

$$= |(-1)^{n+1} b_{n+1} + (-1)^{n+2} b_{n+2} + (-1)^{n+3} b_{n+3} + \cdots + (-1)^m b_m|$$

$$= |(-1)^{n+1}| |b_{n+1} - b_{n+2} + b_{n+3} - \cdots + (-1)^{m-n-1} b_m|$$

$$= (b_{n+1} - b_{n+2}) + (b_{n-3} - b_{n+4}) + \cdots$$

$$= b_{n+1} - (b_{n+2} - b_{n+3}) - (b_{n+4} - b_{n+5}) - \cdots$$

$$\leq b_{n+1} = |a_{n+1}|$$

since the differences in parentheses are non-negative. Thus

$$\left| \sum_{k=n+1}^{m} a_k \right| \leq |a_{n+1}| \tag{3.11}$$

Therefore, by Exercise 16 of Section 3.3, Σa_k converges.

EXAMPLE 3.23 The *alternating harmonic series* $\sum_{k=1}^{\infty} (-1)^{k+1}/k$ converges since the sequence $\{1/k\}$ decreases monotonically to zero. But the series is only conditionally convergent since the harmonic series $\Sigma 1/k$ diverges.

The theorems in this section and the preceding one form a powerful set of tests which help us decide on the convergence of a great variety of series. However, it takes considerable practice to become efficient in using them. The following remarks are intended as bits of advice to keep in mind as you develop, by practice, insights into how to use these tests.

The ratio test is usually indicated when the terms of the series are complicated and when forming ratios of successive terms will lead to significant cancellations. It is especially likely to be helpful when the series resembles a geometric series; thus, it is usually the first test to apply to any power series, $\Sigma c_k (x - x_0)^k$. For example, it is easy to show, by using the ratio test, that the power series $\Sigma 2^k x^k/k$ converges absolutely if $|x| < \frac{1}{2}$ and diverges if $|x| > \frac{1}{2}$. But the ratio test fails to tell us whether the series converges or diverges at the end points $x = \pm \frac{1}{2}$, which is typical of what happens when this test is applied to a power series. When the ratio test fails and especially when the terms of the series are complicated but do not resemble those of a geometric series, then it is likely that a comparison test will be needed. But to apply a comparison test, a familiar series must be found and compared. For example, $\Sigma (2k + 3)/(3k^3 - 5)$ has terms that are much like $2k/3k^3 = (2/3)(1/k^2)$; and since we recall that the familiar p series $\Sigma 1/k^2$ converges, we expect the given series to converge. The integral test is only likely to be useful when the terms are fairly simple; otherwise the corresponding function will be too difficult to integrate. It is most likely to be needed in studying quite unfamiliar series. For example, the terms of $\Sigma 1/(k \log k)$ are bigger than those of the divergent series $\Sigma 1/k$ and smaller than those of the convergent series $\Sigma 1/k^2$.

Thus, no useful comparison comes quickly to mind. Fortunately, the corresponding function $1/(x \log x)$ is not hard to integrate. The alternating series test is indicated simply by the fact that the terms are alternately positive and negative. However, we usually want to know if an alternating series converges absolutely, and so we first apply one of the preceding tests to the series of absolute values; if this series converges, the alternating series test is not needed. Finally, we should note early on whether the terms of the series converge to zero, since this is such an easy test to make. If they do not, the series diverges; if they do, this of course provides no evidence of whether the series converges or diverges.

EXAMPLE 3.24   For what values of x does $\Sigma(-1)^k x^{2k+1}/(2k+1)$ converge?
This is a power series and so we apply the ratio test first:

$$\left|\frac{a_{k+1}}{a_k}\right| = \frac{|(-1)^{k+1}x^{2k+3}/(2k+3)|}{|(-1)^k x^{2k+1}/(2k+1)|} = \frac{|x|^2(2k+1)}{2k+3} \to |x|^2$$

Since the limit $|x|^2$ is less than 1 if $|x| < 1$, and greater than 1 if $|x| > 1$, the series converges absolutely if $|x| < 1$ and diverges if $|x| > 1$. We must use other tests to find out if the series converges at the remaining two points, $x = \pm 1$. At these points, the series of absolute values is $\Sigma 1/(2k+1)$ which is much like the divergent harmonic series $\Sigma 1/k$. In fact,

$$\frac{1/(2k+1)}{1/k} = \frac{k}{2k+1} = \frac{1}{2+1/k} \to \frac{1}{2}$$

and so, by the limit comparison test, $\Sigma 1/(2k+1)$ diverges since $0 < \frac{1}{2} < \infty$. In particular, the series $\Sigma(-1)^k/(2k+1)$, which is the given series at the points $x = \pm 1$, does not converge absolutely. However, by the alternating series test, it does converge since $\{1/(2k+1)\}$ decreases monotonically to zero. Therefore the given power series converges iff $-1 \le x \le 1$; the convergence is conditional at $x = \pm 1$.

The series in the above example is the Taylor series of arctan x at 0 (line 6 of Table 3.2). We have now confirmed the claim in Table 3.2 that this series converges on $[-1, 1]$ and nowhere else. However, despite the suggestive derivation in Example 3.8, we have not yet proved that its sum is arctan x.

EXAMPLE 3.25   Assuming that arctan $x = \Sigma_{k=0}^{\infty}(-1)^k x^{2k+1}/(2k+1)$ when $-1 < x < 1$ (which will be proved in Section 3.8), prove that the same equality holds for $x = 1$.
If $-1 < x < 1$, then

$$\left|\arctan x - \sum_{k=0}^{n}\frac{(-1)^k x^{2k+1}}{2k+1}\right| = \left|\sum_{k=n+1}^{\infty}\frac{(-1)^k x^{2k+1}}{2k+1}\right|$$

$$\le \left|\frac{(-1)^{n+1}x^{2n+3}}{2n+3}\right| \le \frac{1}{2n+3}$$

where the first inequality follows from (3.11). But, since arctan x and x^{2k+1} are continuous, when we let x approach 1 we get

$$\left| \arctan 1 - \sum_{k=0}^{n} \frac{(-1)^k}{2k+1} \right| \le \frac{1}{2n+3}$$

By letting $n \to \infty$, we deduce that $\sum_{k=0}^{\infty}(-1)^k/(2k+1) = \arctan 1 = \pi/4$.

Conditionally convergent series can exhibit very different behavior than finite sums. For example, let s denote the sum of the conditionally convergent alternating harmonic series:

$$s = 1 - \frac{1}{2} + \frac{1}{3} - \frac{1}{4} + \frac{1}{5} - \frac{1}{6} + \frac{1}{7} - \frac{1}{8} \cdots \tag{3.12}$$

and so

$$\frac{s}{2} = \frac{1}{2} - \frac{1}{4} + \frac{1}{6} - \frac{1}{8} \cdots$$

$$= 0 + \frac{1}{2} + 0 - \frac{1}{4} + 0 + \frac{1}{6} + 0 - \frac{1}{8} \cdots$$

since it can be shown that a sum is not changed when extra zero terms are inserted in it. If we add this last equation to (3.12), we get

$$\frac{3s}{2} = 1 + 0 + \frac{1}{3} - \frac{1}{2} + \frac{1}{5} + 0 + \frac{1}{7} - \frac{1}{4} \cdots$$

After removing the zeros, the terms of this sum are just a rearrangement of the terms of the sum (3.12), and so it would seem we have shown that $\frac{3}{2}s = s$, which implies that $s = 0$. But this cannot be true, since the proof of Theorem 3.14 shows that $s \ge b_1 - b_2 = a_1 + a_2 = \frac{1}{2}$; therefore, we have actually shown that the terms of the alternating harmonic series can be rearranged to produce a series with a different sum. So we cannot apply the commutative law of addition too freely to conditionally convergent series. Similarly, the associative and distributive laws also cannot be applied in an unrestricted fashion to conditionally convergent series [see Knopp (1956)].

Theorems 3.12 and 3.13 carry over without change to infinite series of complex numbers, provided the absolute value signs are interpreted as indicating the absolute value of a complex number.

EXERCISES

In Exercises 1–11, find out whether the series is absolutely convergent, conditionally convergent, or divergent.

1. $\sum(-1)^k k^{-3/2}$
2. $\sum(-1)^k k^{-2/3}$
3. $\sum(-1)^k k/(k+1)$
4. $\sum(-1)^k 2^{-1/k}$
5. $\sum(-1)^k k/3^k$
6. $\sum(-1)^k k^2/(1+5^k)$
7. $\sum(-1)^k/\log k$
8. $\sum(-1)^k \sqrt{k}/(k^3+1)$

9. $\Sigma(-1)^k\sqrt{k}/(k+1)$ 10. $\Sigma(-1)^k 2^k/k!$

11. $\Sigma(-1)^k k!/k^k$ *Hint:* Use the fact that $(1 + 1/k)^k \to e$.

12. Give an example of a convergent series Σx_k such that Σx_k^2 diverges.

In Exercises 13–25, find all values of the variable x for which the series converges.

13. $\Sigma x^k/k^2$ 14. $\Sigma k x^k/3^k$

15. $\Sigma(0.1)^k x^k/\sqrt{k}$ 16. $\Sigma 2^k x^k/k5^k$

17. $\Sigma x^k/k^k$ 18. $\Sigma(-1)^k x^{2k}/k!$

19. $\Sigma\sqrt{k}\,(x+1)^k/3^k$ 20. $\Sigma x^{2k+1}/(2k+1)$

21. $\Sigma x^k/\log k$ 22. $\Sigma k^k x^k/k!$ *Hint:* Use Stirling's formula (3.25)

23. $\Sigma k e^{-kx}$

24. $\Sigma 1/(1 + x^k)$

25. $\Sigma x^k(1 - x^k)/k$

26. Assuming that $\log(1 + x) = \Sigma_{k=1}^{\infty}(-1)^{k+1}x^k/k$ when $-1 < x < 1$, prove that the same equality holds for $x = 1$.

27. (Root test)   Suppose the sequence $\{|a_k|^{1/k}\}$ converges to a limit r.

(a) If $r < 1$, prove the series Σa_k converges absolutely.

(b) If $r > 1$, prove the series Σa_k diverges.

(c) Give an example of a convergent series such that $r = 1$.

(d) Give an example of a divergent series such that $r = 1$.

In Exercises 28–30, use the root test to find out if the series converges.

28. $\Sigma 1/2^k$ 29. $\Sigma(4/3)^k$

30. $\Sigma k^k/(2k + 1)^k$

31. Prove the following variation of the ratio test.

(a) If $|a_{k+1}/a_k| \le r$ for all k, where $r < 1$, then Σa_k converges absolutely.

(b) If $|a_{k+1}/a_k| \ge 1$ for all k, then Σa_k diverges.

3.6 Approximate Sum of a Series

In the preceding two sections we explored several methods for finding out whether or not a series converges. The ideas behind these methods can also be used to approximate sums and find error bounds or error estimates. To approximate the sum of the series $\Sigma_{k=1}^{\infty}a_k$, we may simply use one of its partial sums $\Sigma_{k=1}^{n}a_k$. Then, since

$$\sum_{k=1}^{\infty} a_k = \sum_{k=1}^{n} a_k + \sum_{k=n+1}^{\infty} a_k$$

(Theorem 3.4), we can write the error in the form

$$E_n = \left| \sum_{k=1}^{\infty} a_k - \sum_{k=1}^{n} a_k \right| = \left| \sum_{k=n+1}^{\infty} a_k \right|$$

Since the error is itself an infinite series, we will usually only be able to find a bound for it or estimate it.

We assume, at first, that $a_k \geq 0$ for all k and we use the comparison tests of Section 3.4 to find bounds for the error E_n. Suppose, for example, we can find a series Σb_k whose sum we know and whose terms dominate the terms of Σa_k; that is, suppose $a_k \leq b_k$ for all k. Then (see Exercise 10)

$$E_n = \sum_{k=n+1}^{\infty} a_k \leq \sum_{k=n+1}^{\infty} b_k \qquad (3.13)$$

EXAMPLE 3.26   Approximate $\sum_{k=1}^{\infty} 1/k^2$ with an error of at most 0.0001.
We compare with a telescoping series:

$$\frac{1}{k^2} \leq \frac{1}{k(k-1)} = \frac{1}{k-1} - \frac{1}{k}$$

and so

$$E_n \leq \sum_{k=n+1}^{\infty} \left(\frac{1}{k-1} - \frac{1}{k} \right) = \frac{1}{n} \leq 0.0001 \quad \text{if } n \geq 10,000$$

Therefore $\sum_{k=1}^{\infty} 1/k^2$ is within 0.0001 of $\sum_{k=1}^{10,000} 1/k^2 = 1.64483$.

However, inequality (3.13) alone is rarely sufficient for our purposes since there are too few series whose sum we know. More often, we will combine (3.13) with the inequality (3.7), $\sum_{k=n+1}^{m} a_k \leq \int_n^m f(x)\, dx$, which comes from the proof of the integral test. Assume, as we did then, that f is a continuous, non-negative, monotonically decreasing function on $[1, \infty)$ such that $a_k = f(k)$. Then inequality (3.7) implies that

$$E_n = \sum_{k=n+1}^{\infty} a_k \leq \lim_{m \to \infty} \int_n^m f(x)\, dx \qquad (3.14)$$

EXAMPLE 3.27   Approximate $\sum_{k=1}^{\infty} 1/(3k^2 + 1)$ with an error of at most 0.0001.
We could apply inequality (3.14) directly, but $1/(3k^2 + 1)$ is a bit messy to integrate. So we apply (3.13) first. $1/(3k^2 + 1) \leq 1/3k^2$ and so

$$E_n \leq \sum_{k=n+1}^{\infty} \frac{1}{3k^2} \leq \lim_{m \to \infty} \int_n^m \frac{1}{3x^2}\, dx$$

$$= 1/3n \leq 0.0001 \quad \text{if } n \geq 3334$$

Therefore, $\sum_{k=1}^{\infty} 1/(3k^2 + 1)$ is within 0.0001 of $\sum_{k=1}^{3334} 1/(3k^2 + 1) = 0.45633$.

If the terms a_k are not all positive, we can deduce from the proof of Theorem 3.12 (on absolute convergence) that

$$E_n = \left| \sum_{k=n+1}^{\infty} a_k \right| \leq \sum_{k=n+1}^{\infty} |a_k| \qquad (3.15)$$

provided the series Σa_k is absolutely convergent. This error bound can itself be bounded using inequalities (3.13) or (3.14).

In some circumstances, $|a_{n+1}|$ is a reasonable estimate of the error E_n or some multiple of the error. This may seem unlikely, for if the terms a_k are all positive we would be ignoring $\sum_{k=n+2}^{\infty} a_k$, which could contribute substantially to the error. But suppose we modify the hypothesis of the ratio test to $|a_{k+1}/a_k| \leq r_n$ for all $k \geq n+1$ where $r_n < 1$. Then

$$\sum_{k=n+1}^{m} |a_k| = |a_{n+1}| + |a_{n+2}| + \cdots + |a_m|$$

$$= |a_{n+1}| \left[1 + \left|\frac{a_{n+2}}{a_{n+1}}\right| + \left|\frac{a_{n+3}}{a_{n+1}}\right| + \cdots + \left|\frac{a_m}{a_{n+1}}\right| \right]$$

$$\leq |a_{n+1}| \left[1 + r_n + r_n^2 + \cdots + r_n^{m-n-1} \right]$$

$$\leq |a_{n+1}| \sum_{k=0}^{\infty} r_n^k = \frac{|a_{n+1}|}{(1-r_n)}$$

Therefore

Geometric Series

$$E_n \leq \frac{|a_{n+1}|}{1-r_n} \tag{3.16}$$

Two special cases of (3.16) should be noted because they occur quite often. If the ratios $|a_{k+1}/a_k|$ form a decreasing sequence, we can choose $r_n = |a_{n+2}/a_{n+1}|$ and so (3.16) becomes

$$E_n \leq \frac{|a_{n+1}|}{1 - |a_{n+2}/a_{n+1}|} \tag{3.16a}$$

Or, if the ratios $|a_{k+1}/a_k|$ are all bounded above by r, where $r < 1$, then we can choose $r_n = r$ and so (3.16) becomes

$$E_n \leq \frac{|a_{n+1}|}{1-r} \tag{3.16b}$$

Furthermore, whenever the ratios a_{k+1}/a_k are positive and converge to a limit $r < 1$, then $a_{n+1}/(1-r)$ is a good error estimate. Specifically (see Exercise 11),

$$\lim_{n \to \infty} \frac{E_n}{a_{n+1}} = \frac{1}{1-r} \tag{3.16c}$$

EXAMPLE 3.28 Approximate $\sum_{k=1}^{\infty} k/2^k$ with an error of roughly 0.0001.

$$\left|\frac{a_{k+1}}{a_k}\right| = \frac{(k+1)/2^{k+1}}{k/2^k} = \frac{1}{2}\left(\frac{k+1}{k}\right) = \frac{1}{2}\left(1 + \frac{1}{k}\right) \to \frac{1}{2} < 1$$

and so, by (3.16c),

$$E_n = \text{(approximately)} \ \frac{n+1}{2^{n+1}} \left(\frac{1}{1-1/2} \right)$$

$$= \frac{n+1}{2^n} \leq 0.0001 \quad \text{if } n \geq 18$$

Therefore, $\sum_{k=1}^{\infty} k/2^k$ is approximated by $\sum_{k=1}^{18} k/2^k = 1.9999$ with an error of roughly 0.0001.

There is even a set of circumstances in which $|a_{n+1}|$ is actually a bound for the error E_n. If Σa_k is an alternating series in which the terms $\{|a_k|\}$ decrease monotonically, then by inequality (3.11),

$$E_n \le |a_{n+1}| \tag{3.17}$$

EXAMPLE 3.29 Approximate the sum of $\Sigma_{k=1}^{\infty}(-1)^{k+1}/k$ with an error of at most 0.0001.

$$|a_{n+1}| = 1/(n+1) \le 0.0001 \quad \text{if } n \ge 9999$$

Therefore, $\Sigma_{k=1}^{\infty}(-1)^{k+1}/k$ is within 0.0001 of $\Sigma_{k=1}^{9999}(-1)^{k+1}/k = 0.6932$.

However, we may be able to find a smaller bound for the remainder of an alternating series if we consider the positive term series formed by adding two terms at a time.

EXAMPLE 3.30 Approximate the sum of $\Sigma_{k=1}^{\infty}(-1)^{k+1}/k$ with an error of at most 0.0001, by considering the series $\Sigma_{k=1}^{\infty}[1/(2k-1) - 1/2k]$.

$$\frac{1}{2k-1} - \frac{1}{2k} = \frac{1}{2k(2k-1)} \le \frac{1}{(2k-1)^2}$$

and so

$$E_n \le \sum_{k=n+1}^{\infty} \frac{1}{(2k-1)^2} \le \lim_{m \to \infty} \int_n^m \frac{1}{(2x-1)^2} \, dx$$

$$= \frac{1}{2(2n-1)} \le 0.0001 \quad \text{if } n \ge 2501$$

Therefore, $\Sigma_{k=1}^{\infty}(-1)^{k+1}/k$ is within 0.0001 of $\Sigma_{k=1}^{2501}1/2k(2k-1) = 0.69301$. (Note that we used half as many terms as we did in Example 3.29.)

We can use the above techniques, in place of those in Chapter 2 and Sections 1.4 and 1.5, to solve a variety of significant approximation problems. For example, we can use the Maclaurin series for arctan x (with $x = 1$) to approximate the constant π:

$$\sum_{k=0}^{\infty} (-1)^k \frac{x^{2k+1}}{2k+1} = \arctan x$$

and so, by replacing x with $\pi/4$ and multiplying by 4,

$$\sum_{k=0}^{\infty} \frac{4(-1)^k}{2k+1} = \pi$$

However, to approximate π with an error of 0.0001 by applying inequality (3.17) to this series, we would need 20,000 terms. Even if we were to use the method of Example 3.30, we would need 10,000 terms. In Section 3.7 we study several methods for speeding up the rate at which a series converges; that is, we study ways in which we can use fewer terms of a (possibly different) series to get the same degree of accuracy in less time and with less accumulated round-off error. For now, we simply note a special trick for getting better approximations of π. Using trigonometric identities, it can be shown that

$$\frac{\pi}{4} = \arctan \tfrac{1}{2} + \arctan \tfrac{1}{3}$$

$$\frac{\pi}{4} = 4\arctan \tfrac{1}{5} - \arctan \tfrac{1}{239}$$

The first of these is left to Exercise 12; the second, which is called Machin's formula, is a bit messier to derive [see Hatcher (1973) and Wrench (1960)]. Then, to approximate π with an error of 0.0001, we need only 10 terms of the Maclaurin series for $\arctan x$ if we use the first of these two identities, and we need only 5 terms if we use Machin's formula.

Next, we approximate an integral by expressing it as an alternating series and applying inequality (3.17):

$$\sin x = \sum_{k=0}^{\infty} (-1)^k \frac{x^{2k+1}}{(2k+1)!}$$

and so

$$\int_0^1 \sin(x^2)\, dx = \int_0^1 \sum_{k=0}^{\infty} (-1)^k \frac{(x^2)^{2k+1}}{(2k+1)!}\, dx$$

$$= \sum_{k=0}^{\infty} \int_0^1 (-1)^k \frac{x^{4k+2}}{(2k+1)!}\, dx = \sum_{k=0}^{\infty} \frac{(-1)^k}{(4k+3)(2k+1)!}$$

(In Section 3.8 we justify the step in which we brought the summation sign outside the integral.) If we use a partial sum of this last series to approximate $\int_0^1 \sin(x^2)\, dx$ with an error of 0.0001, then, since $E_n \le 1/(4n+7)(2n+3)!$, we only need $n = 2$.

Next, we approximate the function $e^x = \sum_{k=0}^{\infty} x^k/k!$ on the interval $[0, 2]$ by using inequality (3.16). Since the ratios $a_{k+1}/a_k = x/(k+1)$ form a decreasing sequence, we could use version (3.16a) of the ratio test inequality. Instead, we use the simpler error estimate provided by the limit (3.16c):

$$E_n = (\text{approximately})\ \frac{a_{n+1}}{1-r} = \frac{x^{n+1}}{(n+1)!} \le \frac{2^{n+1}}{(n+1)!}$$

$$\le 0.0001 \quad \text{if } n \ge 10.$$

Thus, e^x is within 0.0001 of $\sum_{k=0}^{10} x^k/k!$ on the interval $[0, 2]$.

For our final example, we approximate $\log x$ on the interval $[1, 2]$. First recall that

$$\log(1+x) = \sum_{k=1}^{\infty} (-1)^{k+1} \frac{x^k}{k} \quad \text{if } -1 < x \le 1$$

according to Table 3.2. Although this series converges very slowly, we can transform it to get a representation of the logarithm function which converges much more rapidly.

We replace x by $-x$ and then subtract the two series:

$$\log(1-x) = -\sum_{k=1}^{\infty} \frac{x^k}{k} \quad \text{if } -1 \le x < 1$$

$$\log\left(\frac{1+x}{1-x}\right) = \sum_{k=0}^{\infty} \frac{2x^{2k+1}}{2k+1} \quad \text{if } -1 < x < 1$$

Then we replace $(1+x)/(1-x)$ by x:

$$\log x = \sum_{k=0}^{\infty} \frac{2}{(2k+1)}\left(\frac{x-1}{x+1}\right)^{2k+1} \quad \text{if } x > 0$$

Although we could again use the ratio test inequality (3.16), this time we use the simpler comparison test inequality (3.13). Assuming $1 \le x \le 2$, we get

$$E_n = \sum_{k=n+1}^{\infty} \frac{2}{(2k+1)}\left(\frac{x-1}{x+1}\right)^{2k+1} \le \sum_{k=n+1}^{\infty} \frac{2}{(2n+3)}\left(\frac{x-1}{x+1}\right)^{2k+1}$$

$$= \frac{2}{(2n+3)}\left(\frac{x-1}{x+1}\right)^{2n+3} \sum_{k=0}^{\infty} \left(\frac{x-1}{x+1}\right)^{2k}$$

$$= \frac{2}{(2n+3)}\left(\frac{x-1}{x+1}\right)^{2n+3}\left(\frac{1}{1-\left(\frac{x-1}{x+1}\right)^2}\right)$$

$$\le \frac{2}{2n+3}\left(\frac{1}{3}\right)^{2n+3}\left(\frac{1}{1-\frac{1}{9}}\right) = \frac{9/4}{(2n+3)3^{2n+3}}$$

$$\le 0.0001 \quad \text{if } n \ge 3$$

Thus $\log x$ is within 0.0001 of $\sum_{k=0}^{3} 2[(x-1)/(x+1)]^{2k+1}/(2k+1)$ on the interval $[1,2]$.

EXERCISES

In Exercises 1–9, approximate the sum of the series with an error of

(a) at most 0.1. (A calculator will be helpful.)

(b) at most 0.0001. (Use your program from Section 3.1.)

1. $\displaystyle\sum_{k=1}^{\infty} 1/k^3$

2. $\displaystyle\sum_{k=1}^{\infty} \sqrt{k}/(2k^3+1)$

3. $\displaystyle\sum_{k=2}^{\infty} 1/k(\log k)^5$

4. $\displaystyle\sum_{k=1}^{\infty} (\log k)/k^3$

5. $\displaystyle\sum_{k=1}^{\infty} 2^k/(1+3^k)$

6. $\displaystyle\sum_{k=1}^{\infty} k^2/2^k$

7. $\displaystyle\sum_{k=0}^{\infty} (1+2^k)/k!$

8. $\displaystyle\sum_{k=1}^{\infty} (-1)^k/9k^{3/4}$

9. $\displaystyle\sum_{k=2}^{\infty} (-1)^k/(\log k)^4$

10. If $x_n \le y_n$ for all n, prove that

(a) $\lim_{n \to \infty} x_n \le \lim_{n \to \infty} y_n$ provided both limits exist;

(b) $\displaystyle\sum_{k=1}^{\infty} x_k \le \sum_{k=1}^{\infty} y_k$ provided both sums exist.

11. If $a_k \ge 0$ for all k and $\lim_{k \to \infty} a_{k+1}/a_k = r < 1$, prove that $\lim_{n \to \infty} E_n/a_{n+1} = 1/(1-r)$. *Hint:* Let $r_k = a_{k+1}/a_k$ and consider the sequences $s_n = \mathrm{lub}\{r_n, r_{n+1}, \dots\}$, $t_n = \mathrm{glb}\{r_n, r_{n+1}, \dots\}$.

12. Derive the identity

$$\frac{\pi}{4} = \arctan \tfrac{1}{2} + \arctan \tfrac{1}{3}$$

from the addition formula for the tangent function:

$$\tan(u+v) = \frac{\tan u + \tan v}{1 - \tan u \tan v}$$

13. Approximate $\displaystyle\int_0^1 e^{-x^2}\, dx$ by a partial sum of an infinite series, with an error of at most 0.0001.

14. Approximate $\displaystyle\int_0^\pi (\sin x)/x\, dx$ by a partial sum of an infinite series, with an error of at most 0.0001.

15. Approximate $\cos x$ on the interval $[0, \pi/2]$ by a partial sum of its Maclaurin series, with an error of at most 0.0001.

3.7 Speeding Up Convergence

At the end of the preceding section we saw some isolated tricks for speeding up the rate at which a series converges. Now we explore more systematic methods. In all of these methods we begin with a series Σa_k and replace it with a series $\Sigma a_k'$ which converges more rapidly and has the same sum as Σa_k (or has a sum which is related to that of Σa_k in some known way).

There is an especially easy method for speeding up the convergence of an alternating series Σa_k. As usual, we assume that the terms a_k alternate in sign and that the sequence $\{b_k\} = \{|a_k|\}$ decreases monotonically to zero. In addition, we assume that the sequence of differences $\{b_k - b_{k+1}\}$ decreases monotonically. Recall that the partial sums $s_n = \Sigma_{k=1}^n a_k$ are alternately above and below the sum of the series. This suggests that $a_{n+1}/2$ might be a good estimate of the error and that adding it to s_n might drastically reduce the error. Thus, if we let

$$E_n' = \left| \sum_{k=1}^{\infty} a_k - \left(\sum_{k=1}^n a_k + \frac{a_{n+1}}{2} \right) \right| \tag{3.18}$$

denote the error due to this new approximation, we hope to find a smaller bound than that given by the inequality (3.17). In fact,

$$E_n' = \left| \sum_{k=n+1}^{\infty} a_k - \frac{a_{n+1}}{2} \right|$$

$$= \left(b_{n+1} - b_{n+2} \right) + \left(b_{n+3} - b_{n+4} \right) + \cdots \quad - \tfrac{1}{2} b_{n+1}$$

$$= \tfrac{1}{2} \left(b_{n+1} - b_{n+2} \right) - \tfrac{1}{2} \left(b_{n+2} - b_{n+3} \right) + \tfrac{1}{2} \left(b_{n+3} - b_{n+2} \right) - \cdots .$$

But since $\Sigma(-1)^k (b_k - b_{k+1})/2$ is an alternating series whose terms, in absolute value, decrease monotonically to zero, we can apply inequality (3.17) to the above remainder. Thus

$$E_n' \le \tfrac{1}{2} \left(b_{n+1} - b_{n+2} \right) = \tfrac{1}{2} |a_{n+1} + a_{n+2}| \tag{3.19}$$

Furthermore, the last of the above expressions for E_n' can be used to rewrite the original alternating series. Specifically, by letting $n = 0$, we see that

$$\sum_{k=1}^{\infty} (-1)^{k+1} b_k = \tfrac{1}{2} b_1 + \tfrac{1}{2} \sum_{k=1}^{\infty} (-1)^{k+1} (b_k - b_{k+1}) \tag{3.20}$$

provided both $\{b_k\}$ and $\{b_k - b_{k+1}\}$ decrease monotonically to zero. We may therefore interpret E_n' as the error which arises when the sum of the series on the right side of (3.20) is approximated by its nth partial sum. As the next example suggests, we expect this new series to converge much more rapidly than the original one. We will refer to the method summarized in (3.18) and (3.19) as the *accelerated alternating series* method. However, it is just a special case of a more general method, called the *Euler transform* method, in which the technique of forming the successive differences $b_k - b_{k+1}$ is performed repeatedly [see Johnsonbaugh (1979)].

EXAMPLE 3.31   Use the accelerated alternating series method to approximate $\Sigma_{k=1}^{\infty} (-1)^{k+1}/k$ with an error of at most 0.0001.
    By inequality (3.19)

$$E_n' \le \frac{1}{2} \left| \frac{1}{n+1} - \frac{1}{n+2} \right| = \frac{1}{2(n+1)(n+2)} \le \frac{1}{2(n+1)^2}$$

$$\le 0.0001 \quad \text{if } n \ge 70$$

Therefore, $\Sigma_{k=1}^{\infty}(-1)^{k+1}/k$ is within 0.0001 of $\Sigma_{k=1}^{70}(-1)^{k+1}/k + 1/142 = 0.69310$.

Next we improve on the integral inequality (3.14), again by trying to make a simple estimate of the error. As in the integral test, we assume that f is a continuous, non-negative, decreasing function such that $\int_1^{\infty} f(x)\, dx$ is finite, and we let $a_k = f(k)$. We also assume that f' is continuous, negative, and decreasing in absolute value. (These hypotheses are assured if $f > 0$, $f' < 0$, and $f'' > 0$.) As Figure 3.4 suggests, $\int_{n+1/2}^{\infty} f(x)\, dx$ might be a good estimate of the error. In fact, if we let

$$E_n' = \left| \sum_{k=1}^{\infty} a_k - \left(\sum_{k=1}^{n} a_k + \int_{n+1/2}^{\infty} f(x)\, dx \right) \right| \tag{3.21}$$

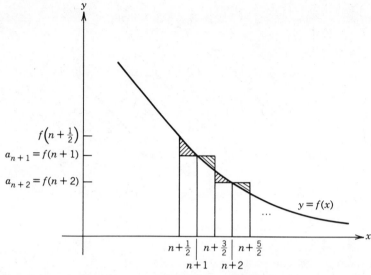

Figure 3.4 $\displaystyle\int_{n+\frac{1}{2}}^{\infty} f(x)\,dx - \sum_{k=n+1}^{\infty} f(k)$

then

$$E_n' = \left| \sum_{k=n+1}^{\infty} a_k - \int_{n+1/2}^{\infty} f(x)\,dx \right|$$

$$= \left| \left(\frac{a_{n+1}}{2} - \int_{n+1/2}^{n+1} f(x)\,dx \right) - \left(\int_{n+1}^{n+3/2} f(x)\,dx - \frac{a_{n+1}}{2} \right) + \cdots \right|$$

and so E_n' is the absolute value of the remainder of an alternating series in which each of the terms in parentheses corresponds to one of the shaded areas in Figure 3.4. If these areas decrease monotonically to zero, then inequality (3.17) would imply that

$$E_n' \le \tfrac{1}{2}\big[f(n + \tfrac{1}{2}) - f(n + 1) \big] \tag{3.22}$$

since the area of the first shaded region is bounded by the area of the rectangle which extends from $n + \frac{1}{2}$ to $n + 1$ horizontally and from $f(n + 1)$ to $f(n + \frac{1}{2})$ vertically.

We will prove that these areas do decrease monotonically to zero and, in the process, derive a special case of a formula due to Euler and Maclaurin [see Boas (1977)]. Let $p(x) = x - [x] - \frac{1}{2}$, where $[x]$ denotes the greatest integer less than or equal to x (Figure 3.5). We first prove that

$$\sum_{k=n+1}^{m} f(k) = \int_{n+1/2}^{m+1/2} f(x)\,dx + \int_{n+1/2}^{m+1/2} p(x) f'(x)\,dx \tag{3.23}$$

by performing on integration by parts on the last integral in (3.23). Before we do, however, we break up this integral into integrals over each of the intervals $[n + \frac{1}{2}, n + 1], [n + 1, n + 2], [n + 2, n + 3], \ldots, [m - 1, m], [m, m + \frac{1}{2}]$, since p can be made into a continuous function on each of these intervals by changing its value at

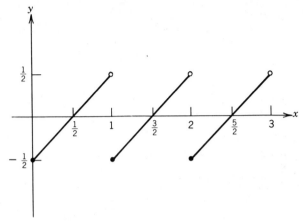

Figure 3.5 $p(x) = x - [x] - \frac{1}{2}$

each right-hand integer endpoint to $\frac{1}{2}$. Thus,

$$\int_k^{k+1} p(x)f'(x)\, dx = p(x)f(x)\big|_k^{k+1} - \int_k^{k+1} p'(x)f(x)\, dx$$

$$= \tfrac{1}{2}f(k+1) + \tfrac{1}{2}f(k) - \int_k^{k+1} f(x)\, dx$$

and so

$$\int_{n+1/2}^{m+1/2} p(x)f'(x)\, dx$$

$$= \int_{n+1/2}^{n+1} p(x)f'(x)\, dx + \sum_{k=n+1}^{m-1} \int_k^{k+1} p(x)f'(x)\, dx + \int_m^{m+1/2} p(x)f'(x)\, dx$$

$$= \tfrac{1}{2}f(n+1) - \int_{n+1/2}^{n+1} f(x)\, dx + \sum_{k=n+1}^{m-1} \left[\tfrac{1}{2}f(k+1) + \tfrac{1}{2}f(k) - \int_k^{k+1} f(x)\, dx \right]$$

$$+ \tfrac{1}{2}f(m) - \int_m^{m+1/2} f(x)\, dx$$

$$= \sum_{k=n+1}^{m} f(k) - \int_{n+1/2}^{m+1/2} f(x)\, dx$$

which proves (3.23). Then, after letting $m \to \infty$, we have the following special case of the Euler–Maclaurin formula:

$$\sum_{k=n+1}^{\infty} f(k) = \int_{n+1/2}^{\infty} f(x)\, dx + \int_{n+1/2}^{\infty} p(x)f'(x)\, dx \qquad (3.24)$$

Therefore, by (3.21),

$$E'_n = \left| \sum_{k=n+1}^{\infty} f(k) - \int_{n+1/2}^{\infty} f(x)\,dx \right|$$

$$= \left| \int_{n+1/2}^{\infty} p(x)f'(x)\,dx \right|$$

$$= \left| \int_{n+1/2}^{n+1} p(x)f'(x)\,dx + \int_{n+1}^{n+3/2} p(x)f'(x)\,dx + \int_{n+3/2}^{n+2} p(x)f'(x)\,dx + \cdots \right|$$

But the terms of this infinite series are alternately negative and positive, since $p(x) \geq 0$ on $[k + \frac{1}{2}, k + 1]$, $p(x) \leq 0$ on $(k, k + \frac{1}{2}]$, and $f'(x) < 0$ everywhere. Furthermore, the first term of the series is greater, in absolute value, than the second term, since the graph of $y = p(x)$ on the interval $(n + \frac{1}{2}, n + \frac{3}{2})$ is symmetric about the point $(n + 1, 0)$ and since $f'(x)$ is decreasing in absolute value. Arguing in the same way for each pair of successive terms of the series, we conclude that the terms decrease monotonically in absolute value. Therefore, by inequality (3.17),

$$E'_n \leq \left| \int_{n+1/2}^{n+1} p(x)f'(x)\,dx \right|$$

$$\leq \int_{n+1/2}^{n+1} |p(x)f'(x)|\,dx$$

$$\leq \int_{n+1/2}^{n+1} \tfrac{1}{2}|f'(x)|\,dx$$

$$= \tfrac{1}{2}\left[f\left(n + \tfrac{1}{2}\right) - f(n + 1) \right]$$

confirming inequality (3.22). We will refer to the method summarized in (3.21) and (3.22) as the *accelerated integral method*.

EXAMPLE 3.32 Use the accelerated integral method to approximate $\sum_{k=1}^{\infty} 1/k^2$ with an error of at most 0.0001.

By inequality (3.22), with $f(x) = 1/x^2$,

$$E'_n \leq \frac{1}{2}\left[\frac{1}{(n + 1/2)^2} - \frac{1}{(n + 1)^2} \right] = \frac{n^2 + 2n + 1 - n^2 - n - 1/4}{2(n + 1/2)^2(n + 1)^2}$$

$$= \frac{n + 3/4}{2(n + 1/2)^2(n + 1)^2} \leq \frac{n + 1}{2(n + 1/2)^3(n + 1)} = \frac{1}{2(n + 1/2)^3}$$

$$\leq 0.0001 \quad \text{if } n \geq 17$$

Therefore, $\sum_{k=1}^{\infty} 1/k^2$ is within 0.0001 of $\sum_{k=1}^{17} 1/k^2 + \int_{17.5}^{\infty} x^{-2}\,dx = \sum_{k=1}^{17} 1/k^2 + 1/17.5 = 1.64495$.

Finally, we apply several ideas of this chapter to the derivation of *Stirling's formula*, a powerful formula for approximating $n!$. We are looking for a sequence x_n

such that $n!/x_n \to 1$, where x_n is much easier to compute than $n!$. But note that $n!/x_n \to 1$ is equivalent to $\log(n!) - \log x_n \to 0$, and also note that

$$\log(n!) = \sum_{k=1}^{n} \log k$$

So instead of approximating $n!$, we approximate $\sum_{k=1}^{n} \log k$, which can be done with the aid of (3.23) with $f(x) = \log x$. We drop the term $\int_{1/2}^{n+1/2} p(x) f'(x)\, dx$, since it is negligible compared to the others; this can be seen from the inequality (3.22), which is valid for $f(x) = \log x$ even though f' is positive, not negative. Thus:

$$\sum_{k=1}^{n} \log k = \text{(approximately)} \int_{1/2}^{n+1/2} \log x\, dx$$

$$= x \log x - x \Big|_{1/2}^{n+1/2}$$

$$= \left(n + \tfrac{1}{2}\right) \log\left(n + \tfrac{1}{2}\right) - \tfrac{1}{2} \log \tfrac{1}{2} - n$$

Furthermore, it is not hard to show (Exercise 9) that $\left(n + \tfrac{1}{2}\right)\left[\log\left(n + \tfrac{1}{2}\right) - \log n\right] \to \tfrac{1}{2}$. Therefore, if we neglect additive constants, we have

$$\log(n!) = \text{(approximately)} \left(n + \tfrac{1}{2}\right) \log n - n$$

To find out how good this approximation is, we let

$$a_n = \log(n!) - \left(n + \tfrac{1}{2}\right) \log n + n$$

and try to find $\lim_{n \to \infty} a_n$. We show that this limit exists by applying the comparison test to the telescoping series $\Sigma(a_k - a_{k+1})$.

$$|a_k - a_{k+1}| = \log(k!) - \log(k+1)! - \left(k + \tfrac{1}{2}\right) \log k$$

$$+ \left(k + \tfrac{3}{2}\right) \log(k+1) - 1$$

$$= \left(k + \tfrac{1}{2}\right)\left[\log(k+1) - \log k\right] - 1$$

$$= \left(k + \tfrac{1}{2}\right) \log\left(1 + \frac{1}{k}\right) - 1$$

$$\leq \left(k + \tfrac{1}{2}\right)\left(\frac{1}{k} - \frac{1}{2k^2} + \frac{1}{3k^3}\right) - 1$$

$$= 1 + \frac{1}{2k} - \frac{1}{2k} - \frac{1}{4k^2} + \frac{1}{3k^2} + \frac{1}{6k^3} - 1$$

$$= \frac{1}{12k^2} + \frac{1}{6k^3} \leq \frac{1}{12k^2} + \frac{1}{6k^2} = \frac{1}{4k^2}$$

where the first inequality comes from the fact that the partial sums of an alternating series are alternately above and below the sum. Specifically,

$$\log(1 + x) = \sum_{k=1}^{\infty} (-1)^k \frac{x^k}{k} \leq \sum_{k=1}^{3} (-1)^k \frac{x^k}{k}$$

Since $\Sigma 1/4k^2$ converges, so does $\Sigma(a_k - a_{k+1})$. But the nth partial sum of this telescoping series is $a_1 - a_{n+1}$, from which we conclude that $\lim_{n \to \infty} a_n$ exists. Although we do not yet know the value of $\lim_{n \to \infty} a_n$, we can at least conclude that

$$e^{a_n} = e^{\log(n!) - (n+\frac{1}{2})\log n + n}$$

$$= e^{\log(n!)} e^{-n \log n} e^{-\frac{1}{2}\log n} e^n$$

$$= \frac{n!}{n^n \sqrt{n}\, e^{-n}}$$

and that

$$\lim_{n \to \infty} \frac{n!}{n^n \sqrt{n}\, e^{-n}} = \lim_{n \to \infty} e^{a_n} = L$$

for some L. We find L with the aid of Wallis' formula (Exercise 18 of Section 3.3). If b_n denotes $n!/n^n \sqrt{n}\, e^{-n}$, then

$$\frac{b_n^2}{b_{2n}} = \frac{(n!)^2 (2n)^{2n} \sqrt{2n}\, e^{-2n}}{(2n)!(n^n \sqrt{n}\, e^{-n})^2} = \frac{(n!)^2 2^{2n} \sqrt{2n}}{(2n)! n}$$

$$= \frac{(n!)^2 2^{2n} \sqrt{2n}}{1 \cdot 3 \cdot 5 \cdots (2n-1) n! 2^n n}$$

$$= \frac{2 \cdot 4 \cdot 6 \cdots (2n) \sqrt{2n}}{1 \cdot 3 \cdot 5 \cdots (2n-1) n}$$

$$= \left[\frac{2 \cdot 2 \cdot 4 \cdot 4 \cdots (2n)(2n)}{1 \cdot 3 \cdot 3 \cdot 5 \cdots (2n-1)(2n+1)} \right]^{1/2} \frac{\sqrt{2n}\sqrt{2n+1}}{n}$$

which, by Wallis' formula, has the limit $2\sqrt{\pi/2} = \sqrt{2\pi}$. And, since $b_n^2/b_{2n} \to L^2/L = L$, it follows that $L = \sqrt{2\pi}$. We have now derived Stirling's formula:

$$\lim_{n \to \infty} \frac{n!}{n^n \sqrt{2\pi n}\, e^{-n}} = 1 \tag{3.25}$$

EXERCISES

In Exercises 1–4, approximate the sum of the series (using the methods of this section) with an error of
  (a) at most 0.1 (A calculator will be helpful);
  (b) at most 0.0001 (Use your program from Section 3.1).

1. $\displaystyle\sum_{k=0}^{\infty} 4(-1)^k/(2k+1)$

2. $\displaystyle\sum_{k=2}^{\infty} (-1)^k/\log k$

3. $\displaystyle\sum_{k=1}^{\infty} (2k + \sqrt{k})/k^3$

4. $\displaystyle\sum_{k=1}^{\infty} (\log k)/k^2$

In *Kummer's acceleration method*, one writes

$$\sum a_k = \sum b_k + \sum (a_k - b_k)$$

where $\sum a_k$ is a series whose sum is to be approximated and where $\sum b_k$ is a series whose sum is known and whose terms are so close to the terms a_k that the series $\Sigma(a_k - b_k)$ converges much faster than Σa_k. For example,

$$\sum_{k=1}^{\infty} \frac{1}{k^2} = \sum_{k=1}^{\infty} \frac{1}{k(k+1)} + \sum_{k=1}^{\infty} \left(\frac{1}{k^2} - \frac{1}{k(k+1)} \right)$$

$$= 1 + \sum_{k=1}^{\infty} \frac{1}{k^2(k+1)}$$

and $\Sigma 1/k^2(k+1)$ converges much faster than $\Sigma 1/k^2$. So it is advantageous to approximate $\Sigma_{k=1}^{\infty}1/k^2$ by $1 + \Sigma_{k=1}^{n}1/k^2(k+1)$. In Exercises 5–7, use Kummer's acceleration method to approximate the sum of the first of the two given series by accelerating its convergence with the aid of the second series. The error is to be (a) at most 0.1 (A calculator will be helpful); (b) at most 0.0001 (Use your program from Section 3.1).

5. $\displaystyle\sum_{k=1}^{\infty} 1/k^2$ $\left(\text{Use } \displaystyle\sum_{k=1}^{\infty} 1/k(k+1) = 1 \right)$

6. $\displaystyle\sum_{k=1}^{\infty} k/(k^3 + 1)$ $\left(\text{Use } \displaystyle\sum_{k=1}^{\infty} 1/k(k+1) = 1 \right)$

7. $\displaystyle\sum_{k=1}^{\infty} 1/k^3$ $\left(\text{Use } \displaystyle\sum_{k=1}^{\infty} 1/k(k+1)(k+2) = \tfrac{1}{4} \right)$

8. Show that the nth partial sum of the series on the right side of (3.20) is the average of the nth and $(n+1)$st partial sums of the series on the left side.

9. Prove that $\lim_{n \to \infty}(n + \tfrac{1}{2})[\log(n + \tfrac{1}{2}) - \log n] = \tfrac{1}{2}$.

10. Find $\displaystyle\lim_{n \to \infty} \frac{(n!)^{1/n}}{n}$

3.8 Uniform Convergence of Series

In this section and the next, we develop the basic theory for applying the operations of calculus to infinite series of functions. The first step is to extend the theory of uniform convergence, introduced in Section 1.6, from sequences to series.

Definition 3.7 Suppose $\Sigma_{k=1}^{\infty} f_k$ is an infinite series of functions which all have the domain D. Let $\{ s_n \}$ denote its sequence of partial sums; thus $s_n(x) = \Sigma_{k=1}^{n} f_k(x)$. We say that the series $\Sigma_{k=1}^{\infty} f_k$ converges *pointwise* to the sum f on D if the sequence $\{ s_n \}$ converges pointwise to f on D. Then we write

$$\sum_{k=1}^{\infty} f_k = f \quad \text{on } D$$

We say that $\Sigma_{k=1}^{\infty} f_k$ converges *uniformly* to the sum f on D if $\{ s_n \}$ converges uniformly to f on D. Then we write

$$\sum_{k=1}^{\infty} f_k = f \quad \text{uniformly on } D$$

Our main result, Theorem 3.15, is an easy extension of Theorems 1.12 and 1.13.

Theorem 3.15 Suppose $\sum_{k=1}^{\infty} f_k$ is a series of continuous functions on an interval D and suppose $\sum_{k=1}^{\infty} f_k = f$ uniformly on D.

(a) Then f is also continuous on D.

(b) If $D = [a, b]$ then $\sum_{k=1}^{\infty} \int_a^b f_k(x)\, dx = \int_a^b \sum_{k=1}^{\infty} f_k(x)\, dx$.

Proof

(a) See Exercise 1.

(b) We are given that the sequence $\{s_n\}$ (where $s_n(x) = \sum_{k=1}^{n} f_k(x)$) converges uniformly to f on $[a, b]$, and we know that each function s_n is continuous on $[a, b]$ since s_n is a finite sum of continuous functions (Theorem A.3a). Hence, by Theorem 1.13,

$$\int_a^b s_n(x)\, dx \to \int_a^b f(x)\, dx = \int_a^b \sum_{k=1}^{\infty} f_k(x)\, dx \qquad (3.26)$$

But, by Theorem A.24a,

$$\int_a^b s_n(x)\, dx = \int_a^b \sum_{k=1}^{n} f_k(x)\, dx = \sum_{k=1}^{n} \int_a^b f_k(x)\, dx$$

which is the nth partial sum of the series $\sum_{k=1}^{\infty} \int_a^b f_k(x)\, dx$. Therefore (3.26) is, in fact, the conclusion we were to prove.

The next theorem, an analog of the comparison test (Theorem 3.10), will be our main tool for proving that a series of functions converges uniformly.

Theorem 3.16 (Weierstrass' M-test) Suppose $\sum f_k$ is a series of functions on the domain D, and suppose $\sum M_k$ is a convergent series of positive numbers such that

$$|f_k(x)| \le M_k \quad \text{for all } x \text{ in } D$$

Then $\sum f_k$ converges uniformly on D.

Proof By the comparison test, $\sum f_k(x)$ converges for each x in D; we denote its sum by $f(x)$. Then, by the triangle inequality, we note that

$$\left| \sum_{k=n+1}^{m} f_k(x) \right| \le \sum_{k=n+1}^{m} |f_k(x)| \le \sum_{k=n+1}^{m} M_k$$

for all $m > n$ and all x in D. And, by Corollary 3.9, for every $\varepsilon > 0$ there exists N such that

$$\sum_{k=n+1}^{m} M_k < \frac{\varepsilon}{2} \quad \text{for all } n \ge N \text{ and all } m > n$$

Hence, for all $n \ge N$ and all $m > n$ and all x in D,

$$\left| \sum_{k=n+1}^{m} f_k(x) \right| \le \sum_{k=n+1}^{m} M_k < \frac{\varepsilon}{2}$$

But, for each fixed value of x and of n, we can choose m (which may depend on x and n) such that $m > n$ and

$$\left| \sum_{k=1}^{m} f_k(x) - f(x) \right| < \frac{\varepsilon}{2}$$

since the series $\Sigma f_k(x)$ converges to $f(x)$. Therefore, for all $n \geq N$ and all x in D,

$$\left| \sum_{k=1}^{n} f_k(x) - f(x) \right| = \left| \left(\sum_{k=1}^{m} f_k(x) - f(x) \right) - \sum_{k=n+1}^{m} f_k(x) \right|$$

$$\leq \left| \sum_{k=1}^{m} f_k(x) - f(x) \right| + \left| \sum_{k=n+1}^{m} f_k(x) \right|$$

$$< \frac{\varepsilon}{2} + \frac{\varepsilon}{2} = \varepsilon$$

EXAMPLE 3.33   Show that $\Sigma \sin(kx)/k^2$ converges uniformly on the whole real line.

$$\left| \frac{\sin(kx)}{k^2} \right| \leq \frac{1}{k^2} \quad \text{for all } x$$

and $\Sigma 1/k^2$ is a convergent p-series; therefore, the M-test guarantees that $\Sigma \sin(kx)/k^2$ converges uniformly on the whole real line. Furthermore, since $\sin(kx)/k^2$ is continuous for each k, so is $\Sigma \sin(kx)/k^2$ (Theorem 3.15a). And, for every interval $[a, b]$, $\int_a^b \Sigma \sin(kx)/k^2 \, dx = \Sigma \int_a^b \sin(kx)/k^2 \, dx$ (Theorem 3.15b).

EXAMPLE 3.34   Rigorously justify the steps of Example 3.8, in which we derived the Maclaurin series for arctan x.

Recall that, if $-1 < x < 1$, then

$$\frac{1}{1-x} = \sum_{k=0}^{\infty} x^k \qquad \text{(the geometric series)}$$

and so, by replacing x with $-x^2$,

$$\frac{1}{1+x^2} = \sum_{k=0}^{\infty} (-1)^k x^{2k}$$

The remaining step is to integrate both sides of this equation. First, let t be any number in the interval $(-1, 1)$. Then, since $\Sigma(t^2)^k$ is a convergent geometric series and since $|(-1)^k x^{2k}| \leq (t^2)^k$ for x in $[-|t|, |t|]$, the M-test assures us that $\Sigma(-1)^k x^{2k}$ converges uniformly on $[-|t|, |t|]$. Therefore, for all t in the interval $(-1, 1)$,

$$\int_0^t \frac{1}{1+x^2} \, dx = \int_0^t \sum_{k=0}^{\infty} (-1)^k x^{2k} \, dx$$

or, by Theorem 3.15b,

$$\arctan t = \sum_{k=0}^{\infty} (-1)^k \frac{t^{2k+1}}{2k+1}$$

Note, however, we have not yet proved that $\sum_{k=0}^{\infty}(-1)^k x^{2k+1}/(2k+1)$ is actually the Maclaurin series of arctan x. This will follow from the theory of power series in the next section.

EXERCISES

1. Prove Theorem 3.15a.

In Exercises 2–9, show that the series converges uniformly on the indicated interval.

2. $\sum \sin(kx)/k^{3/2}$ on $(-\infty, \infty)$

3. $\sum 1/(1 + k^2 x)$ on $[1, \infty)$

4. $\sum x/(x + k^2)$ on $[0, 1]$

5. $\sum x^2/(1 + k^2 x^2)$ on $(-\infty, \infty)$

6. $\sum x^k$ on $[-t, t]$,   where $0 < t < 1$

7. $\sum x^k/(1 + 2^k)$ on $[-t, t]$,   where $0 < t < 2$

8. $\sum x^k/(1 + x^{2k})$ on $[0, t]$,   where $t < 1$

9. $\sum x^k/(1 + x^{2k})$ on $[t, \infty)$,   where $t > 1$

10. Expand $\int_0^{2\pi} \cos(x \cos \theta)/2\pi \, d\theta$ as a Maclaurin series in the variable x. *Hint:* Assume that $\int_0^{2\pi} \cos^{2k}\theta \, d\theta = 2\pi(2k)!/(2^{2k}(k!)^2)$. (The Maclaurin series you are to find is the usual form for the Bessel function $J_0(x)$, which is studied in Chapter 4.)

11. **(a)** Show that $\sum_{k=1}^{\infty} x/k(x + k)$ converges uniformly on $[0, 1]$.

    (b) Show that $\int_0^1 \sum_{k=1}^{\infty} x/k(x + k) \, dx = \lim_{n \to \infty}(\sum_{k=1}^{n} 1/k - \log(n + 1)) = \lim_{n \to \infty}(\sum_{k=1}^{n} 1/k - \log n)$.

    (c) Use (b) to approximate the Euler constant (Exercise 19 of Section 3.4) with an error of at most (i) 0.5; (ii) 0.0001.

12. Suppose the series $\sum_{k=1}^{\infty} f_k'$ converges uniformly on an interval I, and each function f_k' is continuous on I, and $\sum_{k=1}^{\infty} f_k(a)$ converges for some point a in I. Prove that $\sum_{k=1}^{\infty} f_k$ converges on I and

$$\frac{d}{dx}\left(\sum_{k=1}^{\infty} f_k(x) \right) = \sum_{k=1}^{\infty} f_k'(x) \quad \text{for all } x \text{ in } I$$

3.9 Power Series

We especially want to apply the theory of Section 3.8 to Taylor series:

$$\sum_{k=0}^{\infty} f^{(k)}(x_0)(x - x_0)^k/k!$$

It is just as easy, however, to apply it to the seemingly more general power series in which the coefficients of $(x - x_0)^k$ are arbitrary numbers.

Definition 3.8 A *power series about* x_0 is a series of the form

$$\sum_{k=0}^{\infty} c_k(x - x_0)^k$$

First we prove that the set of points at which a power series converges is always an interval centered at x_0.

Theorem 3.17 For every power series $\sum_{k=0}^{\infty} c_k(x - x_0)^k$ there exists R, where $0 \le R \le \infty$, such that

$$\sum_{k=0}^{\infty} c_k(x - x_0)^k \text{ converges absolutely if } |x - x_0| < R$$

$$\sum_{k=0}^{\infty} c_k(x - x_0)^k \text{ diverges if } |x - x_0| > R$$

Proof In this proof and others in this section, we simplify our notation by assuming that $x_0 = 0$. There is no loss of validity in simplifying thus, since the general case of Theorem 3.17 follows from this special case by replacing x with $x - x_0$.

Let $R = \text{lub}\{|x| \,|\, \Sigma c_k x^k \text{ converges}\}$. ($R$ is defined to be ∞ if this least upper bound does not exist.) If $|x| < R$, we can choose y such that $|x| < |y|$ and $\Sigma c_k y^k$ converges. But, by Theorem 3.2, $c_k y^k \to 0$, and so there is a bound M such that $|c_k y^k| \le M$ for all k (Theorem 3.5). Hence

$$|c_k x^k| = |c_k y^k| \left(\frac{|x|}{|y|} \right)^k \le M \left(\frac{|x|}{|y|} \right)^k$$

Since $\Sigma(|x|/|y|)^k$ is a convergent geometric series, $\Sigma c_k x^k$ must therefore converge absolutely (comparison test). On the other hand, if $|x| > R$ then the definition of R implies that $\Sigma c_k x^k$ diverges.

Definition 3.9 The quantity R in Theorem 3.17 is called the *radius of convergence* of the power series $\sum_{k=0}^{\infty} c_k(x - x_0)^k$. The set of points x for which the series converges is called the *interval of convergence*.

By Theorem 3.17, the interval of convergence of a power series with radius of convergence R must have one of the forms

$$(x_0 - R, x_0 + R), [x_0 - R, x_0 + R], (x_0 - R, x_0 + R],$$
$$[x_0 - R, x_0 + R) \quad \text{if } 0 < R < \infty$$
$$[x_0, x_0] \quad \text{if } R = 0$$
$$(-\infty, \infty) \quad \text{if } R = \infty$$

In practice, we can usually find the radius of convergence of a power series from the ratio test and can then use other convergence tests to find out whether $x_0 - R$ and $x_0 + R$ are also in the interval of convergence.

EXAMPLE 3.35 Find the radius and interval of convergence of the power series $\sum_{k=0}^{\infty} (-1)^k x^{2k+1}/(2k + 1)$.

This problem was solved in Example 3.24, in which we used the ratio test to find that $R = 1$ and other convergence tests to find that the interval of convergence is $[-1, 1]$.

EXAMPLE 3.36   Find the radius and interval of convergence of $\sum_{k=0}^{\infty} x^k/k!$.

$$\frac{|a_{k+1}|}{|a_k|} = \frac{|x|^{k+1}/(k+1)!}{|x|^k/k!} = \frac{|x|}{k+1} \to 0$$

Thus $R = \infty$, since the above limit is 0 for all x and $0 < 1$. So the interval of convergence is $(-\infty, \infty)$.

EXAMPLE 3.37   Find the radius and interval of convergence of $\sum_{k=0}^{\infty} k! x^k$.

$$\frac{|a_{k+1}|}{|a_k|} = \frac{(k+1)!|x|^{k+1}}{k!|x|^k} = (k+1)|x| \to \infty \quad \text{if } x \neq 0$$

Thus $R = 0$, since the above limit is ∞ for all $x \neq 0$ and $\infty > 1$. The interval of convergence is $[0,0]$.

An efficient formula for computing the radius of convergence of a power series $\sum c_k(x - x_0)^k$ is

$$R = \lim_{k \to \infty} \left| \frac{c_k}{c_{k+1}} \right| \tag{3.27}$$

provided the limit exists (Exercise 16).

It looks like it should be easy to differentiate and integrate a power series. For example, it looks reasonable to write

$$\frac{d}{dx}\left(c_0 + c_1 x + c_2 x^2 + c_3 x^3 + \cdots \right) = c_1 + 2c_2 x + 3c_3 x^2 + \cdots$$

This is indeed valid; to prove it we apply the theory of uniform convergence from the preceding section.

Lemma 3.18   If R is the radius of convergence of a power series $\sum_{k=0}^{\infty} c_k x^k$, then the series converges uniformly on $[-t, t]$ whenever $0 < t < R$. Furthermore, the sum of this series is a continuous function on the open interval $(-R, R)$.

Proof   Since $|c_k x^k| \leq |c_k| t^k$ for all x in $[-t, t]$, $\sum_{k=0}^{\infty} c_k x^k$ converges uniformly on $[-t, t]$, by the M-test. Therefore, by Theorem 3.15a, the sum is continuous on $[-t, t]$. Thus, the sum is continuous at x whenever $-R < x < R$, since for every such x we can find a number t such that $-t < x < t$ and $0 < t < R$.

Theorem 3.19 $\sum_{k=0}^{\infty} c_k(x - x_0)^k$ and $\sum_{k=1}^{\infty} kc_k(x - x_0)^{k-1}$ have the same radius of convergence. Furthermore, if R is this radius of convergence, then

$$\frac{d}{dx} \sum_{k=0}^{\infty} c_k(x - x_0)^k = \sum_{k=1}^{\infty} kc_k(x - x_0)^{k-1} \tag{3.28}$$

whenever $|x - x_0| < R$ and R is nonzero.

Proof   Again we may assume that $x_0 = 0$. To show that the radius of convergence, R, of the series $\sum kc_k x^{k-1}$ equals the radius of convergence, S, of the series $\sum c_k x^k$, it suffices to prove the following: if $|x| < S$ then $\Sigma kc_k x^{k-1}$ converges absolutely, and if $|x| < R$ then $\Sigma c_k x^k$ converges absolutely. First suppose that $|x| < R$. Then $\Sigma kc_k x^{k-1}$ converges absolutely, which means that $\Sigma |kc_k x^{k-1}|$ converges. But

$$|c_k x^k| \le |kc_k x^k| = |x| \, |kc_k x^{k-1}|$$

and so $\Sigma |c_k x^k|$ also converges (comparison test). Therefore, $R \le S$. Conversely, suppose that $|x| < S$. We then choose y so that $|x| < |y| < S$. Thus, $\Sigma c_k y^k$ converges, from which it follows that the terms $|c_k y^k|$ converge to 0 and so are bounded by some number M (Theorem 3.5). Hence

$$|kc_k x^{k-1}| = \left| c_k y^k \left(\frac{x}{y} \right)^k \frac{k}{x} \right| \le \frac{Mr^k k}{|x|}$$

where $r = |x/y| < 1$. But, by the ratio test, $\Sigma Mr^k k/|x|$ converges, and so $\Sigma |kc_k x^{k-1}|$ also converges. Therefore, $S \le R$. This completes the proof that the two series have the same radius of convergence.

It remains to verify (3.28). If $|x| < R$ then $\Sigma_{k=1}^{\infty} kc_k t^{k-1}$ converges uniformly on $[-|x|, |x|]$ (Lemma 3.18); so, by Theorem 3.15b,

$$\int_0^x \sum_{k=1}^{\infty} kc_k t^{k-1} \, dt = \sum_{k=1}^{\infty} \int_0^x kc_k t^{k-1} \, dt$$

$$= \sum_{k=1}^{\infty} c_k x^k$$

Finally, if we differentiate both sides of this equation with respect to x, we get

$$\sum_{k=1}^{\infty} kc_k x^{k-1} = \frac{d}{dx} \sum_{k=0}^{\infty} c_k x^k$$

The left-hand expression comes from the fundamental theorem of calculus (Theorem A.23a), and the initial value $k = 0$ in the right-hand expression is justified by the fact that $\frac{d}{dx} c_0 = 0$. This completes the proof.

Note that Theorem 3.19 implies that

$$\sum_{k=0}^{\infty} c_k (x - x_0)^k \quad \text{and} \quad \sum_{k=0}^{\infty} c_k (x - x_0)^{k+1}/(k + 1)$$

also have the same radius of convergence R, since the terms in the first series are the derivatives of the corresponding terms in the second series. Furthermore, equation (3.28) implies that, for some constant c,

$$\int \sum_{k=0}^{\infty} c_k (x - x_0)^k \, dx = \sum_{k=0}^{\infty} c_k \frac{(x - x_0)^{k+1}}{k + 1} + c \tag{3.29}$$

whenever $|x - x_0| < R$, since this just says that the derivative of the series on the right is the series inside the integral.

Now that we have finally justified term-by-term differentiation and integration of power series, let's carefully rework some problems of the type considered in Section 3.2 for which we were unable to justify some of the steps.

EXAMPLE 3.38   Using the geometric series $\sum_{k=0}^{\infty} x^k = 1/(1 - x)$, where $|x| < 1$, find the sums of the power series (a) $\sum_{k=0}^{\infty}(k + 1)x^k$, (b) $\sum_{k=0}^{\infty}(k + 2)(k + 1)x^k$, (c) $\sum_{k=0}^{\infty}(x + 3)^{2k}/4^k$, (d) $\sum_{k=1}^{\infty}(-1)^k(x - 1)^k/k$. Also find the interval of convergence of each series.

By Theorem 3.19 applied to the geometric series,

$$\frac{d}{dx}(1 - x)^{-1} = \frac{d}{dx}\sum_{k=0}^{\infty} x^k$$

or

$$-(1 - x)^{-2}\frac{d}{dx}(1 - x) = \sum_{k=1}^{\infty} kx^{k-1}$$

or, after replacing $k - 1$ by j,

$$(1 - x)^{-2} = \sum_{j=0}^{\infty}(j + 1)x^j$$

Thus, the sum of the power series (a) is $(1 - x)^{-2}$. Note that Theorem 3.19 also tells us that the power series (a) has radius of convergence 1 since the geometric series does, but that it gives us no information about the behavior of the series at the two endpoints of its interval of convergence. However, when $x = \pm 1$ the terms of the series do not converge to zero. Therefore the interval of convergence of series (a) is $(-1, 1)$. Next, by Theorem 3.19 applied to part (a) of this example,

$$\frac{d}{dx}(1 - x)^{-2} = \frac{d}{dx}\sum_{k=0}^{\infty}(k + 1)x^k$$

or

$$2(1 - x)^{-3} = \sum_{k=1}^{\infty}(k + 1)kx^{k-1} = \sum_{j=0}^{\infty}(j + 2)(j + 1)x^j$$

Thus, the sum of the power series (b) is $2(1 - x)^{-3}$; its interval of convergence is again $(-1, 1)$. Next, if we replace x by $[(x + 3)/2]^2$ in the geometric series, we get

$$\sum_{k=0}^{\infty}\frac{(x + 3)^{2k}}{4^k} = \frac{1}{1 - [(x + 3)/2]^2} = \frac{-4}{x^2 + 6x + 5}$$

if $|(x + 3)/2| < 1$. So 2 is the radius of convergence of the power series (c) and $(-5, -1)$ is its interval of convergence. To find the sum of the power series (d), we integrate the geometric series; thus, by formula (3.29)

$$\int \frac{1}{1 - x} dx = \sum_{k=0}^{\infty}\frac{x^{k+1}}{k + 1} + c$$

or

$$-\log(1 - x) = \sum_{j=1}^{\infty}\frac{x^j}{j} + c$$

But $c = 0$ since $\log 1 = 0$ and since the sum of $\sum_{j=1}^{\infty} x^j/j$ is zero when $x = 0$.

Finally, we replace x by $1 - x$ and get

$$\sum_{k=1}^{\infty} \frac{(-1)^k (x-1)^k}{k} = -\log\left(1 - (1-x)\right) = -\log x$$

where $|1 - x| < 1$ or $0 < x < 2$. The actual interval of convergence of this power series is $(0, 2]$, since we get the divergent harmonic series at $x = 0$ and the negative of the convergent alternating harmonic series at $x = 2$. Theorem 3.19 does not assure us that the sum of this series is $-\log 2$ at the endpoint $x = 2$, but it is true; the technique used in Example 3.25 will confirm it.

We still do not know, however, that the four power series in Example 3.38 are actually the Taylor series of the functions they sum to. The next theorem assures us that this is always the case.

Theorem 3.20   If $R > 0$ and if $f(x) = \sum_{k=0}^{\infty} c_k (x - x_0)^k$ whenever $|x - x_0| < R$, then f has derivatives of all orders on the interval $(x_0 - R, x_0 + R)$ and

$$c_k = \frac{f^{(k)}(x_0)}{k!} \quad \text{for } k = 0, 1, 2, \dots \tag{3.30}$$

Proof   By repeated use of (3.28), we get

$$f^{(n)}(x) = \sum_{k=n}^{\infty} k(k-1) \cdots (k - n + 1) c_k (x - x_0)^{k-n}$$

So if we let $x = x_0$ in this series, every term becomes zero except for the $k = n$ term. Thus,

$$f^{(n)}(x_0) = n(n-1) \cdots (n - n + 1) c_n = n! c_n$$

which proves (3.30).

Corollary 3.21   If $R > 0$ and if $\sum_{k=0}^{\infty} c_k (x - x_0)^k = \sum_{k=0}^{\infty} d_k (x - x_0)^k$ whenever $|x - x_0| < R$, then $c_k = d_k$ for $k \geq 0$.

Proof   Let $f(x)$ denote the sum of $\sum_{k=0}^{\infty} c_k (x - x_0)^k$ and $g(x)$ denote the sum of $\sum_{k=0}^{\infty} d_k (x - x_0)^k$. Then, by Theorem 3.20, $c_k = f^{(k)}(x_0)/k!$ and $d_k = g^{(k)}(x_0)/k!$ for all k. But we are given that $f(x) = g(x)$ whenever $|x - x_0| < R$, and so f and g have the same kth derivatives; therefore, $c_k = d_k$.

Finally, we return to Table 3.2 to see if we can now confirm all the facts it contains about Taylor series of familiar calculus functions. Recall that Taylor's remainder theorem was the only tool we needed in showing that e^x, $\sin x$, and $\cos x$ are the sums of their Maclaurin series. Also recall that, according to Examples 3.25 and 3.34,

$$\arctan x = \sum_{k=0}^{\infty} (-1)^k \frac{x^{2k+1}}{2k + 1} \quad \text{when } -1 \leq x \leq 1$$

Thanks to Theorem 3.20, we can also conclude that the above power series is actually the Maclaurin series for $\arctan x$ and that $\sum_{k=0}^{\infty} x^k$ is the Maclaurin series for $1/(1 - x)$. Similar techniques can be used to confirm the Maclaurin series for $\log(1 + x)$.

To confirm the binomial series expansion for $(1 + x)^p$ we use a completely different technique. It is not hard to check that

$$1 + \sum_{k=1}^{\infty} p(p - 1) \cdots (p - k + 1)\frac{x^k}{k!} \tag{3.31}$$

is the Maclaurin series for $(1 + x)^p$ and that, by the ratio test, its radius of convergence is 1. It remains only to show that $(1 + x)^p$ is the sum of the series (3.31).

EXAMPLE 3.39   Show that if $f(x)$ denotes the sum of the series (3.31), then $f(x) = (1 + x)^p$ whenever $-1 < x < 1$.

Note that $y = (1 + x)^p$ satisfies the differential equation $y' = py/(1 + x)$ for $x > -1$ and also satisfies the initial condition $y(0) = 1$. The function $y = f(x)$ satisfies $y(0) = 1$ too. We show that $y = f(x)$ also satisfies the differential equation $y' = py/(1 + x)$:

$$y'(1 + x) = \sum_{k=1}^{\infty} kp(p - 1) \cdots (p - k + 1)\frac{x^{k-1}}{k!}(1 + x)$$

$$= \sum_{k=1}^{\infty} p(p - 1) \cdots (p - k + 1)\frac{x^{k-1}}{(k - 1)!}$$

$$+ \sum_{k=1}^{\infty} kp(p - 1) \cdots (p - k + 1)\frac{x^k}{k!}$$

$$= p + \sum_{k=1}^{\infty} p(p - 1) \cdots (p - k)\frac{x^k}{k!}$$

$$+ \sum_{k=1}^{\infty} kp(p - 1) \cdots (p - k + 1)\frac{x^k}{k!}$$

$$= p + \sum_{k=1}^{\infty} p(p - 1) \cdots (p - k + 1)[p - k + k]\frac{x^k}{k!}$$

$$= py$$

But the equation $y'(1 + x) = py$ is separable. Thus

$$\int \frac{dy}{y} = \int \frac{p}{1 + x}\, dx$$

or

$$\log y = p \log(1 + x) + c$$

or

$$y = ae^{p \log(1+x)} \quad \text{where } a = e^c$$

$$= a(1 + x)^p$$

Since $y = f(x)$ and $f(0) = 1$, it follows that $a = 1$; therefore $f(x) = (1 + x)^p$. (Note that if we were to use the uniqueness theorem for differential equations, which is proved in the next section, we would not need to solve the differential equation to reach the conclusion that $f(x) = (1 + x)^p$.)

In Chapter 4, we will turn around the reasoning of Example 3.39 to convert it into a method for solving differential equations. Specifically, we will seek power series solutions by substituting a power series with unknown coefficients into a differential equation and then solving for the coefficients.

Power series also make sense when x is a complex variable and the coefficients c_k are complex. Theorem 3.17 and its proof are still valid as stated, provided the absolute value signs are interpreted as indicating the absolute value of a complex number. Theorem 3.19 is also valid if x is a complex variable, but a different proof is required since the theory of the derivative and integral for functions of a complex variable is different from that for functions of a real variable.

EXERCISES

In Exercises 1–4, find (a) the radius of convergence of the given power series; (b) the interval of convergence of the given power series.

1. $\displaystyle\sum 2^k x^k / k^3$

2. $\displaystyle\sum x^k / \sqrt{k}\, 3^k$

3. $\displaystyle\sum 3^k (x - 2)^k / k!$

4. $\displaystyle\sum k(x + 3)^k / 2^k$

In Exercises 5–10, find the sum of the series by performing operations of algebra or calculus on one of the Taylor series in Table 3.2.

5. $\displaystyle\sum_{k=0}^{\infty} (x - 1)^{3k+1}$

6. $\displaystyle\sum_{k=0}^{\infty} (k + 1)2^k x^k$

7. $\displaystyle\sum_{k=0}^{\infty} x^{2k+1} / (2k + 1)$

8. $\displaystyle\sum_{k=0}^{\infty} x^{2k+1} / (2k + 1)!$

9. $\displaystyle\sum_{k=1}^{\infty} (-1)^{k-1} 2k x^{2k-1} / (2k - 1)!$

10. $\displaystyle x + \sum_{k=1}^{\infty} \left(\frac{-1}{2}\right)\left(\frac{-3}{2}\right) \cdots \left(\frac{1}{2} - k\right) \frac{x^{2k+1}}{(2k + 1)k!}$

In Exercises 11–15, find the sum of the power series by the method of Example 3.39. That is, find the radius of convergence of the series, show that the sum satisfies the given initial-value problem, and solve the initial-value problem.

11. $\displaystyle\sum_{k=0}^{\infty} x^k / k!$; $y' = y$, $y(0) = 1$

12. $\displaystyle\sum_{k=0}^{\infty} (-1)^k x^{2k} / (2k)!$; $y'' + y = 0$, $y(0) = 1$, $y'(0) = 0$

13. $\displaystyle\sum_{k=0}^{\infty} (-1)^k x^{2k+1} / (2k + 1)!$; $y'' + y = 0$, $y(0) = 0$, $y'(0) = 1$

14. $\displaystyle\sum_{k=0}^{\infty} x^{2k}/(2k)!$; $y'' - y = 0$, $y(0) = 1$, $y'(0) = 0$

15. $\displaystyle\sum_{k=0}^{\infty} x^{2k+1}/(2k+1)!$; $y'' - y = 0$, $y(0) = 0$, $y'(0) = 1$

16. If $\lim_{k \to \infty} |c_k/c_{k+1}| = R$, prove that the power series $\Sigma c_k(x - x_0)^k$ has radius of convergence R.

17. If $\lim_{k \to \infty} 1/\sqrt[k]{|c_k|} = R$, prove that the power series $\Sigma c_k(x - x_0)^k$ has radius of convergence R. *Hint:* Use Exercise 27 of Section 3.5.

3.10 A Fundamental Theorem on Differential Equations

In this section we use the deeper theory of Section 3.8 to prove one of the most fundamental facts about differential equations—that an initial-value problem for a first-order equation has a unique solution. Useful background material for our discussion is contained in Section D.1. Existence and uniqueness of solutions is discussed there, and a theorem much like the main one of this section is stated and discussed. Also, a number of terms are defined in Section D.1 and not defined again here. Two that will be especially important in the main theorem of this section are "continuity" for a function of two variables and the "Lipschitz condition" for a function of two variables.

The initial-value problem to be discussed in this section is

$$\frac{dy}{dx} = F(x, y), \qquad y(a) = y_0, \quad \text{where } a \le x \le b \qquad (3.32)$$

and F is continuous in both its variables together. Recall that a function g is a solution of this initial-value problem iff $g(a) = y_0$, g is differentiable at every point of the interval $[a, b]$, and

$$g'(x) = F(x, g(x)) \quad \text{for all } x \text{ in } [a, b] \qquad (3.33)$$

The first fact we establish is that the initial-value problem (3.32) can be rewritten as an integral equation.

Lemma 3.22 The function g is a solution of the initial-value problem (3.32) iff g is continuous at every point of $[a, b]$ and g satisfies

$$g(x) = y_0 + \int_a^x F(t, g(t)) \, dt \quad \text{for all } x \text{ in } [a, b] \qquad (3.34)$$

Proof First we assume that g is continuous and satisfies (3.34). Then $g(a) = y_0$, since the integral of any function from a to a is zero. Furthermore, by the fundamental theorem of calculus (Theorem A.23a), g is differentiable and $g'(x) = F(x, g(x))$. This completes the first half of the proof, except that we neglected to verify a key hypothesis of Theorem A.23 which is that the integrand is continuous. We leave to Exercise 1 the proof that, since F and g are continuous, so is the composite function h, where $h(t) = F(t, g(t))$.

For the second half of the proof, we assume that g is a solution of the initial-value problem (3.32). But, by integrating both sides of (3.33), we get

$$\int_a^x F(t, g(t)) \, dt = \int_a^x g'(t) \, dt$$

$$= g(x) - g(a) = g(x) - y_0$$

which proves that g satisfies (3.34).

This lemma gives us an alternative to proving that the initial-value problem (3.32) has a unique solution; instead, we may prove the equivalent statement that (3.34) has a unique solution. This will be easier. The method for proving the existence of a solution is a celebrated one and goes by the name of "Picard's method." In it, we construct the function g as the limit of the sequence of functions $\{g_n\}$ defined recursively by

$$g_0(x) = y_0$$

$$g_{n+1}(x) = y_0 + \int_a^x F(t, g_n(t)) \, dt \quad \text{for all } n \ge 0 \tag{3.35}$$

EXAMPLE 3.41   Find the Picard sequence (3.35) for the initial-value problem $y' = y$, $y(0) = 1$.

$$g_0(x) = 1 \quad \text{and} \quad g_{n+1}(x) = 1 + \int_0^x g_n(t) \, dt$$

So

$$g_1(x) = 1 + \int_0^x 1 \, dt = 1 + x$$

$$g_2(x) = 1 + \int_0^x (1 + t) \, dt = 1 + x + \frac{x^2}{2!}$$

$$g_3(x) = 1 + \int_0^x \left(1 + t + \frac{t^2}{2!}\right) dt = 1 + x + \frac{x^2}{2!} + \frac{x^3}{3!}$$

$$\vdots$$

$$g_n(x) = 1 + x + \frac{x^2}{2!} + \cdots + \frac{x^n}{n!}$$

Note that $g_n(x)$ is the nth partial sum of $\sum_{k=0}^{\infty} x^k/k!$, and so the approximate solutions $g_n(x)$ converge to e^x, the exact solution of the given initial-value problem.

To make the proof of the existence–uniqueness theorem as simple as possible, we will assume that the function F satisfies a Lipschitz condition on a much larger domain than in Theorem D.1. As a result, the theorem we prove will not be as widely applicable. On the other hand, we will be able to prove a stronger conclusion than that of Theorem D.1; the solutions whose existence we prove will be defined on the whole width of the domain of F, not just on some small subinterval.

Theorem 3.23 (Existence–uniqueness)   Suppose that D is an infinite vertical strip of the form

$$D = \{(x, y) | a \leq x \leq b \quad \text{and} \quad -\infty < y < \infty\}$$

and F is a function that is continuous at every point of D and satisfies a Lipschitz condition on D. Then the initial-value problem

$$\frac{dy}{dx} = F(x, y), \qquad y(a) = y_0$$

has a unique solution on the interval $[a, b]$. (The theorem is also valid on intervals of the form $[b, a]$; see Exercise 6.)

Proof   We start by proving that the solution, if it exists, is unique. We suppose, therefore, that g and h are both solutions of the initial-value problem on the interval $[a, b]$; we must show that $g = h$ on $[a, b]$. By Lemma 3.22,

$$g(x) = y_0 + \int_a^x F(t, g(t)) \, dt$$

and

$$h(x) = y_0 + \int_a^x F(t, h(t)) \, dt$$

for all x in $[a, b]$. Therefore, when we subtract these equations, we get

$$
\begin{aligned}
|g(x) - h(x)| &= \left| \int_a^x [F(t, g(t)) - F(t, h(t))] \, dt \right| \\
&\leq \int_a^x |F(t, g(t)) - F(t, h(t))| \, dt \\
&\leq \int_a^x L |g(t) - h(t)| \, dt \\
&\leq LM(x - a)
\end{aligned}
\tag{3.36}
$$

where L is a Lipschitz constant for F and M denotes the maximum of $|g - h|$ on $[a, b]$. Then, by repeated use of (3.36):

$$
\begin{aligned}
|g(x) - h(x)| &\leq \int_a^x L^2 M(t - a) \, dt \\
&= \frac{L^2 M(x - a)^2}{2!} \\
|g(x) - h(x)| &\leq \int_a^x \frac{L^3 M(t - a)^2}{2!} \, dt \\
&= \frac{L^3 M(x - a)^3}{3!} \\
&\quad\vdots \\
|g(x) - h(x)| &\leq \frac{L^k M(x - a)^k}{k!}
\end{aligned}
$$

But $L^k(x - a)^k / k! \to 0$ as $k \to \infty$, by Example 1.5; therefore, $g = h$ on $[a, b]$.

Now we begin the proof that a solution of the initial-value problem exists. We define the sequence $\{g_n\}$ by Picard's method (3.35). Our goal is to show that this sequence converges uniformly on $[a, b]$ to a limit g and then that g satisfies the integral equation (3.34). The way in which we propose to prove that $\{g_n\}$ converges uniformly is to rewrite g_n as

$$g_n = g_0 + (g_1 - g_0) + (g_2 - g_1) + \cdots + (g_n - g_{n-1})$$

and then apply the M-test to the infinite series $g_0 + \sum_{k=1}^{\infty}(g_k - g_{k-1})$, whose sum is the limit of the sequence $\{g_n\}$. For each $k > 0$, we use the Picard formula (3.35) to write

$$
\begin{aligned}
|g_{k+1}(x) - g_k(x)| &= \left| \int_a^x F(t, g_k(t))\, dt - \int_a^x F(t, g_{k-1}(t))\, dt \right| \\
&= \left| \int_a^x [F(t, g_k(t)) - F(t, g_{k-1}(t))]\, dt \right| \\
&\le \int_a^x |F(t, g_k(t)) - F(t, g_{k-1}(t))|\, dt \\
&\le \int_a^x L|g_k(t) - g_{k-1}(t)|\, dt \qquad (3.37)
\end{aligned}
$$

for all x in $[a, b]$. Next, let M be the maximum value of $\{|F(t, y_0)|\,|\,a \le t \le b\}$, which exists by Theorem A.9. So

$$
\begin{aligned}
|g_1(x) - g_0(x)| &= \left| \int_a^x F(t, g_0(t))\, dt \right| \\
&= \left| \int_a^x F(t, y_0)\, dt \right| \\
&\le \int_a^x M\, dt = M(x - a)
\end{aligned}
$$

Then, by repeated use of (3.37),

$$
\begin{aligned}
|g_2(x) - g_1(x)| &\le \int_a^x LM(t - a)\, dt \\
&= \frac{LM(x - a)^2}{2!} \\
|g_3(x) - g_2(x)| &\le \int_a^x \frac{L^2 M(t - a)^2}{2!}\, dt \\
&= \frac{L^2 M(x - a)^3}{3!} \\
&\ \ \vdots \\
|g_k(x) - g_{k-1}(x)| &\le \frac{L^{k-1} M(x - a)^k}{k!} \\
&\le \frac{L^{k-1} M(b - a)^k}{k!}
\end{aligned}
$$

By the ratio test, the series of constants $\Sigma L^{k-1}M(b-a)^k/k!$ converges, and so, by the M-test, the series $g_0 + \Sigma_{k=1}^{\infty}(g_k - g_{k-1})$ converges uniformly on $[a, b]$ to a function g. Therefore $g_n \to g$ uniformly on $[a, b]$. In particular, Theorem 1.12 implies that g is continuous since, by the fundamental theorem of calculus, each g_n is differentiable and thus continuous.

Finally, we show that g satisfies the integral equation (3.34) and, therefore, the initial-value problem. To accomplish this, we show that $F(x, g_n(x)) \to F(x, g(x))$ uniformly on $[a, b]$. For every $\varepsilon > 0$, we choose N such that

$$|g_n(x) - g(x)| < \frac{\varepsilon}{L}$$

for all $n \geq N$ and all x in $[a, b]$. Then

$$|F(x, g_n(x)) - F(x, g(x))| \leq L|g_n(x) - g(x)| < L\frac{\varepsilon}{L} = \varepsilon$$

for all $n \geq N$ and all x in $[a, b]$. Therefore, $F(x, g_n(x)) \to F(x, g(x))$ uniformly on $[a, b]$, and so, by Theorem 1.13,

$$\int_a^x F(t, g_n(t))\, dt \to \int_a^x F(t, g(t))\, dt$$

If we then add y_0, we get

$$g_{n+1}(x) \to y_0 + \int_a^x F(t, g(t))\, dt$$

for all x in $[a, b]$. But, since $g_{n+1}(x) \to g(x)$ and since sequences have unique limits,

$$g(x) = y_0 + \int_a^x F(t, g(t))\, dt$$

Therefore, by Lemma 3.22, g is a solution of the initial-value problem; this completes the proof.

The Picard sequence $\{g_n\}$ could also be used as a basis for a numerical method of solution for the initial-value problem (3.32). Since the functions g_n converge to the solution, each may be regarded as an approximation to the solution. However, they usually do not converge rapidly enough for such a method to match the effectiveness of the methods in Section 2.8.

EXERCISES

1. Define a function h by $h(t) = F(t, g(t))$. If F is continuous at (x, y) and g is continuous at x, prove that h is continuous at x.

In Exercises 2–5, find the first three terms of the Picard sequence (3.35) for the given initial-value problem.

2. $y' = 3y$, $y(0) = 5$

3. $y' = 2xy$, $y(0) = 1$

4. $y' = 2 + \sin x + y$, $y(0) = 0$

5. $y' = x^2 + y^2,$ $y(0) = 1$

6. Prove that Theorem 3.23 is also valid on intervals of the form $[b, a]$. *Hint:* Apply Theorem 3.23 to the initial-value problem

$$\frac{dy}{dx} = -F(-x, y), \qquad y(-a) = y_0, \quad \text{where } -a \le x \le -b$$

7. Find the nth term of the Picard sequence (3.35) for the initial-value problem $y' = 2xy$, $y(0) = 1$. Show that this nth term equals the nth partial sum of the Maclaurin series of the unique solution $y = \exp(x^2)$.

8. Mimic part of the proof of Theorem 3.23 to show that

$$|g_n(x) - g(x)| \le \frac{L^n M(x - a)^{n+1}}{(n + 1)!}$$

where M is the maximum of $\{|F(t, g(t))| \,|\, a \le t \le b\}$

CHAPTER 4

Series and differential equations

One of the first steps in a mathematical analysis of a scientific problem is to describe the problem (or, more precisely, an idealization of the problem) in mathematical terms. The most widely and successfully used concept in mathematical descriptions of scientific problems has been the differential equation, even though success has often been impeded by the fact that differential equations can be extremely difficult to analyze. These difficulties, rather than thwarting progress, have instead stimulated more research and better understanding. As a result, the subject of differential equations contains a wealth of deep theories and subtle methods of analysis. Some of the greatest contributions to this subject have been based on concepts of convergence and methods of approximation.

In Sections 2.7 and 2.8 we studied methods for producing numerical approximations to solutions of differential equations; in this chapter we study methods of solution and analysis which make use of infinite series. To illustrate the kinds of differential equations we want to study, the kinds of scientific problems which give rise to these equations, and the kinds of questions we might want to ask about the equations, we present several different mathematical descriptions of a single physical phenomenon: the motion of a simple pendulum (Figure 4.1). The presentation is intended to be representative in several senses. Differential equations that describe other phenomena often arise from analogous arguments; the fact that we derive not one but several alternative mathematical descriptions of a single phenomenon is typical; and the types of equations we derive are some of the most commonly occurring ones: second order, often linear (but not always), often with constant coefficients (but not always).

We assume that the pendulum bob at the point B has mass m and that the rod which attaches the bob to the pivot point P has a fixed length r and no appreciable mass. Let s denote the directed distance of the bob from its lowest point and θ the angle of deflection of the rod from the vertical; take the counterclockwise direction to be positive. Since the downward force on the bob is the gravitational force mg, the component of this force in the direction of increasing s is $-mg \sin \theta$. Thus, by Newton's second law of motion, we have

$$m \frac{d^2 s}{dt^2} = -mg \sin \theta$$

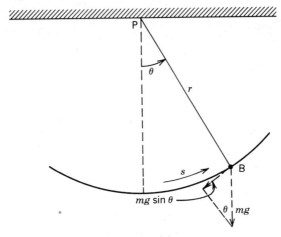

Figure 4.1 Pendulum

where t denotes time. Canceling m and using the fact that $s = r\theta$, we get the differential equation

$$r\frac{d^2\theta}{dt} + g\sin\theta = 0 \tag{4.1}$$

This differential equation describing the motion of a simple pendulum should, in principle, enable us to describe the angle θ as a function of time and to answer such questions as the following. How long is one full period of the pendulum? Does the period change as time increases? Does the period depend on the amplitude or velocity of the pendulum? However, (4.1) is not easy to analyze. Specifically, $\sin\theta$ is not a linear function of θ, and so the powerful theory of second-order linear equations in Sections D.3 and D.4 is not applicable. Therefore, one way to make the equation simpler to solve is to replace $\sin\theta$ by the first nonzero term of its Maclaurin series expansion

$$\sin\theta = \theta - \frac{\theta^3}{3!} + \frac{\theta^5}{5!} - \cdots \tag{4.2}$$

which leads to

$$r\frac{d^2\theta}{dt^2} + g\theta = 0 \tag{4.3}$$

This is a homogeneous, linear differential equation with constant coefficients. Its characteristic equation is $rw^2 + g = 0$, which has the roots $w = \pm i\sqrt{g/r}$; so, by Theorem D.6, its solutions are

$$\theta = c_1\cos\left(\sqrt{\frac{g}{r}}\,t\right) + c_2\sin\left(\sqrt{\frac{g}{r}}\,t\right)$$

If we also impose the initial conditions $\theta(0) = \theta_0$ and $\theta'(0) = 0$, then $\theta_0 = c_1$ and $0 = c_2$; hence

$$\theta = \theta_0\cos\left(\sqrt{\frac{g}{r}}\,t\right) \tag{4.4}$$

Therefore, the period of the pendulum is $2\pi\sqrt{r/g}$, a number that depends only on the length r of the pendulum. Of course, our mathematical description of physical reality was quite crude and we'do not expect that these conclusions would be borne out by experiments, except in an approximate sense. Furthermore, since we replaced $\sin\theta$ by θ, we can only expect the solutions (4.4) to be a reasonable description of the motion of a pendulum that makes small swings.

We expect to find an even better approximation to the behavior of the pendulum if we retain two nonzero terms of the Maclaurin expansion (4.2). In fact, the equation

$$r\frac{d^2\theta}{dt^2} + g\theta - g\frac{\theta^3}{3!} = 0 \tag{4.5}$$

has been used when the angle of deflection θ gets as large as $\pi/4$ radians. Unfortunately, θ^3 is again not a linear function of θ; still, we hope that (4.5) is not quite as difficult to analyze as (4.1).

Many other modifications of the basic pendulum equation (4.1) could be considered. For example, we could weaken or eliminate our assumptions that the rod has no mass and is rigid. Or, as we will actually do, we may allow its length r to vary. In place of Newton's second law, we use the analogous law for torque:

$$N = \frac{dL}{dt}$$

where L is angular momentum. The torque on the rod is its length r times the force $-mg\sin\theta$, and its angular momentum is its length times the momentum $mr\theta'$. Thus

$$-rmg\sin\theta = \frac{d}{dt}\left(mr^2\frac{d\theta}{dt}\right)$$

or

$$mr^2\frac{d^2\theta}{dt^2} + 2mr\frac{dr}{dt}\frac{d\theta}{dt} + mrg\sin\theta = 0$$

or

$$r\frac{d^2\theta}{dt^2} + 2\frac{dr}{dt}\frac{d\theta}{dt} + g\sin\theta = 0 \;\text{.}$$

Then, if we add the assumptions that the pendulum is lengthening at a constant rate v and that $\sin\theta$ is nearly θ, the differential equation becomes

$$(r_0 + vt)\frac{d^2\theta}{dt^2} + 2v\frac{d\theta}{dt} + g\theta = 0 \tag{4.6}$$

where r_0 is the value of the length r when $t = 0$. This differential equation is linear, just as (4.3) is, but the nonconstant coefficient $r_0 + vt$ makes it much more difficult to solve.

The first four sections of this chapter are devoted to a study of second-order linear differential equations with variable coefficients. In particular, equation (4.6) is studied in detail in Section 4.3. In Section 4.5 we briefly study nonlinear equations such as (4.1) and (4.5). Throughout the chapter, familiarity with the theory of second-order differential equations in Sections D3. and D4. is assumed.

4.1 Power Series Solutions

According to Theorem 3.23 or D.1, a first-order initial-value problem

$$\frac{dy}{dx} = F(x, y), \qquad y(a) = y_0$$

has a unique solution y. We can think of this fact as giving us a way of *defining* functions. For example, the natural logarithm function can be defined as the unique solution of

$$\frac{dy}{dx} = \frac{1}{x}, \qquad y(1) = 0, \quad \text{where } x > 0$$

and the exponential function e^x can be defined as the unique solution of

$$\frac{dy}{dx} = y, \qquad y(0) = 1, \quad \text{where } -\infty < x < \infty$$

Properties of such functions can be derived directly from the differential equations which define them. We can even derive series expansions of a function directly from its defining differential equation. For example, the above definition of e^x implies that $y'' = (y')' = y' = y$, $y''' = (y'')' = y' = y, \ldots, y^{(k)} = y^{(k-1)} = y$. Hence, $y^{(k)}(0) = y(0) = 1$. Therefore, the Maclaurin series of $y = e^x$ is the familiar one:

$$\sum_{k=0}^{\infty} \frac{y^{(k)}(0)x^k}{k!} = \sum_{k=0}^{\infty} \frac{x^k}{k!}$$

We would especially like to look at the solutions of second-order linear differential equations with variable coefficients

$$y'' + p(x)y' + q(x)y = r(x) \tag{4.7}$$

from this point of view. According to Theorem D.3, equation (4.7) has a solution that is unique if we impose initial conditions of the form

$$y(x_0) = y_0, \qquad y'(x_0) = y_1 \tag{4.8}$$

Our goal is to find this unique solution, and our initial method for finding it is suggested by the above discussion. The initial-value problem (4.7) and (4.8) *defines* a unique solution y, which we hope has a Taylor series at x_0 that converges to y on some interval. If it does, this solution has the form

$$y = \sum_{k=0}^{\infty} c_k(x - x_0)^k \tag{4.9}$$

and we should be able to find the infinitely many unknown coefficients c_k by substituting the expression (4.9) into (4.7) and solving for c_k. For many equations of the form (4.7), this method works; it is called the *method of substitution by power series with undetermined coefficients*.

In all our discussions, we assume that the differential equation (4.7) is homogeneous (i.e., $r = 0$), even though, as several of the exercises suggest, the method also applies to many nonhomogeneous equations. To find all the solutions of a homogeneous equation of the form (4.7), we need only find two linearly independent solutions and then take all their linear combinations (Theorem D.5). Therefore, all our effort will be put into finding two linearly independent solutions.

EXAMPLE 4.1   Find two power series about zero which represent linearly independent solutions of

$$y'' - 2xy' - 2y = 0$$

Let $y = \sum_{k=0}^{\infty} c_k x^k$. Then $y' = \sum_{k=1}^{\infty} c_k k x^{k-1}$ and $y'' = \sum_{k=2}^{\infty} c_k k(k-1)x^{k-2}$. When we substitute these expressions into the given differential equation, we get

$$\sum_{k=2}^{\infty} c_k k(k-1)x^{k-2} - \sum_{k=1}^{\infty} 2c_k k x^k - \sum_{k=0}^{\infty} 2c_k x^k = 0$$

Or, if we replace k by $j + 2$ in the first sum and move the other two sums to the right side of the equation, and if we use the fact that $\sum_{k=1}^{\infty} c_k k x^k = \sum_{k=0}^{\infty} c_k k x^k$ (since the $k = 0$ term is zero), we get

$$\sum_{j=0}^{\infty} c_{j+2}(j+2)(j+1)x^j = \sum_{k=0}^{\infty} \left[2c_k k x^k + 2c_k x^k \right]$$

or

$$\sum_{k=0}^{\infty} c_{k+2}(k+2)(k+1)x^k = \sum_{k=0}^{\infty} 2c_k(k+1)x^k$$

But, since corresponding coefficients of power series with the same sum are equal (Corollary 3.21), it follows that

$$c_{k+2}(k+2)(k+1) = 2c_k(k+1) \quad \text{for } k \geq 0$$

or

$$c_{k+2} = \frac{2c_k}{k+2} \quad \text{for } k \geq 0 \qquad (4.10)$$

Unfortunately, the difference equation (4.10) also does not have constant coefficients, and so it cannot be solved by the methods of Section 1.3. Instead, we will try to find a pattern for the terms c_k by inspecting the first few terms. Any choice of the initial values c_0 and c_1 will lead to a unique solution of (4.10). In particular, if we define y_1 to be the sum of the power series $\sum_{k=0}^{\infty} c_k x^k$ where $c_0 = 1$, $c_1 = 0$ and the remaining c_k are given by (4.10), and if we define y_2 similarly but with $c_0 = 0$ and $c_1 = 1$, then y_1 and y_2 will be linearly independent solutions of the differential equation. First, we let $c_0 = 1$ and $c_1 = 0$. Then

$$c_2 = \frac{2c_0}{2} = 1 \qquad\qquad c_3 = \frac{2c_1}{3} = 0$$

$$c_4 = \frac{2c_2}{4} = \frac{1}{2}$$

$$c_6 = \frac{2c_4}{6} = \frac{1}{2 \cdot 3}$$

$$\vdots \qquad\qquad\qquad \vdots$$

$$c_{2k} = \frac{1}{2 \cdot 3 \cdots k} = \frac{1}{k!} \qquad c_{2k+1} = 0 \quad \text{for } k \geq 0$$

Therefore, one solution of the given differential equation is

$$y_1 = c_0 + c_2 x^2 + c_4 x^4 + c_6 x^6 + \cdots$$

$$= \sum_{k=0}^{\infty} c_{2k} x^{2k} = \sum_{k=0}^{\infty} \frac{1}{k!} (x^2)^k = \exp(x^2)$$

Normally, however, we cannot expect a power series solution to have a recognizable sum.

Next, we let $c_0 = 0$ and $c_1 = 1$. Then

$$c_3 = \frac{2c_1}{3} = \frac{2}{3} \qquad\qquad c_2 = \frac{2c_0}{2} = 0$$

$$c_5 = \frac{2c_3}{5} = \frac{2^2}{3 \cdot 5}$$

$$\vdots \qquad\qquad\qquad \vdots$$

$$c_{2k+1} = \frac{2^k}{1 \cdot 3 \cdot 5 \cdots (2k+1)} \qquad c_{2k} = 0 \quad \text{for } k \geq 0$$

Therefore, another solution of the given differential equation is

$$y_2 = c_1 x + c_3 x^3 + c_5 x^5 + \cdots$$

$$= \sum_{k=0}^{\infty} c_{2k+1} x^{2k+1} = \sum_{k=0}^{\infty} \frac{2^k x^{2k+1}}{1 \cdot 3 \cdot 5 \cdots (2k+1)}$$

Furthermore, the general solution of the given differential equation is the set of all linear combinations of y_1 and y_2:

$$y = d_1 y_1 + d_2 y_2 = d_1 \exp(x^2) + d_2 \sum_{k=0}^{\infty} \frac{2^k x^{2k+1}}{1 \cdot 3 \cdot 5 \cdots (2k+1)}$$

We have not yet asked ourselves whether the power series solutions we find by this method of undetermined coefficients actually converge and, if so, what their intervals of convergence are. We can compute the radius of convergence of the power series y_2 in Example 4.1 directly from the ratio test:

$$\left| \frac{a_{k+1}}{a_k} \right| = \frac{2^{k+1} |x|^{2k+3}}{1 \cdot 3 \cdot 5 \cdots (2k+1)(2k+3)} \bigg/ \frac{2^k |x|^{2k+1}}{1 \cdot 3 \cdot 5 \cdots (2k+1)} = \frac{2|x|^2}{2k+3} \to 0$$

Therefore the power series y_2 converges for all x.

As we will soon see, however, this last computation is often unnecessary. For there is a quite general theorem which not only guarantees that certain differential equations have power series solutions but also gives a lower bound for their radii of convergence. The theorem simply requires that the coefficient functions be "analytic."

Definition 4.1   A function f is *analytic* at a point x_0 if f can be written as the sum of a power series of the form $\sum_{k=0}^{\infty} c_k (x - x_0)^k$ on some open interval containing x_0.

For example, e^x, $\cos x$, $\sin x$, $1/(1-x)$, and $\log(1+x)$ are all analytic at 0 since each of these functions is the sum of its Maclaurin series on some open interval containing 0. The following facts will help us recognize analytic functions; we omit the proofs [see Smith (1971)].

If $f(x) = \sum_{k=0}^{\infty} c_k(x - x_0)^k$ on some open interval I, then f is analytic at every point in I.

If f and g are analytic at x_0, so are $f + g$, $f - g$, fg, and f/g (except that $g(x_0)$ must be different from zero for f/g to be analytic at x_0).

So it follows that e^x, $\cos x$, and $\sin x$ are analytic at every point on the real line and that $1/(1-x)$ and $\log(1+x)$ are analytic at every point of the interval $(-1,1)$. Also, since a polynomial is a power series at zero with only finitely many nonzero terms and thus has radius of convergence equal to ∞, it is analytic at every point on the real line. It therefore follows that the quotient of two polynomials is analytic at every point where the denominator is different from zero. However, there is a subtle point to notice here. Although we have just seen that $1/(1+x^2)$, for example, must be analytic at every point of the real line, this fact does not imply that the radius of convergence of every Taylor series of the function is ∞. In fact, the radius of convergence of its Maclaurin series is only 1, not ∞. Deeper insight into the behavior of analytic functions can only come from a study of complex numbers and functions of a complex variable. In our example, the denominator $1 + x^2$ equals zero when $x = \pm\sqrt{-1}$, two points that are at a distance 1 from 0; a theorem about complex variables implies that this distance 1 must be the radius of convergence of the Maclaurin series of $1/(1+x^2)$.

Now we state an extension of Theorem D.6 to homogeneous, linear differential equations with variable coefficients.

Theorem 4.1 Suppose that the functions p and q are analytic at x_0. Then the differential equation

$$y'' + p(x)y' + q(x)y = 0$$

has two linearly independent solutions of the form $y = \sum_{k=0}^{\infty} c_k(x - x_0)^k$. Furthermore, the radius of convergence of any such series is at least as large as the smaller of the two radii of convergence of the power series expansions for p and q at x_0.

Although the theory in Chapter 3 is sufficient to enable us to construct a proof of Theorem 4.1, we will not do so—the argument is rather long [see Birkhoff and Rota (1978)]. We continue this section with two more examples of the method of undetermined coefficients.

EXAMPLE 4.2 Find two power series about 0 which represent linearly independent solutions of

$$y'' + \frac{3x}{1+x^2}y' + \frac{1}{1+x^2}y = 0$$

The coefficient functions $3x/(1 + x^2)$ and $1/(1 + x^2)$ are analytic at zero, as can be demonstrated by replacing x with $-x^2$ in the geometric series and then multiplying by $3x$. In particular, both power series have radius of convergence 1; therefore, by Theorem 4.1, the differential equation has two linearly independent power series solutions whose radii of convergence are at least 1. The task of finding these solutions is made easier if we first multiply the equation by $1 + x^2$:

$$(1 + x^2)y'' + 3xy' + y = 0$$

If $y = \sum_{k=0}^{\infty} c_k x^k$, then $y' = \sum_{k=1}^{\infty} c_k k x^{k-1}$ and $y'' = \sum_{k=2}^{\infty} c_k k(k - 1)x^{k-2}$; and so the differential equation becomes

$$(1 + x^2) \sum_{k=2}^{\infty} c_k k(k - 1)x^{k-2} + \sum_{k=1}^{\infty} 3c_k k x^k + \sum_{k=0}^{\infty} c_k x^k = 0$$

$$\sum_{k=2}^{\infty} c_k k(k - 1)x^{k-2} + \sum_{k=2}^{\infty} c_k k(k - 1)x^k + \sum_{k=1}^{\infty} 3c_k k x^k + \sum_{k=0}^{\infty} c_k x^k = 0$$

$$\sum_{k=0}^{\infty} c_{k+2}(k + 2)(k + 1)x^k = - \sum_{k=0}^{\infty} c_k [k(k - 1) + 3k + 1] x^k$$

$$= - \sum_{k=0}^{\infty} c_k(k^2 + 2k + 1)x^k$$

which implies that

$$c_{k+2} = - \frac{(k^2 + 2k + 1)}{(k + 2)(k + 1)} c_k = - \frac{(k + 1)^2}{(k + 2)(k + 1)} c_k$$

$$= - \frac{(k + 1)}{k + 2} c_k \quad \text{for } k \geq 0$$

As in Example 4.1, we first choose $c_0 = 1$ and $c_1 = 0$. Then

$$c_2 = \frac{-c_0}{2} = \frac{-1}{2} \qquad c_{2k+1} = 0$$

$$c_4 = \frac{-3c_2}{4} = \frac{3}{2 \cdot 4}$$

$$c_6 = \frac{-5c_4}{6} = \frac{-3 \cdot 5}{2 \cdot 4 \cdot 6}$$

$$\vdots$$

$$c_{2k} = (-1)^k \frac{1 \cdot 3 \cdot 5 \cdots (2k - 1)}{2 \cdot 4 \cdot 6 \cdots (2k)}$$

$$= (-1)^k \frac{1 \cdot 3 \cdot 5 \cdots (2k - 1)}{2^k k!} \quad \text{for } k \geq 1$$

Therefore, one solution of the given differential equation is

$$y_1 = 1 + \sum_{k=1}^{\infty} (-1)^k \frac{1 \cdot 3 \cdot 5 \cdots (2k - 1)x^{2k}}{2^k k!}$$

$2 \cdot 4 \cdot 6 \cdots 2k = 2^k k!$

If $c_0 = 0$ and $c_1 = 1$, then

$$c_3 = \frac{-2c_1}{3} = \frac{-2}{3} \qquad\qquad c_{2k} = 0$$

$$c_5 = \frac{-4c_3}{5} = \frac{2 \cdot 4}{3 \cdot 5}$$

$$c_7 = \frac{-6c_5}{7} = \frac{-2 \cdot 4 \cdot 6}{3 \cdot 5 \cdot 7}$$

$$\vdots$$

$$c_{2k+1} = (-1)^k \frac{2 \cdot 4 \cdot 6 \cdots (2k)}{1 \cdot 3 \cdot 5 \cdots (2k+1)} = \frac{(-1)^k 2^k (k!)}{1 \cdot 3 \cdot 5 \cdots (2k+1)} \qquad \text{for } k \geq 0$$

Therefore, another solution of the differential equation is

$$y_2 = \sum_{k=0}^{\infty} \frac{(-1)^k 2^k (k!) x^{2k+1}}{1 \cdot 3 \cdot 5 \cdots (2k+1)}$$

EXAMPLE 4.3 Find two power series about 0 which represent linearly independent solutions of

$$y'' - xy = 0$$

(This is called Airy's equation; it occurs in the theory of diffraction.)

When we substitute $y = \sum_{k=0}^{\infty} c_k x^k$ into the differential equation, we get

$$\sum_{k=2}^{\infty} c_k k(k-1) x^{k-2} = \sum_{k=0}^{\infty} c_k x^{k+1}$$

Or, if we let $k = j + 3$ in the left-hand sum,

$$\sum_{j=-1}^{\infty} c_{j+3}(j+3)(j+2) x^{j+1} = \sum_{k=0}^{\infty} c_k x^{k+1}$$

or

$$2c_2 + \sum_{k=0}^{\infty} c_{k+3}(k+3)(k+2) x^{k+1} = \sum_{k=0}^{\infty} c_k x^{k+1}$$

There is no constant term in the right-hand sum, which is the same as saying that the constant term is zero. Thus,

$$c_2 = 0 \quad \text{and} \quad c_{k+3} = \frac{c_k}{(k+2)(k+3)} \qquad \text{for } k \geq 0$$

In particular,

$$c_5 = \frac{c_2}{4 \cdot 5} = 0, \quad c_8 = \frac{c_5}{7 \cdot 8} = 0, \ldots, c_{3k+2} = 0$$

So this time our choice of $c_0 = 1$ and $c_1 = 0$ gives us nonzero values for the coefficients c_{3k}:

$$c_3 = \frac{c_0}{2 \cdot 3} = \frac{1}{2 \cdot 3} \qquad\qquad c_{3k+1} = 0 \quad \text{for } k \geq 0$$

$$c_6 = \frac{c_3}{5 \cdot 6} = \frac{1}{2 \cdot 3 \cdot 5 \cdot 6}$$

$$c_9 = \frac{c_6}{8 \cdot 9} = \frac{1}{2 \cdot 3 \cdot 5 \cdot 6 \cdot 8 \cdot 9}$$

$$\vdots$$

Now redefine k to skip past 0 terms. ➔

$$c_{3k} = \frac{1}{2 \cdot 3 \cdot 5 \cdot 6 \cdots (3k-1)(3k)} \qquad \text{for } k \geq 1$$

Therefore, one solution of Airy's equation is

$$y_1 = c_0 + c_3 x^3 + c_6 x^6 + c_9 x^9 + \cdots$$

$$= \sum_{k=0}^{\infty} c_{3k} x^{3k} = 1 + \sum_{k=1}^{\infty} \frac{x^{3k}}{2 \cdot 3 \cdot 5 \cdot 6 \cdots (3k-1)(3k)}$$

If $c_0 = 0$ and $c_1 = 1$, then

$$c_4 = \frac{c_1}{3 \cdot 4} = \frac{1}{3 \cdot 4} \qquad\qquad c_{3k} = 0 \quad \text{for } k \geq 0$$

$$c_7 = \frac{c_4}{6 \cdot 7} = \frac{1}{3 \cdot 4 \cdot 6 \cdot 7}$$

$$\vdots$$

$$c_{3k+1} = \frac{1}{3 \cdot 4 \cdot 6 \cdot 7 \cdots (3k)(3k+1)} \qquad \text{for } k \geq 1$$

Therefore, another solution of Airy's equation is

$$y_2 = c_1 x + c_4 x^4 + c_7 x^7 + \cdots$$

$$= \sum_{k=1}^{\infty} c_{3k+1} x^{3k+1} = x + \sum_{k=1}^{\infty} \frac{x^{3k+1}}{3 \cdot 4 \cdot 6 \cdot 7 \cdots (3k)(3k+1)}$$

We conclude this section with a brief discussion of how the sum of a series solution can be approximated using the methods of Sections 3.6 and 3.7. To keep the discussion as widely applicable as possible, we do not assume that the difference equation satisfied by the coefficients can be solved explicitly. Even without this assumption, we can often compute both the interval of convergence of the series and a bound or estimate for the error of the approximate sum. However, our discussion concerns just a single example; we do not discuss the breadth of applicability of the techniques.

EXAMPLE 4.4   Approximate the solution of

$$(1 + x^2)y'' + 3xy' + y = 0, \qquad y(0) = 1, \quad y'(0) = 0$$

with an error of about 0.0001.

The given differential equation is the one we solved in Example 4.2. When we substituted $y = \sum_{k=0}^{\infty} c_k x^k$ into this equation, we got

$$c_{k+2} = -\frac{(k+1)}{k+2} c_k \quad \text{for } k \geq 0$$

The initial conditions $y(0) = 1$, $y'(0) = 0$ imply that $c_0 = 1$ and $c_1 = 0$; hence $c_{2k+1} = 0$, which implies that $y = \sum_{k=0}^{\infty} c_{2k} x^{2k}$. Also

$$c_{2k+2} = -\frac{(2k+1)}{2k+2} c_{2k} \quad \text{for } k \geq 0$$

Furthermore, we can express the terms $a_k = c_{2k} x^{2k}$ of the series solution recursively:

$$a_{k+1} = c_{2k+2} x^{2k+2} = -\frac{(2k+1)}{2k+2} c_{2k} x^{2k} x^2$$

$$= -\frac{(2k+1)}{2k+2} a_k x^2 \tag{4.11}$$

This result makes it easy to apply the ratio test to the series solution:

$$\left| \frac{a_{k+1}}{a_k} \right| = \left| \frac{(2k+1)x^2}{2k+2} \right| \to x^2 \tag{4.12}$$

which is less that 1 iff $|x| < 1$. It is more difficult to find out whether the series converges at the endpoints $x = \pm 1$, but the alternating series test does apply. By (4.11) with $x = \pm 1$, we see that the terms decrease in absolute value:

$$|a_{k+1}| = \frac{(2k+1)}{2k+2} |a_k| < |a_k|$$

To show that the absolute values $|a_k|$ converge to zero, we take the logarithm of both sides of this last equation:

$$\log |a_{k+1}| = \log\left(\frac{2k+1}{2k+2} \right) + \log |a_k|$$

$$= \log\left(1 - \frac{1}{2k+2} \right) + \log |a_k|$$

$$< \frac{-1}{2k+2} + \log |a_k|$$

where the inequality arises from the fact that

$$\log(1-x) = -\sum_{k=1}^{\infty} \frac{x^k}{k} = -x - \frac{x^2}{2} - \frac{x^3}{3} - \cdots < -x \quad \text{when } 0 < x < 1$$

But the inequality implies that the terms $-\log |a_k|$ increase more rapidly than the partial sums of the divergent series $\sum 1/(2k+2)$. Hence, $-\log |a_k| \to \infty$, and so $|a_k| \to 0$. We have therefore shown that the interval of convergence of the power series solution is $[-1, 1]$.

To approximate the sum of the series at points x where $|x| < 1$, we can use the ratio test error estimate (3.16c):

$$E_n = \left| \sum_{k=0}^{\infty} a_k - \sum_{k=0}^{n} a_k \right| \leq \sum_{k=n+1}^{\infty} |a_k|$$

$$= \text{(approximately)} \left| \frac{a_{n+1}}{1-r} \right|$$

where $r = x^2$, according to the limit (4.12). Note that $|a_{n+1}| = |c_{n+2} x^{2n+2}| \leq |x|^{2n+2}$, which shows that the error term converges to zero fairly rapidly for points x that are not too close to $x = \pm 1$. But for values of x at or near $x = \pm 1$, it is better to use the accelerated alternating series method of approximation (3.19):

$$\left| \sum_{k=0}^{\infty} a_k - \left(\sum_{k=0}^{n} a_k + \frac{a_{n+1}}{2} \right) \right| \leq \frac{|a_{n+1} + a_{n+2}|}{2}$$

Table 4.1 gives the approximation to the solution y at $x = 0, 0.1, 0.2, \ldots, 0.9, 1.0$ obtained by using the accelerated alternating series method. These approximations were produced by a computer program based on Algorithm 4.1. The function *first* referred to in step 5 describes a_0, which in this example has the constant value 1. The function f referred to in steps 6 and 13 describes the $(k+1)$st term a_{k+1} in terms of a_k, x, and k; in this example, it has the values

$$f(\text{term}, x, k) = -\frac{(2k+1)(\text{term})x^2}{2k+2}$$

given by formula (4.11).

Algorithm 4.1 *Accelerated alternating series method*
1. **Input** *xfirst, xlast, nxvalues, epsilon, maxn*
2. **Repeat steps** 3–14 **with** $i \Leftarrow 0, \ldots, nxvalues$
   3. *sum* $\Leftarrow 0$
   4. $x \Leftarrow xfirst + i(xlast - xfirst) / nxvalues$
   5. *term* $\Leftarrow first(x)$
   6. *nextterm* $\Leftarrow f(term, x, 0)$
   7. $k \Leftarrow -1$
   8. **Repeat steps** 9–13 **until** $k \geq 0$ **and** $|term + nextterm| / 2 < epsilon$
      9. $k \Leftarrow k + 1$
     10. **If** $k > maxn$ **then**
        10a. **Print** x, "No convergence"
        10b. **End**
     11. *sum* $\Leftarrow sum + term$
     12. *term* $\Leftarrow nextterm$
     13. *nextterm* $\Leftarrow f(term, x, k + 1)$
  14. **Print** x, *sum* + *term* / 2
  15. **End**

Table 4.1 The Solution of $(1 + x^2)y'' + 3xy' + y = 0$,
$y(0) = 1, y'(0) = 0$

x	y	x	y
0.1	0.9950	0.6	0.8575
0.2	0.9806	0.7	0.8192
0.3	0.9578	0.8	0.7809
0.4	0.9285	0.9	0.7433
0.5	0.8944	1.0	0.7071

For series whose terms do not alternate in sign, the approximation $\sum_{k=0}^{n} a_k$ and the ratio test error estimate may be used instead. Algorithm 4.1 should then be modified as follows: Eliminate steps 6 and 12, replace the variable *nextterm* in step 13 by the variable *term*, change the second half of the exit test in step 8 to $|term|/(1 - r(x))$ where $r(x)$ is the function that describes the limit of $|a_{k+1}|/|a_k|$, and eliminate "$+ term/2$" from step 14.

EXERCISES

In Exercises 1–9, (a) find two power series about 0 which represent linearly independent solutions of the given differential equation; (b) find the radii of convergence of your two solutions.

1. $y'' + xy' + y = 0$

2. $(1 - x^2)y'' - 5xy' - 3y = 0$

3. $(1 - 2x)y'' + 3y' = 0$

4. $y'' + x^2y' + 2xy = 0$

5. $(2 + x^2)y'' - xy' - 3y = 0$

6. $y'' - x^3y = 0$

7. $y'' - 2xy' + py = 0$,   where p is a constant
   (Hermite's equation)

8. $(1 - x^2)y'' - 2xy' + p(p + 1)y = 0$,   where p is a constant
   (Legendre's equation)

9. $(1 - x^2)y'' - xy' + p^2y = 0$,   where p is a constant
   (Tchebycheff's equation)

10. Show that one of your solutions to Hermite's equation (Exercise 7) is a polynomial of degree n, if $p = 2n$ and n is a non-negative integer. (This polynomial, when multiplied by an appropriate constant, is called the nth Hermite polynomial.)

11. Show that one of your solutions to Legendre's equation (Exercise 8) is a polynomial of degree n, if $p = n$ and n is a non-negative integer. (This polynomial, when multiplied by an appropriate constant, is called the nth Legendre polynomial.)

12. Show that one of your solutions to Tchebycheff's equation (Exercise 9) is a polynomial of degree n, if $p = n$ and n is a non-negative integer. (This polynomial, when multiplied by an appropriate constant, is called the nth Tchebycheff polynomial.)

13. Find the first three nonzero terms of the unique solution of the form $y = \sum_{k=0}^{\infty} c_k x^k$ of

$$y'' + (\sin x)y = 0, \qquad y(0) = 1, \quad y'(0) = 0$$

Hint: Replace sin x by its Maclaurin series and assume that two power series can be multiplied as if they were polynomials.

14. Find the first two nonzero terms of the unique solution of the form $y = \sum_{k=0}^{\infty} c_k x^k$ of

$$y' = \sqrt{1 - y^2}, \qquad y(0) = 0$$

Hint: Square both sides to get rid of the square root symbol; then assume that a power series can be squared as if it were a polynomial.

15. Find the first three nonzero terms of the unique solution of the form $y = \sum_{k=0}^{\infty} c_k x^k$ of

$$y'' + x^3 y' + 3x^2 y = e^x, \qquad y(0) = 0, \quad y'(0) = 0$$

Hint: Replace e^x by its Maclaurin series.

In Exercises 16–18, find a lower bound for the radius of convergence of any solution of the form $y = \sum c_k x^k$ of the given differential equation. Do not solve the differential equation; use Theorem 4.1.

16. $y'' + 4y' + 6xy = 0$ 17. $(1 + x^3)y'' + 4xy' + y = 0$

18. $(1 + x)y'' + 4(\sin x)(1 + x)y' - 3y = 0$

19. Approximate the solution of

$$(1 - x^2)y'' - 5xy' - 3y = 0, \qquad y(0) = 1, \quad y'(0) = 0$$

with an error of about 0.0001. *Suggestions:* Find a difference equation for the coefficients of the series solution, and use it to find the interval of convergence of the series. Approximate the solution at $x = 0.1, 0.2, \ldots, 0.9$ by writing a computer program that implements the modification of Algorithm 4.1 discussed in the last paragraph of the section.

20. Approximate the solution of

$$y'' - xy = 0, \qquad y(0) = 1, \quad y'(0) = 0$$

with an error of about 0.0001. *Suggestions:* Find a difference equation for the coefficients of the series solution, and use it to find the interval of convergence of the series. Approximate the solution at $x = -5, -4, \ldots, 4, 5$ by writing a computer program that implements the modification of Algorithm 4.1 discussed in the last paragraph of the section.

21. Approximate the solution of

$$(4 + x^2)y'' - xy' + 4y = 0, \qquad y(0) = 1, \quad y'(0) = 0$$

with an error of about 0.0001. *Suggestions:* Find a difference equation for the coefficients of the series solution, and use it to find the interval of convergence of the series. Approximate the solution at $x = 0.2, 0.4, \ldots, 1.8, 2.0$ by writing a computer program that implements Algorithm 4.1.

4.2 Regular Singular Points, Part 1

In this section and the next two, we continue our study of second-order, linear, homogeneous differential equations:

$$y'' + p(x)y' + q(x)y = 0 \tag{4.13}$$

Now, however, rather than assuming that the coefficients p and q are analytic at x_0, we will assume that they are analytic on intervals of the form (x_0, a) or (a, x_0) but not at x_0 itself. As a result, even though we will still try to express the solutions of (4.13) in terms of powers of $(x - x_0)$, these solutions will not always be analytic at x_0.

To get some insight into the kinds of solutions we can expect, we first solve a class of differential equations that is much like the class of constant coefficient equations; the method is a modification of the characteristic equation method used in Theorem D.6. The equation

$$x^2 y'' + xPy' + Q y = 0 \tag{4.14}$$

(where P and Q are constants) is called the *Cauchy–Euler* differential equation. It can be put in the form (4.13) by dividing through by x^2. Then we see that $p(x) = P/x$ and $q(x) = Q/x^2$ are analytic on $(0, \infty)$ and $(-\infty, 0)$, but not at 0, unless their numerators are zero. We solve (4.14) by transforming it into a constant coefficient differential equation as follows. Let $u(t) = y(e^t)$. Then $y(x) = u(\log x)$, since $x = e^t$ implies $t = \log x$; hence

$$y' = \frac{dy}{dx} = \frac{du}{dt}\frac{dt}{dx} = \frac{1}{x}u'$$

$$y'' = \frac{d}{dx}\left(\frac{1}{x}u'\right) = \frac{1}{x}\frac{du'}{dt}\frac{dt}{dx} - \frac{1}{x^2}u' = \frac{1}{x^2}u'' - \frac{1}{x^2}u'$$

So (4.14) becomes the constant coefficient equation

$$u'' - u' + Pu' + Qu = 0 \tag{4.15}$$

whose characteristic equation is

$$w^2 + (P - 1)w + Q = 0 \tag{4.16}$$

If w is a root of this quadratic equation, then $u(t) = e^{wt}$ is a solution of (4.15). Therefore

$$y(x) = u(\log x) = e^{w \log x} = x^w$$

must be a solution of (4.14). More generally, Theorem D.6 implies (see Exercise 7):

Theorem 4.2

(a) If the quadratic equation (4.16) has two distinct real roots w_1 and w_2, then the solutions of the Cauchy–Euler differential equation (4.14) are

$$y = c_1 x^{w_1} + c_2 x^{w_2} \quad \text{on } x > 0$$

(b) If (4.16) has a double root w, then the solutions of (4.14) are

$$y = c_1 x^w + c_2 x^w \log x \quad \text{on } x > 0$$

(c) If (4.16) has two distinct complex roots $a \pm ib$, then the solutions of (4.14) are

$$y = c_1 x^a \cos(b \log x) + c_2 x^a \sin(b \log x) \quad \text{on } x > 0$$

Note that, unless the exponent w happens to be a non-negative integer, the solution $y = x^w$ is not analytic at 0. But if $w \geq 2$, it can be shown that $y = x^w$ is twice differentiable at $x = 0$ and is therefore a solution of (4.14) at that point. The logarithmic terms are, of course, undefined at $x = 0$. If we replace x by $-x$ in the above three formulas, we get solutions of (4.14) that are valid on the interval $(-\infty, 0)$. Therefore, (4.14) has solutions which are analytic at every point of the real line except zero, and sometimes the solution is also valid at $x = 0$. Furthermore, if we generalize (4.14) to

$$(x - x_0)^2 y'' + (x - x_0)Py' + Qy = 0 \tag{4.17}$$

we can adapt Theorem 4.2 to this setting by replacing x with $x - x_0$ in the three forms of the solution.

EXAMPLE 4.5   Find the general solution of $x^2 y'' + 5xy' + 4y = 0$, on the interval $(0, \infty)$.

The characteristic equation (4.16) is $w^2 + 4w + 4 = 0$, which has the double root $w = -2$. Therefore, by Theorem 4.2b,

$$y = c_1 x^{-2} + c_2 x^{-2} \log x \quad \text{for } x > 0$$

Furthermore, if we replace x by $|x|$, we obtain a solution that is valid for all $x \neq 0$.

Our goal in this section is to solve the following generalization of the Cauchy–Euler equation (4.17):

$$(x - x_0)^2 y'' + (x - x_0)P(x)y' + Q(x)y = 0 \tag{4.18}$$

where $P(x)$ and $Q(x)$ are analytic at x_0. If we put this equation in our earlier form for a second-order, linear, homogeneous differential equation

$$y'' + p(x)y' + q(x)y = 0 \tag{4.19}$$

we see that $p(x) = P(x)/(x - x_0)$ and $q(x) = Q(x)/(x - x_0)^2$. So the power series method of Section 4.1 does not apply, unless $P(x_0) = 0$, $Q(x_0) = 0$, and $Q'(x_0) = 0$. The following definition will help us distinguish between cases of the differential equation (4.19) to which the method of Section 4.1 applies, cases to which the method of this section applies, and cases which we will not treat at all.

Definition 4.2   We say that x_0 is an *ordinary point* of the differential equation (4.19) if $p(x)$ and $q(x)$ are analytic at x_0; otherwise, we say that x_0 is a *singular point* of (4.19). In the latter case, we say that x_0 is a *regular singular point* of (4.19) if

$$p(x) = \frac{P(x)}{x - x_0} \quad \text{and} \quad q(x) = \frac{Q(x)}{(x - x_0)^2}$$

where $P(x)$ and $Q(x)$ are analytic at x_0; otherwise, we say that x_0 is an *irregular singular point* of (4.19).

Thus, the method of power series in Section 4.1 applies to differential equations of the form (4.19) in which x_0 is an ordinary point, and it produces solutions which are analytic at x_0. The method of this section, on the other hand, applies to differential equations of the form (4.19) in which x_0 is a regular singular point, and it produces solutions which are analytic on intervals of the form (x_0, a) and (a, x_0). If x_0 is an irregular singular point of (4.19), we will not attempt to find solutions that are valid near x_0.

EXAMPLE 4.6   Find all the ordinary, regular singular, and irregular singular points of

$$(x^4 - x^2)y'' + (2x + 1)y' + x^2(x + 1)y = 0$$

Note that

$$p(x) = \frac{2x + 1}{x^4 - x^2} = \frac{2x + 1}{x^2(x - 1)(x + 1)}$$

and

$$q(x) = \frac{x^2(x + 1)}{x^4 - x^2} = \frac{1}{x - 1}$$

Thus, 0, 1, and -1 are singular points and all other values of x are ordinary points. Furthermore, 1 is a regular singular point since

$$(x - 1)p(x) = \frac{2x + 1}{x^2(x + 1)} \quad \text{and} \quad (x - 1)^2 q(x) = x - 1$$

are analytic at 1; and -1 is a regular singular point since

$$(x + 1)p(x) = \frac{2x + 1}{x^2(x - 1)} \quad \text{and} \quad (x + 1)^2 q(x) = \frac{(x + 1)^2}{x - 1}$$

are analytic at -1. But 0 is an irregular singular point since $xp(x)$ is not analytic at 0.

Some of the most important differential equations in applied mathematics have the form (4.19). Table 4.2 lists the best known ones. The first column gives the name of the mathematician or scientist who is usually acknowledged as being the first to analyze the equation. For example, Bessel began the analysis of the equation $x^2y'' + xy' + (x^2 - p^2)y = 0$, which is therefore known as "Bessel's equation." Also, certain solutions of Bessel's equation are known as "Bessel functions," and corresponding names are given to certain solutions of the other equations. Although the solutions of these seven equations have been studied and used extensively and are therefore considered to be well-known, most of them cannot be expressed explicitly by formulas that use only the elementary functions of calculus. Instead, they are defined and analyzed by more indirect means. For example, we can think of them as being defined by the differential equations for which they are solutions or by certain infinite series whose sums they are.

The method we use to solve equations of the type (4.18), which includes several of the equations in Table 4.2, is known as *Frobenius' method*. It can be thought of as a combination of the power series method of Section 4.1 and the method of Theorem 4.2

Table 4.2 Some Classical Differential Equations

Name	Equation	Singular points
Airy	$y'' - xy = 0$	None
Hermite	$y'' - 2xy' + 2py = 0$	None
Tchebycheff	$(1 - x^2)y'' - xy' + p^2 y = 0$	± 1
Legendre	$(1 - x^2)y'' - 2xy' + p(p + 1)y = 0$	± 1
Bessel	$x^2 y'' + xy' + (x^2 - p^2)y = 0$	0
Laguerre	$xy'' + (1 - x)y' + py = 0$	0
Gauss	$x(1 - x)y'' + [c - (a + b + 1)x]y' - aby = 0$	0, 1
	(Also called the hypergeometric equation)	

for the Cauchy–Euler equation. If 0 is a regular singular point of the differential equation, we start by looking for solutions in which a power series is multiplied by a power function x^w:

$$y = x^w \sum_{k=0}^{\infty} c_k x^k = \sum_{k=0}^{\infty} c_k x^{k+w} \tag{4.20}$$

Then (see Exercise 8)

$$y' = \sum_{k=0}^{\infty} c_k(k + w)x^{k+w-1}$$

$$y'' = \sum_{k=0}^{\infty} c_k(k + w)(k + w - 1)x^{k+w-2}$$

why $k=0$

More generally, if x_0 is a regular singular point, we replace x by $x - x_0$ in (4.20).

EXAMPLE 4.7   Find two linearly independent functions of the form (4.20) that are solutions of

$$2x^2 y'' + (x - x^2)y' - y = 0, \qquad x > 0$$

Since

$$xp(x) = \frac{x(x - x^2)}{2x^2} = \frac{1 - x}{2} \quad \text{and} \quad x^2 q(x) = \frac{x^2(-1)}{2x^2} = \frac{-1}{2}$$

it follows that $xp(x)$ and $x^2 q(x)$ are both analytic at 0, but neither p nor q is. Thus, 0 is a regular singular point of the given differential equation, and so we should try Frobenius' method. After substituting Frobenius' solution (4.20) into the given differential equation, we have

$$\sum_{k=0}^{\infty} 2c_k(k+w)(k+w-1)x^{k+w} + \sum_{k=0}^{\infty} c_k(k+w)x^{k+w}$$

$$- \sum_{k=0}^{\infty} c_k(k+w)x^{k+w+1} - \sum_{k=0}^{\infty} c_k x^{k+w} = 0$$

or, after collecting together the first, second, and fourth terms,

$$\sum_{k=0}^{\infty} c_k[2(k+w)(k+w-1)+(k+w)-1]x^{k+w} = \sum_{k=1}^{\infty} c_{k-1}(k+w-1)x^{k+w}$$

Hence, by equating corresponding coefficients and canceling the common factor $k + w - 1$, we have

$$c_k[2(k+w)+1] = c_{k-1} \quad \text{for } k \geq 1$$
$$c_0[2w(w-1)+w-1] = 0 \qquad (\text{the } k = 0 \text{ term}) \tag{4.21}$$

We certainly do not want to solve this last equation by setting $c_0 = 0$, for then (4.21) would force all the remaining coefficients c_k to be 0. So

$$0 = 2w(w-1) + w - 1$$
$$= (2w+1)(w-1) \tag{4.22}$$

and therefore $w = 1$ or $w = -\frac{1}{2}$. We will get a Frobenius solution (4.20) for each of these two values of w. First we choose $w = 1$. Then the difference equation (4.21) becomes

$$c_k = \frac{c_{k-1}}{2(k+1)+1} = \frac{c_{k-1}}{2k+3} \quad \text{for } k \geq 1$$

Furthermore, if we let $c_0 = 1$, we get

$$c_1 = \frac{1}{5}, \quad c_2 = \frac{c_1}{7} = \frac{1}{5 \cdot 7}, \quad c_3 = \frac{c_2}{9} = \frac{1}{5 \cdot 7 \cdot 9}, \cdots$$

In general,

$$c_k = \frac{1}{5 \cdot 7 \cdots (2k+3)} \quad \text{for } k \geq 1$$

and so

$$y_1 = x\left(1 + \sum_{k=1}^{\infty} \frac{x^k}{5 \cdot 7 \cdots (2k+3)}\right)$$

When $w = -\frac{1}{2}$, the difference equation (4.21) becomes

$$c_k = \frac{c_{k-1}}{2(k-\frac{1}{2})+1} = \frac{c_{k-1}}{2k} \quad \text{for } k \geq 1$$

Furthermore, if we let $c_0 = 1$, we get

$$c_1 = \frac{1}{2}, \quad c_2 = \frac{c_1}{2 \cdot 2} = \frac{1}{2^2 \cdot 2}, \quad c_3 = \frac{c_2}{2 \cdot 3} = \frac{1}{2^3 \cdot 3!}, \quad c_4 = \frac{c_3}{2 \cdot 4} = \frac{1}{2^4 \cdot 4!}$$

In general,

$$c_k = \frac{1}{k! 2^k} \quad \text{for } k \geq 0$$

and so

$$y_2 = x^{-1/2} \sum_{k=0}^{\infty} \frac{x^k}{k! 2^k} = x^{-1/2} \sum_{k=0}^{\infty} \frac{(x/2)^k}{k!} = x^{-1/2} e^{x/2}$$

Therefore, the general solution of the given differential equation is

$$y = c_1 x \left(1 + \sum_{k=1}^{\infty} \frac{x^k}{5 \cdot 7 \cdots (2k+3)} \right) + c_2 x^{-1/2} e^{x/2}$$

An application of the ratio test shows that the first of our two solutions is defined and analytic for all x. The second solution is defined and analytic only for $x > 0$. However, if we replace the factor $x^{-1/2}$ by $|x|^{-1/2}$, we have a solution that is valid for all $x \neq 0$.

Now let us try to carry out this same computation for an arbitrary differential equation of the form (4.18). We want to see if Frobenius' method always yields a quadratic equation such as (4.22) for the unknown exponents w and if we always get a difference equation that can be solved for the unknown coefficients c_k.

As usual, we consider only the case where $x_0 = 0$:

$$x^2 y'' + x P(x) y' + Q(x) y = 0$$

And we assume that $P(x)$ and $Q(x)$ are analytic at 0. Our first step is to write $P(x)$ as a power series

$$P(x) = \sum_{k=0}^{\infty} P_k x^k = P_0 + x \sum_{k=1}^{\infty} P_k x^{k-1} = P_0 + x P_1(x)$$

where P_1 is also analytic at 0. Similarly, $Q(x) = Q_0 + x Q_1(x)$ where Q_1 is analytic at 0. So if we replace y by $\sum_{k=0}^{\infty} c_k x^{k+w}$ in the differential equation, we get

$$\sum_{k=0}^{\infty} c_k (k+w)(k+w-1) x^{k+w} + \sum_{k=0}^{\infty} c_k (k+w)(P_0 + x P_1(x)) x^{k+w}$$

$$+ \sum_{k=0}^{\infty} c_k (Q_0 + x Q_1(x)) x^{k+w} = 0$$

or

$$\sum_{k=0}^{\infty} c_k \left[(k+w)(k+w-1) + (k+w)P_0 + Q_0 \right] x^{k+w}$$

$$= - \sum_{k=0}^{\infty} c_k \left[(k+w)P_1(x) + Q_1(x) \right] x^{k+w+1} \tag{4.23}$$

Since the coefficient $(k+w)P_1(x) + Q_1(x)$ is analytic at zero, we may expand it as a power series in x. Let us imagine doing so and then multiplying this series by x^{k+w+1}

and rearranging the terms of the right side of (4.23) in increasing powers of x. We will not write out the result of these manipulations explicitly; nevertheless, we can perceive some of the consequences of performing these manipulations and then equating corresponding coefficients of (4.23). For example, there is no x^w term in the right side of the equation, since the term with the lowest power of x is $-c_0 x^{w+1}$ times the constant term of the power series expansion of $wP_1(x) + Q_1(x)$. Therefore, the coefficient of x^w in the left side must be zero:

$$c_0[w(w-1) + wP_0 + Q_0] = 0$$

Or, since we do not want $c_0 = 0$,

$$w^2 + (P_0 - 1)w + Q_0 = 0 \qquad (4.24)$$

This generalization of the characteristic equation (4.16) and of (4.22) in Example 4.7 is called the *indicial equation* of the differential equation (4.18); its solutions provide the values of the exponent w in Frobenius' solutions $y = \sum_{k=0}^{\infty} c_k x^{k+w}$.

Now let us imagine equating corresponding coefficients of x^{k+w} for $k > 0$ in (4.23). This produces a difference equation for c_k in which the left side is c_k times the coefficient

$$\mathscr{C}_k = (k+w)(k+w-1) + (k+w)P_0 + Q_0$$

and the right side is an expression involving only $c_0, c_1, \ldots, c_{k-1}$. Hence, if $\mathscr{C}_k$ is different from zero when $k > 0$, we can divide the difference equation by $\mathscr{C}_k$ and thus solve for c_k. To see whether $\mathscr{C}_k$ is different from zero, we rewrite it with the aid of the indicial equation (4.24):

$$\mathscr{C}_k = k^2 + kw + k(w-1) + w(w-1) + kP_0 + wP_0 + Q_0$$
$$= k^2 + kw + k(w-1) + kP_0$$
$$= k(k + 2w - 1 + P_0)$$

Furthermore, if w_1 and w_2 denote the roots of the indicial equation (4.24), then $w_1 + w_2 = 1 - P_0$ since the sum of the roots of a quadratic equation always equals the negative of the coefficient of the linear term. Therefore

$$\mathscr{C}_k = k(k + 2w - (w_1 + w_2))$$

We assume now that w_1 and w_2 are real and that $w_1 \geq w_2$. So, if we let $w = w_1$, then

$$\mathscr{C}_k = k(k + w_1 - w_2)$$

which is a positive number whenever $k > 0$. Thus, the difference equation can be solved for c_k and we have Frobenius' solution $y = x^{w_1}\sum_{k=0}^{\infty} c_k x^k$. But if we let $w = w_2$, then

$$\mathscr{C}_k = k(k + w_2 - w_1) = k(k - (w_1 - w_2))$$

which, if $w_1 - w_2$ is an integer, equals zero when $k = w_1 - w_2$. Thus, although we get a second Frobenius solution $y = x^{w_2}\sum_{k=0}^{\infty} c_k x^k$ if $w_1 - w_2$ is not an integer, Frobenius' method seems to break down in the special case when the solutions of the indicial equation differ by an integer. Sometimes, however, we will be lucky. It can happen that the right side of the difference equation for c_k

$$\mathscr{C}_k c_k = (\text{expression in } c_0, c_1, \ldots, c_{k-1})$$

is zero when $\mathscr{C}_k$ is zero. If it is, we can again solve the difference equation for c_k and thus find a second Frobenius solution. But even if this hope fails, it still turns out that there is a second solution, one with a logarithmic term as in Theorem 4.2b.

The above discussion hints at, but does not prove, the following theorem. While not completely general, it does describe the solutions of many important differential equations of the form (4.18) [see Birkhoff and Rota (1978) for a proof].

Theorem 4.3 Suppose that P and Q are functions that are analytic at 0. Let R be the minimum of the radii of convergence of the Maclaurin series of P and Q, and let w_1 and w_2 be the roots of the indicial equation

$$w^2 + (P_0 - 1)w + Q_0 = 0$$

where $P_0 = P(0)$ and $Q_0 = Q(0)$. Assume that w_1 and w_2 are real and that $w_1 \geq w_2$. Then the differential equation

$$x^2 y'' + x P(x) y' + Q(x) y = 0$$

has a nonzero solution y_1 on the interval $(0, R)$ of the form

$$y_1 = \sum_{k=0}^{\infty} c_k x^{k+w_1}$$

On the same interval $(0, R)$, there is also a solution y_2 that is linearly independent of y_1 and has one of the following three forms:

(a) If $w_1 - w_2$ is not an integer, then

$$y_2 = \sum_{k=0}^{\infty} d_k x^{k+w_2}$$

(b) If $w_1 - w_2 = 0$, then

$$y_2 = y_1 \log x + \sum_{k=1}^{\infty} d_k x^{k+w_2}$$

(c) If $w_1 - w_2$ is a positive integer, then

$$y_2 = C y_1 \log x + \sum_{k=0}^{\infty} d_k x^{k+w_2} \quad (\text{where } C \text{ could be zero})$$

EXAMPLE 4.8 Find a nonzero solution of the form $\sum c_k x^{k+w}$ of

$$x^2 y'' - (x^2 + x) y' + y = 0, \qquad x > 0$$

$P(x) = -(x + 1)$ and $Q(x) = 1$. Thus, $P_0 = -1$ and $Q_0 = 1$, and so the indicial equation is $w^2 - 2w + 1 = 0$, which has the double root $w = 1$. So we substitute the series $\sum_{k=0}^{\infty} c_k x^{k+1}$ into the given differential equation:

$$\sum_{k=0}^{\infty} c_k(k+1)kx^{k+1} - \sum_{k=0}^{\infty} c_k(k+1)x^{k+2} - \sum_{k=0}^{\infty} c_k(k+1)x^{k+1}$$

$$+ \sum_{k=0}^{\infty} c_k x^{k+1} = 0$$

or

$$\sum_{k=0}^{\infty} c_k[(k+1)k - (k+1) + 1]x^{k+1} = \sum_{k=1}^{\infty} c_{k-1}kx^{k+1}$$

and so

$$c_k[k^2 + k - k - 1 + 1] = c_{k-1}k \quad \text{or} \quad c_k = c_{k-1}/k \quad \text{for } k \geq 1$$

Hence, if $c_0 = 1$ then

$$c_1 = 1, \qquad c_2 = \frac{c_1}{2} = \frac{1}{2}, \qquad c_3 = \frac{c_2}{3} = \frac{1}{3!}, \ldots, c_k = \frac{1}{k!}$$

Therefore, $y_1 = \sum_{k=0}^{\infty} x^{k+1}/k! = x\sum_{k=0}^{\infty} x^k/k! = xe^x$ is one solution of the given differential equation.

According to Theorem 4.3b, there is also a solution of the form

$$y_2 = xe^x \log x + \sum_{k=1}^{\infty} d_k x^{k+1}, \qquad x > 0$$

We consider such solutions in Section 4.4.

EXERCISES

In Exercises 1–5, find all the solutions of the given Cauchy–Euler equation on the interval $(0, \infty)$.

1. $x^2 y'' - 2y = 0$
2. $2x^2 y'' - 5xy' + 3y = 0$
3. $x^2 y'' - xy' + y = 0$
4. $x^2 y'' + xy' + 4y = 0$
5. $x^2 y'' - 5xy' + 13y = 0$
6. Derive the characteristic equation (4.16) directly from the Cauchy–Euler equation (4.14) by substituting x^w for y.
7. (a) If $u(t) = te^{wt}$ and $y(x) = u(\log x)$, show that $y(x) = x^w \log x$.
   (b) If $u(t) = e^{at} \cos(bt)$ and $y(x) = u(\log x)$, show that $y(x) = x^a \cos(b \log x)$.
8. Using Theorem 3.19, verify that

$$\frac{d}{dx}\left(x^w \sum_{k=0}^{\infty} c_k x^k\right) = \sum_{k=0}^{\infty} c_k(k+w)x^{k+w-1}$$

In Exercises 9–11, find all the ordinary, regular singular, and irregular singular points of the given differential equation.

9. $(x - 1)^2 y'' - (x^2 - x) y' + y = 0$

10. $x^3(1 - x^2) y'' + (2x - 3) y' + xy = 0$

11. $(2x + 1)(x - 2)^2 y'' + (x + 2) y' = 0$

In Exercises 12–18, (a) show that 0 is a regular singular point of the given differential equation; (b) find the indicial equation and its roots; (c) find a nonzero solution of the form $\sum c_k x^{k+w}$; (d) if the indicial equation has two distinct real roots which do not differ by an integer, find another solution of this form that is linearly independent of the first.

12. $2xy'' + y' + xy = 0$

13. $2x^2 y'' - 3xy' + (3 - x) y = 0$

14. $x^2 y'' + xy' + (x^2 - \frac{1}{9}) y = 0$ (Bessel's equation with $p = \frac{1}{3}$)

15. $x^2 y'' + xy' + x^2 y = 0$ (Bessel's equation with $p = 0$)

16. $x^2 y'' + xy' + (x^2 - 1) y = 0$ (Bessel's equation with $p = 1$)

17. $x(1 - x) y'' + (\frac{1}{2} - 3x) y' - y = 0$ (Gauss' equation with $a = 1$, $b = 1$, $c = \frac{1}{2}$)

18. $xy'' + (1 - x) y' + py = 0$, p a constant   (Laguerre's equation)

19. Show that your solution to Laguerre's equation (Exercise 18) is a polynomial of degree n, if $p = n$ and n is a non-negative integer. (This polynomial, when multiplied by an appropriate constant, is called the nth Laguerre polynomial.)

20. Use Frobenius' method to find all solutions of $y'' + p^2 y = 0$.

21. Show that the change of variables $u(x) = \sqrt{x}\, y(x)$ transforms the differential equation $u'' + u = 0$ into Bessel's equation with $p = \frac{1}{2}$. Use this fact to solve Bessel's equation with $p = \frac{1}{2}$.

22. In Bessel's equation, if p has the form $n + \frac{1}{2}$ for some positive integer n, find two linearly independent solutions of the form $\sum c_k x^{k+w}$.

4.3 Bessel Functions: An Application

In the introduction to this chapter we derived a differential equation for the motion of a lengthening pendulum:

$$(r_0 + vt) \frac{d^2\theta}{dt^2} + 2v \frac{d\theta}{dt} + g\theta = 0 \tag{4.25}$$

where r_0 is the initial length of the pendulum and v is the constant rate at which it is lengthening. In this section we analyze (4.25); the presentation is adapted from *Mathematical Methods in the Physical Sciences* by Mary L. Boas (New York: John Wiley & Sons, Inc., 1966, pp. 565–570).

We can simplify (4.25) a little by changing the independent variable to x, where

$$x = \frac{r_0}{v} + t = \frac{r_0 + vt}{v}$$

Since $\dfrac{dx}{dt} = 1$, it follows that $\dfrac{d\theta}{dt} = \dfrac{d\theta}{dx}$, and so (4.25) becomes

$$vx \frac{d^2\theta}{dx^2} + 2v \frac{d\theta}{dx} + g\theta = 0$$

Then we multiply the equation by x/v; also, we rename x back to t since this new variable still denotes time:

$$t^2 \frac{d^2\theta}{dt^2} + 2t \frac{d\theta}{dt} + \frac{gt\theta}{v} = 0 \tag{4.26}$$

We want to study the behavior of the angle θ for all values of t, from small to arbitrarily large. So if we are to express θ as an infinite series, we would want it to converge for all positive values of t. We have a choice: Since every $t_0 > 0$ is an ordinary point of the differential equation (4.26), we could expand θ in a power series about such a t_0. But then the singularity at zero would cause the radius of convergence of such a series to be finite. On the other hand, $t_0 = 0$ is a regular singular point, which is more difficult to work with. But, by Theorem 4.3, all the series solutions about $t_0 = 0$ converge on the entire interval $(0, \infty)$, since $P(t) = 2$ and $Q(t) = gt/v$ are power series about $t_0 = 0$ with infinite radii of convergence. We therefore choose Frobenius' method and its extensions.

We do not, however, have to carry out Frobenius' method for (4.26). This equation happens to belong to a well-known class of differential equations that can be transformed into Bessel's equation. Specifically, every equation of the form

$$t^2 u'' + (1 - 2a)tu' + \left[b^2c^2t^{2c} + a^2 - p^2c^2 \right] u = 0 \tag{4.27}$$

can be transformed into Bessel's equation "of order p"

$$x^2 y'' + xy' + \left(x^2 - p^2 \right) y = 0 \tag{4.28}$$

by means of the change of variables

$$u(t) = t^a y(x), \qquad x = bt^c \tag{4.29}$$

(The verification is left to Exercise 1.) Our pendulum equation (4.26) has the form (4.27), where

$$a = \frac{-1}{2} \qquad b = 2\sqrt{\frac{g}{v}} \qquad c = \frac{1}{2} \qquad p = 1$$

Therefore, it transforms into Bessel's equation of order 1 (i.e., where $p = 1$) by means of the change of variables

$$\theta = \frac{1}{\sqrt{t}} y(x) \quad \text{where } x = b\sqrt{t} \text{ and } b = 2\sqrt{\frac{g}{v}} \tag{4.30}$$

Thus, if we know the general solution y of Bessel's equation of order 1, equation (4.30) gives us the general solution θ of our pendulum equation (4.26).

Bessel's differential equation has been studied extensively, beginning at least as far back as Bessel's own work in 1824 on the elliptic motion of planets. Not only can many useful differential equations (such as those of the form (4.27)) be transformed into Bessel's equation, but the solutions of his equation have many significant properties. So rather than immediately continue our study of the pendulum problem, we take an excursion into Bessel functions with an eye to developing those properties which best illuminate the behavior of the pendulum. In particular, we would like to know whether Bessel functions display any nearly periodic behavior, since that would indicate that the motion of a lengthening pendulum is described by a function not too different from the trigonometric functions (4.4) that describe the motion of a pendulum in our simplest model. Also, we would like to know what happens to the motion after a long elapse of time.

The indicial equation for Bessel's differential equation (4.28) is

Always $P_0 = 1$
$Q_0 = -p^2$

$$w^2 + (1 - 1)w - p^2 = 0$$

and so $w = \pm p$. We may assume, without loss of generality, that $p \geq 0$. So, in the notation of Theorem 4.3, the roots are $w_1 = p$ and $w_2 = -p$; hence, $w_1 - w_2 = 2p$. Therefore, if $2p$ is not an integer, Bessel's equation has linearly independent solutions y_1 and y_2 of the form

$$y_1 = \sum_{k=0}^{\infty} c_k x^{k+p} \quad \text{and} \quad y_2 = \sum_{k=0}^{\infty} c_k x^{k-p}$$

Furthermore, if $2p$ is an integer but p is not, Bessel's equation still has two linearly independent solutions of this form (Exercise 22 of Section 4.2). However, if p is an integer, the solution y_2 has a logarithmic term. Note also that, in the notation of Theorem 4.3,

$$P(x) = 1 \quad \text{and} \quad Q(x) = x^2 - p^2$$

which are both power series about zero with infinite radii of convergence; thus, all the solutions of Bessel's equation can be expressed in terms of series that converge on the interval $(0, \infty)$.

From now on, we consider only Bessel equations of order n, where n is a non-negative integer; these are the Bessel equations that occur most often in applications. If we replace p by n in (4.28) and then let $y = \sum_{k=0}^{\infty} c_k x^{k+n}$, we get

$$\sum_{k=0}^{\infty} c_k(k + n)(k + n - 1)x^{k+n} + \sum_{k=0}^{\infty} c_k(k + n)x^{k+n}$$

$$+ \sum_{k=0}^{\infty} c_k x^{k+n+2} - \sum_{k=0}^{\infty} c_k n^2 x^{k+n} = 0$$

or

$$\sum_{k=0}^{\infty} c_k \left[(k + n)(k + n - 1 + 1) - n^2 \right] x^{k+n} = - \sum_{k=2}^{\infty} c_{k-2} x^{k+n}$$

Therefore

$$c_k(k^2 + 2kn + n^2 - n^2) = \begin{cases} -c_{k-2} & \text{if } k \geq 2 \\ 0 & \text{if } k = 0, 1 \end{cases}$$

Since the coefficient of c_1 is $1 + 2n$, it follows from the above formula that $c_1 = 0$. And, since

$$c_k = \frac{-c_{k-2}}{k(k + 2n)} \quad \text{for } k \geq 2$$

it follows that $c_{2k+1} = 0$ for $k \geq 0$ and that

$$c_2 = \frac{-c_0}{2(2 + 2n)} = \frac{-c_0}{2 \cdot 2(n + 1)}$$

$$c_4 = \frac{-c_2}{4(4 + 2n)} = \frac{c_0}{2 \cdot 4 \cdot 2^2(n + 1)(n + 2)}$$

$$c_6 = \frac{-c_4}{6(6 + 2n)} = \frac{-c_0}{2 \cdot 4 \cdot 6 \cdot 2^3(n + 1)(n + 2)(n + 3)}$$

$$\vdots$$

$$c_{2k} = \frac{(-1)^k c_0}{2 \cdot 4 \cdot 6 \cdots (2k)2^k(n + 1)(n + 2) \cdots (n + k)}$$

$$= \frac{(-1)^k c_0 n!}{k!2^k 2^k(n + k)!}$$

We choose $c_0 = 1/n!2^n$; this choice produces the solution known as J_n, the *Bessel function of order n*:

$$J_n(x) = \sum_{k=0}^{\infty} \frac{(-1)^k}{k!(n + k)!} \left(\frac{x}{2}\right)^{2k+n} \qquad n = 0, 1, 2, \ldots \qquad (4.31)$$

It is easy to check that this series converges for all x.

The following theorem gives a few of the most important properties of J_n; others are left to the exercises.

Theorem 4.4

(a) $\frac{d}{dx}(x^{-n}J_n(x)) = -x^{-n}J_{n+1}(x)$

(b) $\frac{d}{dx}(x^n J_n(x)) = x^n J_{n-1}(x)$

(c) $xJ_{n+1}(x) = nJ_n(x) - xJ_n'(x)$

(d) $xJ_{n-1}(x) = xJ_n'(x) + nJ_n(x)$

Proof (a)

$$-x^n \frac{d}{dx}(x^{-n}J_n(x)) = -x^n \frac{d}{dx} \sum_{k=0}^{\infty} \frac{(-1)^k}{k!(n + k)!} \left(\frac{x}{2}\right)^{2k} \frac{1}{2^n}$$

$$= -x^n \sum_{k=1}^{\infty} \frac{(-1)^k 2k}{k!(n + k)!} \left(\frac{x}{2}\right)^{2k-1} \frac{1}{2^{n+1}}$$

$$= \sum_{k=1}^{\infty} \frac{(-1)^{k-1}}{(k - 1)!(n + k)!} \left(\frac{x}{2}\right)^{2k-1+n}$$

$$= \sum_{k=0}^{\infty} \frac{(-1)^k}{k!(n + k + 1)!} \left(\frac{x}{2}\right)^{2k+1+n}$$

$$= J_{n+1}(x)$$

Part (b) is proved similarly (Exercise 2). The recursion formulas (c) and (d) result from applying the product rule for derivatives to (a) and (b), respectively. For example, by (a),

$$-x^{-n}J_{n+1}(x) = -nx^{-n-1}J_n(x) + x^{-n}J_n'(x)$$

which, after it is multiplied by $-x^{n+1}$, is the formula (c).

Corollary 4.5

(a) Between any two positive roots of $J_n(x) = 0$ there is a root of $J_{n+1}(x) = 0$.
(b) Between any two positive roots of $J_n(x) = 0$ there is a root of $J_{n-1}(x) = 0$.

Proof

(a) We apply the mean value theorem to the function f defined by $f(x) = x^{-n}J_n(x)$. If x_1 and x_2 are distinct positive roots of $J_n(x) = 0$, then $f(x_1) = 0 = f(x_2)$. Hence $f'(z) = 0$ at some point z between x_1 and x_2. Therefore, by Theorem 4.4a, $-z^{-n}J_{n+1}(z) = 0$, which implies that $J_{n+1}(z) = 0$.
(b) This theorem is proved similarly (Exercise 3).

We can prove even more about the zeros of the Bessel functions J_n: There are infinitely many of them and they occur arbitrarily far out on the positive x axis. Furthermore, we can find out roughly how far apart the consecutive zeros are, at least those far out on the x axis. These insights will be gained after transforming Bessel's equation into a form in which there is no first derivative term. According to equation (4.27) and the change of variables (4.29), we should let

$$a = \tfrac{1}{2} \qquad b = c = 1 \quad \text{and} \quad u(x) = \sqrt{x}\, y(x)$$

Then Bessel's equation (4.28) of order n becomes

$$x^2 u'' + \left(x^2 + \tfrac{1}{4} - n^2\right)u = 0$$

or

$$u'' + \left(1 + \frac{1 - 4n^2}{4x^2}\, u\right) = 0 \tag{4.32}$$

For large x, this equation is much like $u'' + u = 0$, whose solutions have the form

$$u = c_1 \cos x + c_2 \sin x = u_0 \cos\left(x - x_0\right)$$

where $c_1 = u_0 \cos(x_0)$ and $c_2 = u_0 \sin(x_0)$. This suggests, but does not prove, that, for large x, the Bessel function J_n satisfies

$$J_n(x) = (\text{approximately}) \; \frac{u_0}{\sqrt{x}} \cos\left(x - x_0\right) \tag{4.33}$$

We will not make precise the sense in which this approximation is valid, nor will we compute the values of x_0 and u_0; but all of this can be done. We will simply use (4.33) as a source for qualitative statements about J_n and for inspiring guesses about other properties of J_n. For example, since the consecutive zeros of the cosine function are π units apart, equation (4.33) suggests that consecutive zeros of J_n might be approximately π units apart, at least those far out on the positive x axis. In fact, the next result shows that we can get precise information from the comparison of (4.32) with $u'' + u = 0$.

Lemma 4.6 Suppose we are given differential equations $y'' + r(x)y = 0$ and $y'' + s(x)y = 0$ whose coefficients r and s are continuous on an open interval (a, b) and satisfy $s(x) > r(x)$ for all x in (a, b). If f and g are solutions of the respective differential equations on this interval and if x_1 and x_2 are consecutive zeros of f in the interval (a, b), then g has a zero between x_1 and x_2.

Proof Let $h(x) = g(x)f'(x) - f(x)g'(x)$. Then

$$h'(x) = g(x)f''(x) + g'(x)f'(x) - f'(x)g'(x) - f(x)g''(x)$$
$$= -r(x)f(x)g(x) + s(x)g(x)f(x)$$
$$= (s(x) - r(x))f(x)g(x)$$

Now let us suppose that g does not have a zero between x_1 and x_2. Then, by the intermediate value theorem, g must be of constant sign on the open interval (x_1, x_2). The function f also has constant sign there, since x_1 and x_2 are *consecutive* zeros of f. We assume that f and g are both positive on (x_1, x_2); other choices of signs lead to similar proofs. The above computation then shows that $h'(x) > 0$ on (x_1, x_2). Hence, by Lemma A.26,

$$0 < \int_{x_1}^{x_2} h'(x)\,dx$$
$$= h(x_2) - h(x_1)$$
$$= g(x_2)f'(x_2) - f(x_2)g'(x_2) - g(x_1)f'(x_1) + f(x_1)g'(x_1)$$
$$= g(x_2)f'(x_2) - g(x_1)f'(x_1)$$

However, $g(x_2) \geq 0$ and $g(x_1) \geq 0$ since g is continuous. Also, $f'(x_1) \geq 0$ and $f'(x_2) \leq 0$; otherwise, we could show that $f(x) < 0$ for some x between x_1 and x_2 (Exercise 9). These facts imply that $g(x_2)f'(x_2) - g(x_1)f'(x_1) \leq 0$, which contradicts our conclusion above. Therefore, g must have a zero between x_1 and x_2.

Theorem 4.7 J_n has infinitely many positive zeros and they occur arbitrarily far out on the positive x axis.

Proof First we prove that the theorem is true for J_0. We may apply Lemma 4.9 to the differential equations

$$u'' + u = 0 \quad \text{and} \quad u'' + \left(1 + \frac{1}{4x^2}\right)u = 0$$

since $1 + 1/4x^2 > 1$ on the interval $(0, \infty)$. Also, we may let $f(x) = \sin x$ and $g(x) = \sqrt{x}\,J_0(x)$, since these functions satisfy the respective differential equations. Therefore, since πn and $\pi(n + 1)$ are consecutive zeros of the sine function for every positive integer n, we have proved the theorem for J_0. But then, by repeated use of

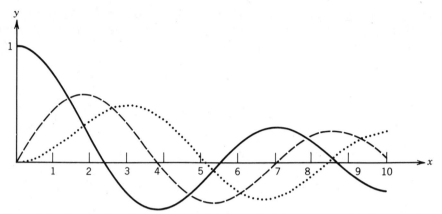

Figure 4.2 Graphs of the Bessel functions J_0 (————), J_1 (– – – – –), and J_2 (· · · · · · · · ·)

Corollary 4.5a, an induction argument shows that the theorem is true for all the remaining Bessel functions J_n.

Figure 4.2 displays graphs of the first few Bessel functions J_n, and Table 4.3 gives numerical approximations to their first few zeros. The figure suggests yet another property of the zeros of J_n: Although J_n has infinitely many positive zeros, no finite interval contains infinitely many of them. We omit the proof of this fact.

We have discussed only one solution, J_n, of Bessel's equation of order n. We know the form of another solution from Theorem 4.3, but the techniques for finding it will not be taken up until the next section. As we will see, the function

$$K_0(x) = J_0(x) \log x - \sum_{k=1}^{\infty} \frac{(-1)^k H_k x^{2k}}{2^{2k}(k!)^2}, \quad x > 0$$

where H_k denotes the kth partial sum of the harmonic series, is another solution of Bessel's equation of order 0. For Bessel's equation of order $n > 0$, a second solution on $(0, \infty)$ is

$$K_n(x) = J_n(x) \log x - \frac{1}{2} \sum_{k=0}^{n-1} \frac{(n-k-1)!}{k!} \left(\frac{x}{2}\right)^{2k-n}$$

$$- \frac{1}{2} \sum_{k=0}^{\infty} \frac{(-1)^k (H_k + H_{n+k})}{k!(n+k)!} \left(\frac{x}{2}\right)^{2k+n}$$

Table 4.3 The First Few Positive Zeros of J_0, J_1, J_2

J_0	J_1	J_2
2.4048	3.8317	5.1356
5.5201	7.0156	8.4172
8.6537	10.1735	11.6198
11.7915	13.3237	14.7960
14.9309	16.4706	17.9598

where H_k is defined as above for $k > 0$ and $H_0 = 0$. Therefore, since J_n and K_n are linearly independent, the general solution of Bessel's equation of order n may be written as

$$y = c_1 J_n(x) + c_2 K_n(x) \tag{4.34}$$

Now let us apply our new knowledge about the solutions of Bessel's equation to our study of the motion of a lengthening pendulum. According to the transformation (4.30) and the solutions (4.34), the solutions of the pendulum equation (4.26) can be written as

$$\theta = \frac{1}{\sqrt{t}} \left(c_1 J_1(b\sqrt{t}) + c_2 K_1(b\sqrt{t}) \right) \tag{4.35}$$

Next, we assume that initial conditions can be chosen so that $c_2 = 0$. We make this assumption partly out of convenience; since we know so little about the function K_1, our analysis will be simpler if we eliminate K_1 from the description of the angle θ. It is not unreasonable, however, to make this simplification, because it can be shown that other initial conditions lead to similar conclusions about the behavior of θ. Thus,

$$\theta = \frac{c_1 J_1(b\sqrt{t})}{\sqrt{t}} \tag{4.36}$$

Furthermore, if we let $x = b\sqrt{t}$, then

$$\theta' = \frac{d\theta}{dt} = \frac{d\theta}{dx}\frac{dx}{dt} = \frac{d}{dx}\frac{c_1 b J_1(x)}{x}\frac{dx}{dt}$$

$$= \frac{-c_1 b J_2(x)}{x}\frac{dx}{dt} \quad \text{by Corollary 4.7a}$$

$$= \frac{-c_1 J_2(b\sqrt{t})}{\sqrt{t}}\frac{b}{2\sqrt{t}}$$

and so

$$\theta' = \frac{-c_1 b J_2(b\sqrt{t})}{2t} \tag{4.37}$$

Finally we are prepared to draw some conclusions about the motion of the pendulum. First, equation (4.33) tells us that, for large x, $J_1(x)$ behaves something like a sinusoidal function with amplitude proportional to $1/\sqrt{x}$. Therefore, according to (4.36), the amplitude of θ is eventually nearly proportional to $1/(\sqrt[4]{t}\sqrt{t}) = t^{-3/4}$. In particular, the angular motion gradually dies out. The angular velocity, however, dies out even more rapidly since θ' is eventually nearly proportional to $1/(\sqrt[4]{t}\,t) = t^{-5/4}$.

More exact information about the motion of the pendulum can be deduced from our table of zeros of J_n. Note that at the top of a swing of the pendulum, $\theta' = 0$ and so $J_2 = 0$ according to (4.37). And at the bottom of a swing, $\theta = 0$ and so $J_1 = 0$ according to equation (4.36). Thus, we can find the "quarter periods" of the pendulum, by which we mean the time that elapses between the top and bottom of a swing, as follows. If $x_1, x_2, x_3, \ldots$ denote the infinitely many positive zeros of J_1 arranged in increasing order and $y_1, y_2, y_3, \ldots$ denote the infinitely many positive

Table 4.4 Quarter Periods of the Pendulum

Upswing	Downswing
11.692	
21.631	22.844
31.520	32.651
41.401	42.501
51.274	52.359

zeros of J_2 arranged in increasing order, then $x = b\sqrt{t}$ implies that $t = x^2/b^2$ and therefore the quarter periods are $(y_1^2 - x_1^2)/b^2$, $(x_2^2 - y_1^2)/b^2$, $(y_2^2 - x_2^2)/b^2$, etc. The first column of Table 4.4 contains the differences $y_i^2 - x_i^2$; the second column contains $x_{i+1}^2 - y_i^2$. These numbers, when multiplied by $1/b^2 = v/4g$, are therefore the quarter periods. Some immediate observations we can make are that the periods are getting longer and at a rate that seems nearly constant. Also, each downswing takes considerably longer than the preceding upswing, but then the next upswing takes slightly less long. These observations suggest, of course, an even more extensive analysis of the function θ and therefore of Bessel functions. But we choose to end our study of the lengthening pendulum here.

EXERCISES

1. Show that if the expression (4.29) is substituted for u in (4.27), then y satisfies Bessel's equation of order p.

2. Prove Theorem 4.4b.

3. Prove Corollary 4.5b.

In Exercises 4–6, transform the given equation into Bessel's equation by finding the constants a, b, c, and p in equation (4.27). Then express the general solution of the given equation in terms of the functions J_n and K_n.

4. $t^2 u'' + (t^2 - \frac{15}{4})u = 0$

5. $t^2 u'' + tu' + (r^2 t^{2s/n} - s^2)u = 0$

6. $u'' + r^2 t^{(1-2n)/n} u = 0$

7. Suppose $\{y_n\}$ is a sequence of functions defined recursively by

$$y_{n+1}(x) = -y_n'(x) + \frac{ny_n(x)}{x}, \qquad x > 0$$

If the function y_0 satisfies Bessel's equation of order 0, prove that y_n satisfies Bessel's equation of order n, for every integer $n \geq 0$.

8. Prove the identities

   (a) $J_{n-1}(x) - J_{n+1}(x) = 2J_n'(x)$

   (b) $J_{n-1}(x) + J_{n+1}(x) = 2nJ_n(x)/x$

   (c) $J_n(-x) = (-1)^n J_n(x)$

   (d) $\int_0^x t^n J_{n-1}(t)\, dt = x^n J_n(x)$

9. Suppose f is defined on the closed interval $[x_1, x_2]$ and is differentiable at x_1 and x_2. If $f(x_1) = f(x_2) = 0$ and if $f(x) > 0$ whenever $x_1 < x < x_2$, prove that $f'(x_1) \geq 0$ and $f'(x_2) \leq 0$. *Hint:* Write out the definitions of $f'(x_1)$ and $f'(x_2)$.

10. The Legendre polynomials can be defined by

$$P_n(x) = \frac{1}{2^n n!} \frac{d^n}{dx^n}(x^2 - 1)^n, \qquad n = 0, 1, 2, \ldots$$

Deduce that

(a) $P_n(x) = \displaystyle\sum_{k=0}^{[n/2]} \frac{(-1)^k (2n - 2k)! x^{n-2k}}{2^n k! (n - k)! (n - 2k)!}$ from the binomial theorem.

(b) $(n + 1)P_{n+1}(x) = (2n + 1)xP_n(x) - nP_{n-1}(x)$ from (a).

(c) $\dfrac{d^n}{dx^n} xf(x) = x \dfrac{d^n}{dx^n} f(x) + n \dfrac{d^{n-1}}{dx^{n-1}} f(x)$, if f is n times differentiable.

(d) $P_{n+1}'(x) = (n + 1)P_n(x) + xP_n'(x)$ from the definition of P_n and (c).

(e) $nP_n(x) = xP_n'(x) - P_{n-1}'(x)$ from (b) and (d).

(f) $(1 - x^2)P_n'(x) = nP_{n-1}(x) - nxP_n(x)$ from (d) and (e).

(g) $((1 - x^2)P_n'(x))' + n(n + 1)P_n(x) = 0$ from (e) and (f).

(h) P_n satisfies Legendre's equation:

$$(1 - x^2)y'' - 2xy' + n(n + 1)y = 0$$

11. In 1733 Daniel Bernoulli studied the motion of a heavy chain hanging from a fixed point and oscillating about its (vertical) equilibrium position. (This was one of the earliest problems that led to a differential equation with a regular singular point.) Bernoulli showed that the motion is described reasonably well by the partial differential equation

$$\frac{\partial^2 y}{\partial t^2} = gx \frac{\partial^2 y}{\partial x^2} + g \frac{\partial y}{\partial x}, \qquad 0 < x < L$$

where t is time, x is the vertical position measured upward from the bottom of the chain, y is the horizontal displacement of the chain, g is the (constant) acceleration due to gravity, and L is the length of the chain (Figure 4.3).

(a) Assume, as Bernoulli did, that there exist solutions of the form $y(x, t) = u(x)\sin(2\pi\omega t)$, in which each point of the chain oscillates sinusoidally in time with the uniform frequency ω. By substituting this expression for y into Bernoulli's partial differential equation, find a second-order, linear, homogeneous differential equation satisfied by u. Transform this equation into a Bessel equation, and use the transformation to write the general solution for u.

(b) Since $u(0)$ is clearly finite, the fact that $|K_n(x)|$ becomes arbitrarily large for x near 0 implies that the coefficient of K_n in your answer to (a) must be zero. Also, $u(L)$ must be zero since the top of the chain is fixed in place. Show that this restriction forces ω to have one of the values

$$\omega = \frac{z\sqrt{g}}{4\pi\sqrt{L}}$$

where z is a positive zero of a particular J_n. (The smallest value of z, Bernoulli suggested, gives rise to the fundamental frequency, ω, of the chain and larger values produce the higher harmonics.)

(c) Find the fundamental frequency of a hanging chain 4 ft long.

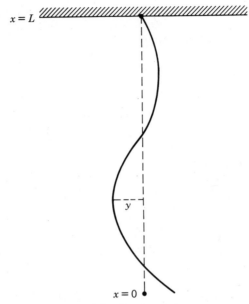

$x = L$

y

$x = 0$

Figure 4.3   Hanging chain

4.4 Regular Singular Points, Part 2

In this section we develop three techniques for computing the solution y_2 in Theorem 4.3. Recall that the theorem describes the form of two linearly independent solutions y_1 and y_2 of the differential equation

$$x^2y'' + xP(x)y' + Q(x)y = 0, \qquad 0 < x < R, \tag{4.38}$$

where P and Q can be represented as sums of power series about 0 with radius of convergence at least R. The indicial equation is then

$$w^2 + (P_0 - 1)w + Q_0 = 0 \tag{4.39}$$

where $P_0 = P(0)$ and $Q_0 = Q(0)$. We assume that (4.39) has real roots w_1 and w_2, where $w_1 \geq w_2$; thus, equation (4.38) has a nonzero solution of the form

$$y_1 = \sum_{k=0}^{\infty} c_k x^{k+w_1} \tag{4.40}$$

Sometimes there is also a solution of the form

$$y_2 = \sum_{k=0}^{\infty} c_k x^{k+w_2}$$

that is linearly independent of y_1; but if not, there must be one of the form

$$y_2 = y_1 \log x + \sum_{k=0}^{\infty} c_k x^{k+w_2} \tag{4.41}$$

It is solutions of this latter form that we now seek. According to Theorem 4.3, the general solution of the differential equation (4.38) always includes solutions of the form (4.41) when $w_1 = w_2$ and might or might not include such solutions when $w_1 - w_2$ is a positive integer.

The most straightforward method for computing solutions of the form (4.41) is simply to substitute an expression of that form into the differential equation (4.38) and then solve for the unknown coefficients c_k. We can do part of the computation abstractly. Let $y_2 = y_1 \log x + u$, where y_1 is a known nonzero solution of the form (4.40). Then

$$y_2' = y_1' \log x + \frac{y_1}{x} + u'$$

$$y_2'' = y_1'' \log x + \frac{2y_1'}{x} - \frac{y_1}{x^2} + u''$$

So the result of substituting y_2 into the differential equation (4.38) is

$$x^2 y_1'' \log x + 2xy_1' - y_1 + x^2 u'' + xP(x) y_1' \log x$$
$$+ P(x) y_1 + xP(x) u' + Q(x) y_1 \log x + Q(x) u = 0$$

Or, since y_1 satisfies the differential equation (4.38),

$$2xy_1' - y_1 + x^2 u'' + P(x) y_1 + xP(x) u' + Q(x) u = 0$$

or

$$x^2 u'' + xP(x) u' + Q(x) u = -2xy_1' + (1 - P(x)) y_1 \tag{4.42}$$

Therefore, to find the solution y_2, it remains only to substitute the expression $u = \sum_{k=0}^{\infty} c_k x^{k+w}$ into the differential equation (4.42) and solve for the unknown coefficients c_k; then the solution y_2 is given by

$$y_2 = y_1 \log x + u \tag{4.43}$$

Since the left side of (4.42) is exactly the left side of the original equation (4.38) (with y replaced by u), the amount of additional computation, beyond that needed to compute y_1, is less than one might first imagine.

EXAMPLE 4.9   Find two linearly independent solutions, represented as series expansions in powers of x, of Bessel's equation of order 0:

$$x^2 y'' + xy' + x^2 y = 0$$

The indicial equation is $w^2 = 0$ and so $w_1 = w_2 = 0$; thus, one of the solutions will have the form (4.41). Although we found a nonzero solution of this equation in the preceding section, it will be instructive to start the problem from the beginning. Also, we will be better able to illustrate the second of our three methods of solution if we keep w general at first. Thus, we substitute $y_1 = \sum_{k=0}^{\infty} c_k x^{k+w}$ into the differential equation:

$$\sum_{k=0}^{\infty} c_k(k+w)(k+w-1)x^{k+w} + \sum_{k=0}^{\infty} c_k(k+w)x^{k+w} + \sum_{k=0}^{\infty} c_k x^{k+w+2} = 0$$

$$\sum_{k=0}^{\infty} c_k(k+w)^2 x^{k+w} = -\sum_{k=2}^{\infty} c_{k-2} x^{k+w} \qquad (4.44)$$

Then, equating corresponding coefficients of x^{k+w}, we get the equations $c_0 w^2 = 0$, $c_1(1+w)^2 = 0$, and

$$c_k = \frac{-c_{k-2}}{(k+w)^2} \quad \text{for } k \geq 2 \qquad (4.45)$$

The first of these yields the indicial equation $w^2 = 0$; the second implies that $c_1 = 0$. Thus, from the difference equation (4.45), we conclude that $c_{2k+1} = 0$ for all $k \geq 0$ and, by letting $w = 0$ and $c_0 = 1$,

$$c_2 = \frac{-1}{2^2}, \qquad c_4 = \frac{-c_2}{4^2} = \frac{1}{2^2 4^2}, \qquad c_6 = \frac{-c_4}{6^2} = \frac{-1}{2^2 4^2 6^2}$$

More generally,

$$c_{2k} = \frac{(-1)^k}{2^2 4^2 \cdots (2k)^2} = \frac{(-1)^k}{(2^k k!)^2}$$

and, therefore, one nonzero solution of the given differential equation is

$$y_1 = \sum_{k=0}^{\infty} \frac{(-1)^k x^{2k}}{2^{2k}(k!)^2}$$

We find a second solution y_2 by using the method summarized in (4.42) and (4.43). Specifically, we solve $x^2 u'' + xu' + x^2 u = -2xy_1'$ for u, where $u = \sum_{k=0}^{\infty} c_k x^{k+w}$. But since u has the same form as y_1, we see from (4.44) that

$$\sum_{k=0}^{\infty} c_k k^2 x^k = -\sum_{k=2}^{\infty} c_{k-2} x^k - 2x \sum_{k=1}^{\infty} \frac{(-1)^k 2kx^{2k-1}}{2^{2k}(k!)^2}$$

$$= -\sum_{k=2}^{\infty} c_{k-2} x^k + \sum_{k=1}^{\infty} \frac{(-1)^{k-1} 2^2 kx^{2k}}{2^{2k}(k!)^2}$$

Again there is no linear term in the right side of the equation, and so $c_1 = 0$. Also, since the last sum on the right contains no odd power of x, we see that $c_{2k+1} = -c_{2k-1}/(2k+1)^2$ for $k \geq 1$; therefore, $c_{2k+1} = 0$ for $k \geq 0$. Finally, when coefficients of even powers of x are equated, the result is

$$c_{2k} = \frac{-c_{2k-2}}{2^2 k^2} + \frac{(-1)^{k-1}}{2^{2k} k(k!)^2} \quad \text{for } k \geq 1$$

If we choose $c_0 = 0$ (as Theorem 4.3b says we may), then

$$c_2 = \frac{1}{2^2}$$

$$c_4 = -\frac{c_2}{2^2 2^2} - \frac{1}{2^4 2(2!)^2} = -\frac{1}{2^4 (2!)^2} - \frac{1}{2^5 (2!)^2} = \frac{-1}{2^4 (2!)^2}\left(1 + \frac{1}{2}\right)$$

$$c_6 = \frac{-c_4}{2^2 3^2} + \frac{1}{2^6 3(3!)^2}$$

$$= \frac{1}{2^6 (3!)^2}\left(1 + \frac{1}{2}\right) + \frac{1}{2^6 (3!)^2 3} = \frac{1}{2^6 (3!)^2}\left(1 + \frac{1}{2} + \frac{1}{3}\right)$$

$$\vdots$$

$$c_{2k} = \frac{(-1)^{k-1} H_k}{2^{2k}(k!)^2} \quad \text{where } H_k = 1 + \frac{1}{2} + \cdots + \frac{1}{k} \quad \text{for } k \geq 1$$

We have thus found the solution

$$y_2 = y_1 \log x + \sum_{k=1}^{\infty} \frac{(-1)^{k-1} H_k x^{2k}}{2^{2k}(k!)^2}$$

$$= \left(\sum_{k=0}^{\infty} \frac{(-1)^k x^{2k}}{2^{2k}(k!)^2}\right) \log x + \sum_{k=1}^{\infty} \frac{(-1)^{k-1} H_k x^{2k}}{2^{2k}(k!)^2}$$

Our second method for finding solutions of the form (4.41) is usually more efficient than the one above; but it takes longer to explain. The explanation, however, will give added insight into why the logarithmic term arises. Observe first that the trial solution in Frobenius' method can be thought of as depending on both x and w:

$$y_w(x) = \sum_{k=0}^{\infty} c_k(w) x^{k+w} \tag{4.46}$$

We have indicated in this equation that the coefficients c_k can be thought of as functions of w since the variable w usually appears in the difference equation for c_k. Equation (4.45) in the preceding example is a case in point. In fact, if we had not replaced w by 0 in that example, we could have expressed the coefficients c_k as explicit functions of w:

$$c_{2k}(w) = \frac{(-1)^k c_0}{(w+2)^2 (w+4)^2 \cdots (w+2k)^2} \quad \text{and} \quad c_{2k+1}(w) = 0$$

Now let us see what happens when we substitute an expression of the form (4.46) into the differential equation (4.38). As usual, we equate corresponding coefficients of x^{k+w} and solve the resulting difference equation for $c_k(w)$. However, just solving the difference equation does not make y_w a solution of the differential equation. The coefficients of x^{k+w} for $k > 0$ will certainly be zero; but, if w is not one of the solutions of the indicial equation, the coefficient of x^w will not be zero. In fact, in our analysis of Frobenius' method in Section 4.2 we found that this coefficient is $c_0(w^2 + (P_0 - 1)w + Q_0)$, where the expression in parentheses is the left side of the indicial equation. In other words, if $y_w(x)$ has the form (4.46), where $c_k(w)$ satisfies the difference equation that results from substituting y_w into the differential equation (4.38), then

$$x^2 y_w'' + xP(x)y_w' + Q(x)y_w = c_0(w)(w^2 + (P_0 - 1)w + Q_0)x^w$$

Or, if we let $c_0(w) = 1$ and denote the solutions of the indicial equation by w_1 and w_2, then

$$x^2 y_w'' + xP(x)y_w' + Q(x)y_w = (w - w_1)(w - w_2)x^w \qquad (4.47)$$

In particular, y_w is not a solution of (4.38) if $w \neq w_1$ and $w \neq w_2$; but y_{w_1} and y_{w_2} are solutions of (4.38), provided the coefficients $c_k(w)$ exist at w_1 and w_2, respectively.

The real value of (4.47), however, is that we can use it to find a second solution of the differential equation (4.38) when Frobenius' method alone does not produce two linearly independent solutions. First consider the case when $w_1 = w_2$. We find another solution of (4.38) by differentiating equation (4.47) *with respect to w* and then replacing w by w_2. (The following computations are intended only as a rough derivation; we will not justify the steps rigorously.) Using primes to indicate differentiation with respect to x and interchanging that differentiation and the differentiation with respect to w, we get

$$x^2\left(\frac{d}{dw}y_w\right)'' + xP(x)\left(\frac{d}{dw}y_w\right)' + Q(x)\left(\frac{d}{dw}y_w\right)$$

$$= \frac{d}{dw}(w - w_2)^2 x^w$$

$$= 2(w - w_2)x^w + (w - w_2)^2 x^w \log x$$

$$= 0 \quad \text{when } w = w_2$$

Therefore, the function $\dfrac{dy_w}{dw}$, when evaluated at $w = w_2$, is a solution of (4.38). Next, we find a more useful way of expressing this solution. By differentiating (4.46) with respect to w and interchanging the differentiation and summation signs, we get

$$\frac{dy_w}{dw} = \sum_{k=0}^{\infty} \frac{d}{dw}\left(c_k(w)x^{k+w}\right)$$

$$= \sum_{k=0}^{\infty} c_k(w)x^{k+w} \log x + \sum_{k=1}^{\infty} c_k'(w)x^{k+w}$$

where the second sum starts at $k = 1$ since we chose $c_0(w)$ to be identically 1. Hence, after replacing w by w_2, we find the solution promised in Theorem 4.3b:

$$y_2 = \sum_{k=0}^{\infty} c_k(w_2)x^{k+w_2} \log x + \sum_{k=1}^{\infty} c_k'(w_2)x^{k+w_2}$$

$$= y_1 \log x + \sum_{k=1}^{\infty} c_k'(w_2)x^{k+w_2} \qquad (4.48)$$

Furthermore, we have found a useful formula, $c_k'(w_2)$, for the unknown coefficients of x^{k+w_2}.

In the case when $w_1 - w_2$ is a positive integer and Frobenius' method does not produce two linearly independent solutions, an argument much like the one above shows that

$$y_2 = \left(\sum_{k=0}^{\infty} \lim_{w \to w_2} \left[(w - w_2)c_k(w) \right] x^{k+w_2} \right) \log x$$
$$+ \sum_{k=0}^{\infty} \lim_{w \to w_2} \frac{d}{dw} \left[(w - w_2)c_k(w) \right] x^{k+w_2} \tag{4.49}$$

It can also be shown that the sum in parentheses is a solution of the form $\sum_{k=0}^{\infty} c_k(w_1)x^{k+w_1}$.

EXAMPLE 4.10 Find two linearly independent solutions, represented as series expansions in powers of x, of

$$x^2y'' + xy' + 2xy = 0$$

As usual, we substitute $y = \sum_{k=0}^{\infty} c_k x^{k+w}$ into the differential equation.

$$\sum_{k=0}^{\infty} c_k(k + w)(k + w - 1)x^{k+w} + \sum_{k=0}^{\infty} c_k(k + w)x^{k+w} + \sum_{k=0}^{\infty} 2c_k x^{k+w+1} = 0$$

or

$$\sum_{k=0}^{\infty} c_k(k + w)^2 x^{k+w} = - \sum_{k=1}^{\infty} 2c_{k-1} x^{k+w}$$

Setting corresponding coefficients equal, we find that $c_0 w^2 = 0$ and

$$c_k = \frac{-2c_{k-1}}{(w + k)^2} \quad \text{for } k \geq 1$$

Hence, the indicial equation is $w^2 = 0$, and so $w_1 = w_2 = 0$. But since we want to express the coefficients c_k as functions of w, we do not replace w by 0 in the difference equation. We let $c_0 = 1$; then

$$c_1 = \frac{-2c_0}{(w + 1)^2} = \frac{-2}{(w + 1)^2}$$

$$c_2 = \frac{-2c_1}{(w + 2)^2} = \frac{2^2}{(w + 1)^2(w + 2)^2}$$

$$\vdots$$

$$c_k = \frac{(-1)^k 2^k}{(w + 1)^2(w + 2)^2 \cdots (w + k)^2} \quad \text{for } k \geq 1$$

Therefore, $c_k(w_1) = c_k(0) = (-1)^k 2^k/(k!)^2$ for $k \geq 0$, and so one solution of the given differential equation is

$$y_1 = \sum_{k=0}^{\infty} c_k(w_1)x^{k+w_1} = \sum_{k=0}^{\infty} \frac{(-1)^k 2^k x^k}{(k!)^2}$$

We find a second solution by using (4.48). It will be easier to compute the derivatives c_k' if we take the logarithm of $|c_k|$ and then differentiate with respect to w:

$$\log |c_k| = \log(2^k) - \log\left[(w+1)^2(w+2)^2 \cdots (w+k)^2\right]$$

$$= \log(2^k) - 2\sum_{j=1}^{k} \log(w+j)$$

and so

$$\frac{c_k'}{c_k} = -2\sum_{j=1}^{k} \frac{1}{w+j}$$

So, when w is replaced by $w_2 = 0$, we get

$$c_k'(0) = \left(-2\sum_{j=1}^{k} \frac{1}{j}\right)c_k(0) = -2H_k \frac{(-1)^k 2^k}{(k!)^2}$$

where H_k again denotes the kth partial sum of the harmonic series. Therefore, a second solution of the given differential equation is

$$y_2 = y_1 \log x + \sum_{k=1}^{\infty} \frac{(-1)^{k+1} 2^{k+1} H_k x^k}{(k!)^2}$$

EXAMPLE 4.11 Find two linearly independent solutions, represented as series expansions in powers of x, of Bessel's equation of order 1:

$$x^2 y'' + xy' + (x^2 - 1)y = 0$$

We substitute $y = \sum_{k=0}^{\infty} c_k x^{k+w}$ into the differential equation and get

$$\sum_{k=0}^{\infty} c_k(k+w)(k+w-1)x^{k+w}$$

$$+ \sum_{k=0}^{\infty} c_k(k+w)x^{k+w} + \sum_{k=0}^{\infty} c_k x^{k+w+2} - \sum_{k=0}^{\infty} c_k x^{k+w} = 0$$

or

$$\sum_{k=0}^{\infty} c_k\left((k+w)(k+w-1) + (k+w) - 1\right)x^{k+w} = -\sum_{k=2}^{\infty} c_{k-2} x^{k+w}$$

The coefficient of x^{k+w} on the left side can be simplified:

$$c_k\left((k+w)(k+w-1) + (k+w) - 1\right) = c_k\left((k+w)(k+w) - 1\right)$$

$$= c_k(k+w-1)(k+w+1)$$

Hence, after setting corresponding coefficients equal, we find that

$$c_0(w - 1)(w + 1) = 0$$

$$c_1 w(w + 2) = 0$$

$$c_k = \frac{-c_{k-2}}{(w + k - 1)(w + k + 1)} \quad \text{for } k \geq 2$$

The first of these equations is the indicial equation, which implies that $w_1 = 1$, $w_2 = -1$. The second implies that $c_1 = 0$. And the difference equation implies that $c_{2k+1} = 0$ for $k \geq 0$ and

$$c_2 = \frac{-c_0}{(w + 1)(w + 3)} = \frac{-1}{(w + 1)(w + 3)} \quad (\text{choosing } c_0 = 1)$$

$$c_4 = \frac{-c_2}{(w + 3)(w + 5)} = \frac{1}{(w + 1)(w + 3)^2(w + 5)}$$

$$c_6 = \frac{-c_4}{(w + 5)(w + 7)} = \frac{-1}{(w + 1)(w + 3)^2(w + 5)^2(w + 7)}$$

$$\vdots$$

$$c_{2k} = \frac{(-1)^k}{(w + 1)(w + 3)^2 \cdots (w + 2k - 1)^2(w + 2k + 1)} \quad \text{for } k \geq 1$$

$$(4.50)$$

Therefore

$$c_{2k}(w_1) = c_{2k}(1) = \frac{(-1)^k}{2 \cdot 4^2 \cdot 6^2 \cdots (2k)^2(2k + 2)}$$

$$= \frac{(-1)^k}{(2 \cdot 4 \cdot 6 \cdots (2k))^2(k + 1)}$$

$$= \frac{(-1)^k}{(2^k k!)^2(k + 1)} = \frac{(-1)^k}{2^{2k}k!(k + 1)!} \quad \text{for } k \geq 0$$

So one solution of the given differential equation is

$$y_1 = \sum_{k=0}^{\infty} c_k(w_1)x^{k+w_1} = \sum_{k=0}^{\infty} \frac{(-1)^k x^{2k+1}}{2^{2k}k!(k + 1)!}$$

which is twice the Bessel function J_1 found in (4.31).

If we were to attempt to use Frobenius' method to find a solution of the form $y = \sum_{k=0}^{\infty} c_k x^{k+w_2} = \sum_{k=0}^{\infty} c_k x^{k-1}$, we would find that $c_0 = 0$ and $c_{2k} = c_2(-1)^{k-1}/2^{2k-2}(k - 1)!k!$ for $k \geq 1$. This produces exactly the solution y_1 we just found. So, instead, we find a second solution by using formula (4.49). We start by using (4.50) to find

$$\lim_{w \to w_2} (w - w_2)c_{2k}(w) = \lim_{w \to -1} (w + 1)c_{2k}(w)$$

$$= \lim_{w \to -1} \frac{(-1)^k}{(w + 3)^2 \cdots (w + 2k - 1)^2(w + 2k + 1)}$$

$$= \frac{(-1)^k}{2^2 4^2 \cdots (2k - 2)^2(2k)}$$

$$= \frac{(-1)^k}{(2 \cdot 4 \cdots 2(k - 1))^2(2k)}$$

$$= \frac{(-1)^k}{\left(2^{k-1}(k - 1)!\right)^2(2k)} = \frac{(-1)^k}{2^{2k-1}(k - 1)!k!}$$

for $k \geq 1$. Also note that $\lim_{w \to w_2} (w - w_2)c_0(w) = 0$ since $c_0(w) = 1$. Thus, the coefficient of $\log x$ in formula (4.49) is

$$\sum_{k=1}^{\infty} \frac{(-1)^k x^{2k-1}}{2^{2k-1}(k - 1)!k!} = \sum_{k=0}^{\infty} \frac{(-1)^{k+1} x^{2k+1}}{2^{2k+1}k!(k + 1)!}$$

which is just the negative of the Bessel function J_1. To find the second term in formula (4.49), we take the logarithm of $|(w - w_2)c_{2k}(w)|$ and then differentiate with respect to w:

$$\log |(w + 1)c_{2k}(w)| = -\log\left[(w + 3)^2 \cdots (w + 2k - 1)^2(w + 2k + 1)\right]$$

$$= -2 \sum_{j=1}^{k-1} \log(w + 2j + 1) - \log(w + 2k + 1)$$

and so

$$\frac{((w + 1)c_{2k}(w))'}{(w + 1)c_{2k}(w)} = -2 \sum_{j=1}^{k-1} \frac{1}{w + 2j + 1} - \frac{1}{w + 2k + 1}$$

So, by letting w approach $w_2 = -1$, we get

$$\lim_{w \to -1} \frac{d}{dw}(w + 1)c_{2k}(w) = \left(-2 \sum_{j=1}^{k-1} \frac{1}{2j} - \frac{1}{2k}\right)\left(\lim_{w \to -1}(w + 1)c_{2k}(w)\right)$$

$$= \left(-H_{k-1} - \frac{1}{2k}\right)\frac{(-1)^k}{2^{2k-1}(k - 1)!k!}$$

where $H_0 = 0$. These computations assume that $k \geq 1$; for $k = 0$, we see that $(w - w_2)c_0(w) = w + 1$ whose derivative is 1. Therefore, a second solution of the differential equation is

$$y_2 = -J_1 \log x + x^{-1} + \sum_{k=1}^{\infty}\left(H_{k-1} + \frac{1}{2k}\right)\frac{(-1)^{k+1} x^{2k-1}}{2^{2k-1}(k - 1)!k!}$$

We discuss only briefly a third method for finding a second solution of the differential equation (4.38). This method, called *reduction of order*, applies to the wider class of second-order, linear, homogeneous differential equations:

$$y'' + p(x)y' + q(x)y = 0 \tag{4.51}$$

The second solution which it produces is not in the form of an infinite series, but rather an integral; for some applications, this is more useful.

The method is to seek a solution y_2 in the form of a product $y_1 u$, where y_1 is a known solution of (4.51) and u is an unknown function. To find u, we simply substitute the product $y = y_1 u$ into the differential equation (4.51). The result (Exercise 11) is

$$u'' + \left(p(x) + \frac{2 y_1'(x)}{y_1(x)} \right) u' = 0 \tag{4.52}$$

Since this is a first-order linear and separable equation in u', it can be solved for u' with the techniques of Appendix D.2. Then u itself is found by a second integration. Examples are left to the exercises.

EXERCISES

In Exercises 1–4, find two linearly independent solutions, represented as series expansions in powers of x, of the given differential equation. Use (a) the method summarized in (4.42) and (4.43), or (b) the method summarized in (4.48) and (4.49).

1. $xy'' + y' - y = 0$ **2.** $x^2 y'' + 3xy' + (1 + x)y = 0$

3. $xy'' - y = 0$ **4.** $x^2 y'' + 2xy' + xy = 0$

5. Use the method summarized in (4.42) and (4.43) to find two linearly independent solutions, represented as series expansions in powers of x, of Bessel's equation of order 1, $x^2 y'' + xy' + (x^2 - 1)y = 0$.

6. Use the method summarized in (4.48) to find two linearly independent solutions, represented as series expansions in powers of x, of Bessel's equation of order 0, $x^2 y'' + xy' + x^2 y = 0$.

7. Use the method summarized in (4.49) to find two linearly independent solutions, represented as series expansions of x, of Bessel's equation of order 2, $x^2 y'' + xy' + (x^2 - 4)y = 0$.

In Exercises 8–11, use the reduction of order method to find a second solution of the given differential equation.

8. $x^2 y'' + 7xy' + 9y = 0$, $y_1 = x^{-3}$

9. $(x - 1)y'' - xy' + y = 0$, $y_1 = e^x$

10. $x^2 y'' - x(x + 2)y' + (x + 2)y = 0$, $y_1 = x$

11. $(1 - x^2)y'' - 2xy' + 2y = 0$, $y_1 = x$ (Legendre's equation with $p = 1$)

12. If y_1 satisfies the differential equation (4.51) and u satisfies the differential equation (4.52), prove that y, u satisfies (4.51).

13.   Recall that one solution of Bessel's equation of order n

$$x^2 y'' + xy' + (x^2 - n^2) y = 0$$

is denoted J_n. Use the reduction of order method to show that there is a second solution of the form

$$J_n(x) \int \frac{dx}{x[J_n(x)]^2}$$

4.5 Asymptotic Series

Up to now, it has been important for our series solutions of differential equations to converge at all points for which the solution is valid. In this section, however, we will find that certain divergent series provide useful representations of solutions. In fact, these representations give both qualitative information about the behavior of the solution and accurate numerical approximations.

Before discussing these representations, we introduce some concepts that make it easier for us to describe the behavior of a function qualitatively.

Definition 4.3   Let f and g be functions defined in a punctured neighborhood of x_0, with $g(x) \neq 0$ in this neighborhood. (A "punctured neighborhood of x_0" is a set of points of the form $(x_0, x_0 + h)$ or $(x_0 - h, x_0)$ or $(x_0 - h, x_0) \cup (x_0, x_0 + h)$.) We say that:

(a) $f(x) = O(g(x))$ as $x \to x_0$ if there is a punctured neighborhood N of x_0 and a constant C such that

$$|f(x)| \leq C|g(x)| \quad \text{for all } x \text{ in } N$$

(b) $f(x) = o(g(x))$ as $x \to x_0$ if for every $\varepsilon > 0$ there exists a punctured neighborhood N of x_0 such that

$$|f(x)| \leq \varepsilon |g(x)| \quad \text{for all } x \text{ in } N$$

If the punctured neighborhood N referred to in (a) or (b) above has the form $(x_0, x_0 + h)$, we write $x \to x_0^+$ in place of $x \to x_0$; and if N has the form $(x_0 - h, x_0)$, we write $x \to x_0^-$ in place of $x \to x_0$.

Roughly speaking, Definition 4.3a says that f grows no more rapidly than g near x_0, and 4.3b says that f grows more slowly than g near x_0. Note also that Definition 4.3b is equivalent to the statement that

$$\lim_{x \to x_0} \frac{f(x)}{g(x)} = 0$$

Many of the results in this book can be restated in the language of "big oh" and "little oh," as the new symbols in Definitions 4.3a and 4.3b are usually read. And, as we will see, it can help our discussion of the qualitative behavior of a function to use this language. Some examples are

$$x = o(1) \qquad \text{as } x \to 0$$

$$\frac{1}{1 + x} = o(1) \qquad \text{as } x \to \infty$$

$$\frac{x^2 \sin x}{1 + x} = O(x) \qquad \text{as } x \to \infty$$

$$e^{-x} = o(x^{-p}) \qquad \text{as } x \to \infty \quad \text{for all } p > 0$$

$$\log x = o(x^p) \qquad \text{as } x \to \infty \quad \text{for all } p > 0$$

$$\log x = o(x^{-p}) \qquad \text{as } x \to 0^+ \quad \text{for all } p > 0$$

$$\sin x - x = O(x^3) \qquad \text{as } x \to 0$$

$$\cos x - 1 = O(x^2) \qquad \text{as } x \to 0$$

$$e^x - \sum_{k=0}^{n} x^k = O(x^{n+1}) \qquad \text{as } x \to 0 \quad \text{for all } n \geq 0$$

As an example of how such statements might be read and interpreted, consider the fifth line of the above list. We can read it as "$\log x$ is little oh of x^p as x approaches positive infinity, for all positive values of p"; and we can interpret it as saying that, for large values of x, $\log x$ grows more slowly than every positive power of x.

Definitions 4.3a and 4.3b also make sense for sequences $\{x_n\}$ and $\{y_n\}$ in place of functions f and g; of course, n can only have the limits $\pm \infty$ and the punctured neighborhoods must have the forms $\{N, N + 1, N + 2, \ldots\}$ and $\{-N, -N - 1, -N - 2, \ldots\}$ for ∞ and $-\infty$, respectively. Some examples are

$$x^n = o(1) \qquad \text{as } n \to \infty \text{ if } |x| < 1$$

$$x^n = o(n!) \qquad \text{as } n \to \infty \text{ for all } x$$

$$F_n = O\left(\frac{1 + \sqrt{5}}{2}\right)^n \qquad \text{as } n \to \infty \text{ where } F_n \text{ is the } n\text{th term of the Fibonacci sequence}$$

$$H_n - \log n - \gamma = o(1) \qquad \text{as } n \to \infty \text{ where } H_n \text{ is the } n\text{th partial sum of the harmonic series and } \gamma \text{ denotes Euler's constant}$$

One place where it would have been especially handy to use the big oh and little oh language was in our discussion in Section 4.3 of the behavior of the Bessel functions $J_n(x)$ for large positive values of x. Three especially significant asymptotic properties of Bessel functions are

$$J_n(x) - \sqrt{\frac{2}{\pi x}} \cos\left(x - \frac{n\pi}{2} - \frac{\pi}{4}\right) = O(x^{-3/5}) \qquad \text{as } x \to \infty$$

$$\frac{x_m - m\pi}{m\pi} = o(1) \qquad \text{as } m \to \infty$$

$$(x_{m+1} - x_m) - \pi = o(1) \qquad \text{as } m \to \infty$$

where x_m denotes the mth positive zero of a fixed Bessel function J_n. This gives us precise, provable statements expressed in an intuitively appealing fashion. The third statement, for example, can be interpreted as saying that the zeros of J_n far out on the positive real axis are approximately π units apart. We will not, however, prove these properties [see Wilf (1978)].

Now we are ready to explain what we mean by a series that represents a function asymptotically even though it may be a divergent series.

Definition 4.4   Let $\{g_n\}$ be a sequence of functions that are defined and never zero on a punctured neighborhood of x_0. $\{g_n\}$ is an *asymptotic sequence* at x_0 if, for all positive integers n,

$$g_{n+1}(x) = o(g_n(x)) \quad \text{as } x \to x_0$$

The most commonly used asymptotic sequences are $\{(x - x_0)^n\}$, which is asymptotic at x_0, and $\{1/x^n\}$, which is asymptotic at ∞.

Definition 4.5   If $\{g_n\}$ is asymptotic at x_0 and f is defined on a punctured neighborhood of x_0, then the series $\sum_{k=0}^{\infty} c_k g_k(x)$ is an *asymptotic series* for f at x_0 with respect to $\{g_n\}$ if, for all $n \geq 0$,

$$f(x) - \sum_{k=0}^{n} c_k g_k(x) = o(g_n(x)) \quad \text{as } x \to x_0 \tag{4.53}$$

The coefficients c_k in Definition 4.5 are uniquely determined by f, $\{g_n\}$, and x_0, since formula (4.53) implies that

$$0 = \lim_{x \to x_0} \frac{f(x) - \sum_{k=0}^{n} c_k g_k(x)}{g_n(x)} = \lim_{x \to x_0} \left(\frac{f(x) - \sum_{k=0}^{n-1} c_k g_k(x)}{g_n(x)} - \frac{c_n g_n(x)}{g_n(x)} \right)$$

and so

$$c_n = \lim_{x \to x_0} \frac{f(x) - \sum_{k=0}^{n-1} c_k g_k(x)}{g_n(x)} \tag{4.54}$$

This uniquely determines c_n recursively. The function f, however, is not uniquely determined by $\{g_n\}$, $\{c_n\}$, and x_0 (see Exercise 1).

One large class of asymptotic series that we are already familiar with are the Taylor series. Suppose f is infinitely differentiable in some open interval containing x_0. Then, even if the Taylor series of f fails to converge to f, we can still prove that $\sum_{k=0}^{\infty} f^{(k)}(x_0)(x - x_0)^k/k!$ is an asymptotic series for f at x_0 with respect to $\{(x - x_0)^n\}$, as follows. By Taylor's remainder formula,

$$f(x) - \sum_{k=0}^{n} \frac{f^{(k)}(x_0)}{k!} (x - x_0)^k = \frac{f^{(n+1)}(z)}{(n+1)!} (x - x_0)^{n+1}$$

for some point z between x_0 and x. Therefore,

$$\lim_{x \to x_0} \frac{f(x) - \sum_{k=0}^{n} f^{(k)}(x_0)(x - x_0)^k / k!}{(x - x_0)^n} = \lim_{x \to x_0} \frac{f^{(n+1)}(z)}{(n + 1)!}(x - x_0) = 0$$

In the next example, we use an asymptotic series representation of a function f to produce highly accurate numerical approximations of f even though the nth term of the series gets large rapidly in absolute value as n increases. We can see how this might happen as follows. If $\sum_{k=0}^{\infty} c_k g_k(x)$ denotes the asymptotic series for f at x_0, then

$$f(x) - \sum_{k=0}^{n} c_k g_k(x) = o(g_n(x)) \quad \text{as } x \to x_0$$

So, for a fixed value of n, if $|g_n(x)|$ is small for values of x near x_0 (which is often the case), then $f(x)$ is approximated closely by $\sum_{k=0}^{n} c_k g_k(x)$ for these values of x. However, the magnitude of the error of this approximation can depend on n and x in a very complicated way. In particular, if the asymptotic series is divergent, the error gets arbitrarily large as n gets large.

EXAMPLE 4.12   Let f be the function defined by

$$f(x) = \int_0^\infty \frac{e^{-t}}{1 + tx}\, dt \quad \text{for } x \geq 0$$

Find the asymptotic series for f at 0 with respect to $\{x^n\}$.

We use the geometric series to rewrite the integrand as follows:

$$\frac{e^{-t}}{1 + tx} = e^{-t}\left(\frac{1}{1 - (-tx)}\right) = e^{-t} \sum_{k=0}^{\infty} (-1)^k t^k x^k = \sum_{k=0}^{\infty} (-1)^k t^k e^{-t} x^k$$

Although this series converges only when $|tx| < 1$, we pretend that it can be integrated term by term over a much wider range:

$$f(x) = \int_0^\infty \frac{e^{-t}}{1 + tx}\, dt = \sum_{k=0}^{\infty} (-1)^k x^k \int_0^\infty t^k e^{-t}\, dt$$

As we now show, this artificial step did produce the asymptotic series for f at 0 with respect to $\{x^n\}$. First we evaluate the coefficient of x^k in the above series; integrating by parts, we get

$$\int_0^\infty t^k e^{-t}\, dt = -t^k e^{-t}\Big|_0^\infty - \int_0^\infty kt^{k-1}(-e^{-t})\, dt$$

$$= k \int_0^\infty t^{k-1} e^{-t}\, dt$$

So, by applying this formula k times, we get

$$\int_0^\infty t^k e^{-t}\, dt = k! \int_0^\infty e^{-t}\, dt = k!$$

Therefore, it is the series

$$\sum_{k=0}^{\infty} (-1)^k k! x^k \tag{4.55}$$

that we wish to show is asymptotic to f at 0. Our technique is to modify the above derivation of the asymptotic series by using finite sums in place of the divergent infinite series. We apply Taylor's remainder formula to the function g defined by $g(u) = 1/(1 - u)$; thus, for some point z between u and 0

$$\frac{1}{1 - u} = g(u) = \sum_{k=0}^{n} \frac{g^{(k)}(0)}{k!} u^k + \frac{g^{(n+1)}(z)}{(n + 1)!} u^{n+1} = \sum_{k=0}^{n} u^k + \frac{u^{n+1}}{(1 - z)^{n+1}}$$

since $g^{(k)}(u) = k!/(1 - u)^{k+1}$. Therefore, replacing u with $-tx$, we get

$$\left| \frac{1}{1 + tx} - \sum_{k=0}^{n} (-1)^k t^k x^k \right| = \frac{|tx|^{n+1}}{|1 - z|^{n+1}} \le t^{n+1} x^{n+1}$$

since $u = -tx \le 0$. Hence

$$\left| f(x) - \sum_{k=0}^{n} (-1)^k k! x^k \right|$$

$$= \left| \int_0^{\infty} \frac{e^{-t}}{1 + tx} \, dt - \sum_{k=0}^{n} (-1)^k x^k \int_0^{\infty} t^k e^{-t} \, dt \right|$$

$$= \left| \int_0^{\infty} e^{-t} \left(\frac{1}{1 + tx} - \sum_{k=0}^{n} (-1)^k x^k t^k \right) dt \right|$$

$$\le \int_0^{\infty} e^{-t} t^{n+1} x^{n+1} \, dt = (n + 1)! x^{n+1} \tag{4.56}$$

and so

$$f(x) - \sum_{k=0}^{n} (-1)^k k! x^k = o(x^n) \quad \text{as } x \to 0$$

This proves that the series (4.55) is asymptotic to f at 0. But we have shown even more. Although the series (4.55) diverges for all $x \ne 0$, the bound (4.56) enables us to use the series to produce highly accurate numerical approximations of $f(x)$ for values of x near 0. If, for example, $0 \le x \le 1/(n + 1)$, then, by (4.56),

$$\left| f(x) - \sum_{k=0}^{n} (-1)^k k! x^k \right| \le \frac{(n + 1)!}{(n + 1)^{n+1}} = \frac{n!}{(n + 1)^n}$$

Suppose we now choose $n = 5$; then $f(x)$ is approximated by $1 - x + 2x^2 - 6x^3 + 24x^4 - 120x^5$ with an error of at most $5!/6^5 \le 0.0155$ if $0 \le x \le \frac{1}{6}$. Note, however, that since the series $\sum_{k=0}^{\infty} (-1)^k k! x^k$ diverges, adding more terms of the

Table 4.5 $s_n = \sum_{k=0}^{n}(-1)^k k!\left(\frac{1}{6}\right)^k$
and $s_n - 0.8716$

n	s_n	Error
0	1.0000	0.1285
1	0.8333	−0.0383
2	0.8889	0.0173
3	0.8611	−0.0105
4	0.8796	0.0080
5	0.8642	−0.0074
6	0.8796	0.0080
7	0.8616	−0.0100
8	0.8856	0.0140
9	0.8496	−0.0220
10	0.9096	0.0380
11	0.7996	−0.0720
12	1.0197	0.1481

series to our approximation eventually produces a much poorer approximation to $f(x)$ rather than a better one. Table 4.5 shows the first several partial sums of the divergent series $\sum_{k=0}^{\infty}(-1)^k k!\left(\frac{1}{6}\right)^k$ and also the errors of these approximations to the exact value of $f\left(\frac{1}{6}\right)$, which happens to be 0.8716 (to four places). The pattern is typical of a divergent asymptotic series; the partial sums get close to the exact value rather quickly but then diverge away rather quickly. Finding the best choice of n is not easy, but we will not study this problem.

In the remainder of this section, we analyze the two nonlinear pendulum equations

$$r\frac{d^2\theta}{dt^2} + g\sin\theta = 0 \tag{4.57}$$

$$r\frac{d^2\theta}{dt^2} + g\theta - \frac{g\theta^3}{3!} = 0 \tag{4.58}$$

that we derived in the introduction to this chapter. In particular, we find an asymptotic series solution of (4.58). Nonlinear differential equations often yield up more information to asymptotic methods than they do to traditional methods such as those in Appendix D and Sections 4.1 through 4.4.

First, however, we try to solve both equations by an elementary method. Since the independent variable t does not occur in either equation, the first of the two methods in Section D.4 is applicable. That is, we can let

$$u = \frac{d\theta}{dt} \quad \text{and so} \quad \frac{d^2\theta}{dt^2} = u\frac{du}{d\theta}$$

Then (4.57) becomes

$$ru\frac{du}{d\theta} + g\sin\theta = 0$$

and so

$$\frac{r}{g}\int u\,du = -\int\sin\theta\,d\theta$$

or

$$\frac{ru^2}{2g} = \cos\theta + c \qquad (4.59)$$

If we also impose the initial conditions

$$\theta = \theta_0 > 0, \qquad \frac{d\theta}{dt} = 0 \quad \text{when } t = 0 \qquad (4.60)$$

we see that $c = -\cos\theta_0$ and so (4.59) becomes

$$\frac{ru^2}{2g} = \cos\theta - \cos\theta_0$$

$$= \left(1 - 2\sin^2\left(\frac{\theta}{2}\right)\right) - \left(1 - 2\sin^2\left(\frac{\theta_0}{2}\right)\right)$$

$$= 2\sin^2\left(\frac{\theta_0}{2}\right) - 2\sin^2\left(\frac{\theta}{2}\right)$$

and so

$$\frac{d\theta}{dt} = u = -2\sqrt{\frac{g}{r}}\sqrt{\sin^2\left(\frac{\theta_0}{2}\right) - \sin^2\left(\frac{\theta}{2}\right)} \qquad (4.61)$$

The negative square root was chosen here since the initial conditions (4.60) represent the state of the pendulum when it has swung as far to the right as possible, and so, for values of t greater than zero but not too large, θ is a decreasing function of t. Finally, we would like to solve the separable equation (4.61). Unfortunately, after separating the variables, we find that the integral with respect to θ cannot be evaluated exactly. Still, we can at least set up the integral and evaluate it numerically. In particular, we can find the period p of the pendulum numerically. So we separate the variables and integrate both sides from $t = 0$ (or $\theta = \theta_0$) to $t = p/4$ (or $\theta = 0$):

$$-\int_0^{p/4} 2\sqrt{\frac{g}{r}}\,dt = \int_{\theta_0}^0 \frac{d\theta}{\sqrt{\sin^2\left(\theta_0/2\right) - \sin^2\left(\theta/2\right)}}$$

or

$$p = 2\sqrt{\frac{r}{g}}\int_0^{\theta_0} \frac{d\theta}{\sqrt{\sin^2\left(\theta_0/2\right) - \sin^2\left(\theta/2\right)}}$$

However, this integral is improper since the integrand tends to infinity as θ approaches θ_0. So to make it easier to evaluate numerically, we convert it into a proper integral by making the substitution

$$\sin\left(\frac{\theta}{2}\right) = k\sin\phi \quad \text{where } k = \sin\left(\frac{\theta_0}{2}\right)$$

Then

$$\theta = 2\arcsin(k\sin\phi)$$

which implies

$$d\theta = \frac{2k\cos\phi\,d\phi}{\sqrt{1 - k^2\sin^2\phi}}$$

and so

$$p = 2\sqrt{\frac{r}{g}}\int_0^{\pi/2}\frac{2k\cos\phi\,d\phi}{\sqrt{1 - k^2\sin^2\phi}\sqrt{k^2 - k^2\sin^2\phi}}$$

$$= 4\sqrt{\frac{r}{g}}\int_0^{\pi/2}\frac{d\phi}{\sqrt{1 - k^2\sin^2\phi}} \tag{4.62}$$

The second column of Table 4.6 gives the value of $p\sqrt{g/r}$ for several values of the initial amplitude θ_0. Note in particular that the period depends on the initial amplitude, which was not the case for the simpler pendulum equation

$$r\frac{d^2\theta}{dt^2} + g\theta = 0 \tag{4.63}$$

When we solved (4.63) in the introduction to the chapter, we found that $p\sqrt{g/r}$ always has the value $2\pi = 6.2831$ (to four places). But since (4.63) was derived from the original pendulum equation (4.57) by replacing $\sin\theta$ with θ, it is not surprising that the periods in Table 4.6 are closest to 2π when θ_0 is small.

We made one hidden assumption in the process of finding the period of the solution θ: We assumed that θ really is a periodic function. That it is periodic, however, follows from (4.59), several of whose solutions are graphed in Figure 4.4. What we have done in Figure 4.4 is to ignore the fact that $u = \theta'$ and instead treat u and θ as variables whose only relationship is that given by (4.59). The graphs of the solutions to (4.59) form what is called the "phase-space portrait" of the differential equation (4.57). The fact that the phase-space portrait consists of closed curves implies

Table 4.6 Three Methods for Finding the Period of a Pendulum with Initial Amplitude θ_0

θ_0	$r\theta'' + g\sin\theta = 0$	$r\theta'' + g\theta - g\theta^3/6 = 0$	$2\pi/(1 - \theta_0^2/16)$
$\pi/16$	6.2984	6.2984	6.2984
$\pi/8$	6.3443	6.3447	6.3443
$\pi/4$	6.5344	6.5413	6.5351
$\pi/3$	6.7430	6.7674	6.7455

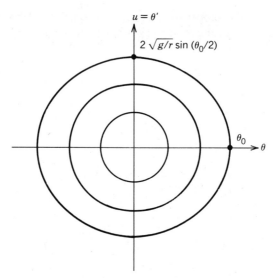

Figure 4.4   Solutions of $ru^2/2g = \cos\theta - \cos\theta_0$

that there is some time $t_0 > 0$ such that $\theta(t_0) = \theta(0)$ and $\theta'(t_0) = \theta'(0)$. But this statement says that the functions $\theta(t)$ and $\theta(t + t_0)$ satisfy the same initial conditions; and it is easy to check that both functions satisfy the differential equation (4.57). Therefore, by the uniqueness theorem for second-order differential equations (Theorem D.3), $\theta(t) = \theta(t + t_0)$. This proves that θ has period t_0.

The same method can be applied to the second of our two nonlinear pendulum equations:

$$r\frac{d^2\theta}{dt^2} + g\theta - \frac{g\theta^3}{3!} = 0, \quad \text{where } \theta = \theta_0 \text{ and } \frac{d\theta}{dt} = 0 \text{ when } t = 0$$

Its phase-space portrait consists of the solutions of

$$\frac{r}{g}u^2 = \frac{\theta^4}{12} - \theta^2 - \frac{\theta_0^4}{12} + \theta_0^2 \tag{4.64}$$

where $u = \dfrac{d\theta}{dt}$; and the period of θ is

$$p = 4\sqrt{\frac{r}{g}} \int_0^{\theta_0} \frac{d\theta}{\sqrt{\left(\theta_0^2 - \theta^2\right)\left(1 - \left(\theta_0^2 + \theta^2\right)/12\right)}} \tag{4.65}$$

$$= 4\sqrt{\frac{r}{g}} \int_0^{\pi/2} \frac{d\phi}{\sqrt{1 - \theta_0^2(1 + \sin^2\phi)/12}} \tag{4.66}$$

where $\theta = \theta_0 \sin\phi$ (see Exercise 2). The third column of Table 4.6 gives some values of $p\sqrt{g/r}$.

To get more information about the solution of the nonlinear pendulum equation (4.58), we apply our asymptotic concepts. First, however, we simplify our notation by making the change of variables

$$x = \sqrt{\frac{g}{r}}\, t \quad \text{and} \quad y = \frac{\theta}{\theta_0} \tag{4.67}$$

Then

$$\frac{d\theta}{dt} = \theta_0 \frac{dy}{dt} = \theta_0 \frac{dy}{dx}\frac{dx}{dt} = \theta_0 \sqrt{\frac{g}{r}}\,\frac{dy}{dx}$$

and

$$\frac{d^2\theta}{dt^2} = \theta_0 \sqrt{\frac{g}{r}}\,\frac{d^2y}{dx^2}\frac{dx}{dt} = \theta_0 \frac{g}{r}\frac{d^2y}{dx^2}$$

and so (4.58) becomes

$$\theta_0 g \frac{d^2y}{dx^2} + \theta_0\, g y = \frac{\theta_0^{\,3} g y^3}{6}$$

or

$$\frac{d^2y}{dx^2} + y = \varepsilon y^3, \quad \text{where } \varepsilon = \frac{\theta_0^{\,2}}{6} \tag{4.68}$$

Also, the initial conditions (4.60) transform into

$$y(0) = 1, \qquad y'(0) = 0 \tag{4.69}$$

The form of (4.68) was chosen not only to emphasize its similarity to the linear equation $y'' + y = 0$, but also to suggest a new type of series solution. Rather than expand y in powers of the independent variable x as we have in previous sections, we try to expand y in powers of ε:

$$y = \sum_{k=0}^{\infty} y_k \varepsilon^k \tag{4.70}$$

where each coefficient y_k is a function of x. Thus, when $\varepsilon = 0$, $y = y_0$; so y_0 ought to be a solution of the linear equation $y'' + y = 0$, and the remaining coefficients y_k ought to be small "perturbations" of y_0. We will not, however, try to prove that the series in formula (4.70) converges; we will be satisfied with an asymptotic series solution.

Our first step in finding the coefficients y_k is to substitute the series (4.70) into (4.68) and equate corresponding powers of ε:

$$\left(y_0'' + y_1''\varepsilon + y_2''\varepsilon^2 + y_3''\varepsilon^3 + \cdots \right) + \left(y_0 + y_1\varepsilon + y_2\varepsilon^2 + y_3\varepsilon^3 + \cdots \right)$$

$$= \varepsilon\left(y_0 + y_1\varepsilon + y_2\varepsilon^2 + \cdots \right)^3$$

Hence

$$\begin{aligned}
y_0'' + y_0 &= 0 \\
y_1'' + y_1 &= y_0^{\,3} \\
y_2'' + y_2 &= 3y_0^{\,2} y_1 \\
y_3'' + y_3 &= 3y_0 y_1^{\,2} + 3y_0^{\,2} y_2
\end{aligned} \tag{4.71}$$

$$\vdots \qquad \vdots \qquad \vdots$$

Furthermore, to guarantee that the series (4.70) also satisfies the initial conditions (4.69), we impose the following initial conditions on the functions y_k:

$$y_0(0) = 1, \; y_0'(0) = 0 \quad \text{and} \quad y_k(0) = y_k'(0) = 0 \quad \text{for } k \geq 1$$

Now we solve for the first few functions y_k in the infinite system of linear differential equations (4.71). It is easy to see that the unique solution of the first equation satisfying the initial conditions $y_0(0) = 1$ and $y_0'(0) = 0$ is

$$y_0(x) = \cos x$$

Therefore the second equation becomes

$$y_1'' + y_1 = \cos^3 x = \tfrac{3}{4} \cos x + \tfrac{1}{4} \cos(3x) \quad \text{where } y_1(0) = y_1'(0) = 0 \quad (4.72)$$

This is harder to solve, but it can be shown that there is a particular solution of the form

$$y_1 = c_1 \cos x + c_2 \sin x + c_3 \cos(3x) + c_4 \sin(3x)$$

Using this fact and the techniques of Section D.4, it can then be shown (Exercise 3) that

$$y_1 = \tfrac{1}{32} \cos x - \tfrac{1}{32} \cos(3x) + \tfrac{3}{8} x \sin x$$

So far, we have shown that the series solution (4.70) begins

$$y = \cos x + \varepsilon\left(\tfrac{1}{32}\right)(\cos x - \cos(3x) + 12x \sin x) + \cdots \quad (4.73)$$

Unfortunately, although the solution y of the pendulum equation (4.68) is periodic, we have found that the function y_1 is not. In fact, the term $x \sin x$ becomes arbitrarily large for large x. This suggests that $y_0 + \varepsilon y_1$ is unlikely to be a satisfactory approximation to y. However, an examination of why this method of finding a series solution failed will suggest a modification that succeeds. The underlying cause of the failure is that y_0 and the other periodic terms of (4.73) have period 2π, whereas y itself has a slightly longer period. We should get better approximations if we can force all the functions y_k to have exactly the same period as y. But to do this, we need some additional flexibility—some other quantity that can be varied. The idea that succeeds is to also express the frequency (i.e., the reciprocal of the period) as an asymptotic series with respect to $\{\varepsilon^n\}$. To help us do this, we make yet another change of the independent variable:

$$u = \omega x \quad \text{where} \quad \frac{2\pi}{\omega} \text{ is the period of } y \quad (4.74)$$

Thus

$$\frac{dy}{dx} = \frac{dy}{du}\frac{du}{dx} = \omega \frac{dy}{du} \quad \text{and} \quad \frac{d^2y}{dx^2} = \omega^2 \frac{d^2y}{du^2}$$

and so the pendulum equation (4.68) becomes

$$\omega^2 \frac{d^2y}{du^2} + y = \varepsilon y^3 \quad \text{where } y(0) = 1 \quad \text{and} \quad y'(0) = 0 \quad (4.75)$$

Our modification of the failed method (4.70) is to seek constants ω_k and functions y_k

such that

$$\omega = \sum_{k=0}^{\infty} \omega_k \varepsilon^k \quad \text{and} \quad y = \sum_{k=0}^{\infty} y_k \varepsilon^k \tag{4.76}$$

satisfy (4.75). As before, the series are understood to be merely asymptotic representations. We also assume that $\omega_0 = 1$, since the period of the solution y of equation (4.68) is $2\pi/1$ when $\varepsilon = 0$. Now we are ready to substitute both series (4.76) into the differential equation (4.75):

$$\left(1 + \omega_1\varepsilon + \omega_2\varepsilon^2 + \cdots\right)^2 \left(y_0'' + y_1''\varepsilon + y_2''\varepsilon^2 + \cdots\right)$$

$$+ \left(y_0 + y_1\varepsilon + y_2\varepsilon^2 + \cdots\right) = \varepsilon\left(y_0 + y_1\varepsilon + y_2\varepsilon^2 + \cdots\right)^3$$

Then, equating corresponding coefficients of ε, we get

$$y_0'' + y_0 = 0 \qquad\qquad\qquad\qquad y_0(0) = 1, \quad y_0'(0) = 0$$

$$y_1'' + y_1 = -2\omega_1 y_0'' + y_0^3 \qquad\qquad y_1(0) = 0, \quad y_1'(0) = 0$$

$$y_2'' + y_2 = -\left(\omega_1^2 + 2\omega_2\right)y_0'' - 2\omega_1 y_1'' + 3y_0^2 y_1 \qquad y_2(0) = 0, \quad y_2'(0) = 0$$

$$\vdots \qquad \vdots \qquad \vdots \tag{4.77}$$

The solution of the first equation of the system (4.77) is

$$y_0 = \cos u$$

Therefore, the second equation becomes

$$y_1'' + y_1 = 2\omega_1 \cos u + \cos^3 u = \left(2\omega_1 + \tfrac{3}{4}\right)\cos u + \tfrac{1}{4}\cos(3u) \tag{4.78}$$

It can be shown that this has a particular solution of the form

$$y_1 = c_1 \cos u + c_2 \sin u + c_3 u \cos u + c_4 u \sin u + c_5 \cos(3u) + c_6 \sin(3u)$$

which still has some nonperiodic terms. However (and this is the key point), if we choose ω_1 so that the coefficient of $\cos u$ in (4.78) is zero, then there is a particular solution for which the coefficients c_3 and c_4 are zero. So, by applying the methods of Section D.4 to (4.78) with $\omega_1 = -\tfrac{3}{8}$ and the initial conditions $y_1(0) = y_1'(0) = 0$, we get the periodic solution

$$y_1 = \tfrac{1}{32}\left(\cos u - \cos(3u)\right)$$

(Exercise 4). Therefore, by using (4.74) and (4.67) to change the variables back again, we have

$$y = \cos u + \varepsilon\left(\tfrac{1}{32}\right)\left[\cos u - \cos(3u)\right] + \cdots$$

$$= \cos(\omega x) + \varepsilon\left(\tfrac{1}{32}\right)\left[\cos(\omega x) - \cos(3\omega x)\right] + \cdots$$

and so

$$\theta = \theta_0\left(\cos\left(\omega\sqrt{\tfrac{g}{r}}\,t\right) + \frac{\theta_0^2}{192}\left[\cos\left(\omega\sqrt{\tfrac{g}{r}}\,t\right) - \cos\left(3\omega\sqrt{\tfrac{g}{r}}\,t\right)\right] + \cdots\right)$$

$$\tag{4.79}$$

Also

$$\omega = \omega_0 + \omega_1 \varepsilon + \omega_2 \varepsilon^2 + \cdots$$

$$= 1 - \tfrac{3}{8}\varepsilon + \cdots = 1 - \tfrac{3}{8}\left(\frac{\theta_0{}^2}{6}\right) + \cdots = 1 - \frac{\theta_0{}^2}{16} + \cdots \qquad (4.80)$$

We claim, but do not prove here, that (4.79) and (4.80) are indeed the beginnings of asymptotic series for θ and ω, respectively. We also claim that the coefficients ω_k can be chosen so that the functions y_k are all periodic with the same period as y. As a result, equations (4.79) and (4.80) are useful representations of θ and ω. In the fourth column of Table 4.6, for example, we give some values of the period $2\pi/\omega$, where ω is approximated by $1 - \theta_0{}^2/16$. As we can see, using just two terms of the asymptotic expansion of ω gives a remarkably good approximation of the period, even for fairly large initial amplitudes. Similarly, using just the first two terms of (4.79) gives a good approximation to θ, provided $\theta_0{}^2/6$ is sufficiently small.

We can summarize the above method of analyzing the differential equation (4.68) as follows. We make the change of variable (4.74) and then substitute the two asymptotic series (4.76) into the differential equation, choosing $\omega_0 = 1$ and choosing ω_k for $k \geq 1$ so that the functions y_k are periodic. This method of analyzing the periodic solutions of a differential equation, especially this manner of choosing the constants ω_k, is called *Lindstedt's procedure*; the class of differential equations to which it is usually applied are those of the form

$$y'' + y = F(y, y')$$

For a more thorough discussion of Lindstedt's procedure and other methods for finding asymptotic solutions of nonlinear differential equations, see Nayfeh (1973).

EXERCISES

1. Show that $\sum_{k=0}^{\infty} x^k$ is the asymptotic series for $1/(1 - x) + \exp(-1/x^2)$ at 0 with respect to $\{x^n\}$.

2. **(a)** Derive the phase-space equation (4.64) for the differential equation (4.58) with the initial conditions (4.60).

   (b) Then derive formulas (4.65) and (4.66) for the period of the solution of this initial-value problem.

3. Solve the initial-value problem (4.72).

4. Solve the differential equation (4.78) with $\omega_1 = -\tfrac{3}{8}$ and the initial conditions $y_1(0) = y_1'(0) = 0$.

In Exercises 5–8, verify the given big oh or little oh statement.

5. $x^2 \sin x/(1 + x) = O(x)$ as $x \to \infty$

6. $\log x = o(x^{-p})$ as $x \to 0^+$ for all $p > 0$

7. $\sin x - x = O(x^3)$ as $x \to 0$

8. $F_n = O\big((1 + \sqrt{5})^n/2^n\big)$ as $n \to \infty$

9. Suppose the sequence $\{g_n\}$ is asymptotic at 0, and suppose the series $\sum_{k=0}^{\infty} c_k g_k(x)$ is asymptotic to f at 0 with respect to $\{g_n\}$. Show that the sequence $\{g_n(1/x)\}$ is asymptotic at ∞ and that the series $\sum_{k=0}^{\infty} c_k g_k(1/x)$ is asymptotic to $f(1/x)$ at ∞ with respect to $\{g_n(1/x)\}$.

10. Let f be the function defined by

$$f(x) = \int_0^{\infty} \frac{e^{-xt}}{1 + t^2} \, dt \quad \text{for } x > 0$$

(a) Find the asymptotic series for f at ∞ with respect to $\{1/x^n\}$. *Hint:* Use the geometric series to rewrite the integrand and then pretend that the resulting series can be integrated term by term; also use the fact that

$$\int_0^{\infty} t^k e^{-xt} \, dt = k! x^{-k-1}$$

(b) Use Taylor's remainder formula to prove that the series you found in part (a) is indeed asymptotic to f at ∞ with respect to $\{1/x^n\}$.

11. Let f be the function defined by

$$f(x) = \exp(x^2) \int_x^{\infty} \exp(-t^2) \, dt \quad \text{for } x > 0$$

Find the asymptotic series for f at ∞ with respect to $\{1/x^n\}$. *Hint:* Write the integrand as $(t \exp(-t^2)) t^{-1}$ and integrate by parts.

12. Apply the Lindstedt procedure to find the first three terms of the asymptotic series solution (4.76) of the initial-value problem

$$\frac{d^2 y}{dx^2} + y = \varepsilon y^2, \qquad y(0) = 1, \quad y'(0) = 1$$

Hints: The functions y_1 and y_2 will be periodic if ω_1 is chosen to be 0 and ω_2 is chosen to be $-\frac{5}{12}$. In the system of differential equations analogous to (4.77), the equations for y_1 and y_2 then have particular periodic solutions of the form

$$y_1 = c_1 + c_2 \cos(2u) + c_3 \sin(2u)$$

$$y_2 = c_1 + c_2 \cos(2u) + c_3 \sin(2u) + c_4 \cos(3u) + c_5 \sin(3u)$$

where $u = \omega x$ and $2\pi/\omega$ is the period of y.

CHAPTER 5

Fourier series and
least squares

$\mathbf{A}$ great many natural phenomena display periodic behavior. A planet rotates and returns to its original orientation; the tide in a bay comes in and goes out again; the number of hours of daylight in a day repeats the following year; the prongs of a tuning fork vibrate back and forth; a weight suspended from a spring bobs up and down; the current in an electrical circuit surges back and forth. Of course, no naturally occurring phenomenon repeats itself exactly. Still, to solve many real-world problems associated with such phenomena, it is often reasonable to assume that the behavior is perfectly periodic and to use periodic functions to represent associated numeric quantities.

In mathematics, a function f is said to have *period* p (where p is a positive real number) if $f(x + p) = f(x)$ for all real numbers x. For example, $\sin x$ and $\cos x$ have period 2π, and $\sin(2\pi x/p)$ has period p. A simple physical setting in which periodic behavior can be characterized by periodic functions consists of an object rotating on a circular path (see Figure 5.1). For example, if the object moves with constant angular velocity v, its position is a periodic function of time t. For then $\theta = vt$, if we assume that $\theta = 0$ when $t = 0$, and so $x = r\cos\theta = r\cos(vt)$ and $y = r\sin\theta = r\sin(vt)$. Thus, x and y are periodic functions with period $2\pi/v$. The reciprocal of the period, $v/2\pi$, is called the *frequency* of the object's position; it represents the number of complete revolutions the object makes when the time t increases by one unit. For example, if the object is rotating at $\pi/2$ radians per second, the period of the object's position is 4 and its frequency is $\frac{1}{4} = 0.25$ cycles per second. Some of the many other natural uses of the periodic functions $r\cos(vt)$ and $r\sin(vt)$ are to describe the side to side motion of a swinging pendulum [see (4.4)] and the vertical motion of a weight bobbing on a spring (see Example D.9).

A vibrating string is a much more complex physical system displaying periodic behavior. The problem of analyzing the motion of a vibrating string proved so difficult in the eighteenth and nineteenth centuries that mathematicians proposed and debated a variety of competing solutions. The debates were fruitful; not only were useful solutions of the problem produced, but the methods and ideas developed for solving this problem grew to have wide applicability. In particular, the vibrating string problem helped inspire the idea of a Fourier series.

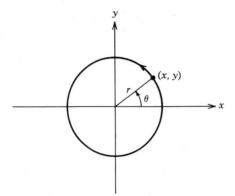

Figure 5.1   An object rotating on the circle $x^2 + y^2 = r^2$

In 1747 d'Alembert derived the following equation of motion for a taut vibrating string, where t is time, x is the coordinate along the string when it is at equilibrium, and $y = y(t, x)$ is the deflection of the string from its straight-line equilibrium position (see Figure 5.2):

$$\frac{\partial^2 y}{\partial x^2} = \frac{1}{c^2} \frac{\partial^2 y}{\partial t^2} \tag{5.1}$$

The constant c, we will discover later, is related to the speed at which the string vibrates. Equation (5.1) is only reasonable, however, if the deflections from the equilibrium position are relatively small; so we should regard Figure 5.2 as a greatly exaggerated illustration of a vibrating string photographed at an instant in time.

D'Alembert also proposed a collection of solutions of (5.2):

$$y = F(x + ct) + G(x - ct) \tag{5.2}$$

It is not hard to verify that these functions are all solutions of (5.1), provided F and G are twice differentiable (Exercise 3 of Section 5.1). It is harder to derive (5.2); we omit the derivation because D'Alembert's approach is not the one we will pursue in depth.

It is instructive, however, to see what effect a typical set of initial and boundary conditions has on D'Alembert's solutions. We suppose that the string has a finite length L, that it is started in motion not by being imparted a velocity in the y direction, but simply by being displaced in the y direction, and that its two ends are tied down. These suppositions can be represented symbolically by the "initial conditions"

$$\frac{\partial y}{\partial t}(0, x) = 0, \qquad y(0, x) = f(x) \quad \text{whenever } 0 \le x \le L \tag{5.3}$$

and the "boundary conditions"

$$y(t, 0) = 0, \qquad y(t, L) = 0 \quad \text{whenever } t \ge 0 \tag{5.4}$$

Figure 5.2   A vibrating string

where $t = 0$ represents the initial time, $y = f(x)$ represents the initial displacement, and $x = 0$ and $x = L$ represent the ends of the string. For convenience, we assume that the functions F and G in (5.2) are defined on the entire real line. We begin by applying the first of the initial conditions (5.3) to the solution (5.2):

$$\frac{\partial y}{\partial t} = F'(x + ct)\frac{\partial}{\partial t}(x + ct) + G'(x - ct)\frac{\partial}{\partial t}(x - ct)$$

$$= cF'(x + ct) - cG'(x - ct)$$

So

$$0 = \frac{\partial y}{\partial t}(0, x) = cF'(x) - cG'(x)$$

Hence

$$G'(x) = F'(x)$$

and so

$$G(x) = F(x) + d \quad \text{for some constant } d$$

Therefore, D'Alembert's solutions (5.2) may be written

$$y = F(x + ct) + F(x - ct) + d$$

But the constant d can be absorbed by simply replacing the function F by $F + d/2$; so we may as well write

$$y = F(x + ct) + F(x - ct)$$

Next we apply the second of the initial conditions (5.3):

$$f(x) = y(0, x) = F(x) + F(x) = 2F(x)$$

and so

$$F(x) = \frac{f(x)}{2}$$

We may regard this last equation as holding for all x, not just for x between 0 and L, by simply defining f outside the interval $[0, L]$ to be $2F$. Therefore, the solutions (5.2) may be written

$$y = \tfrac{1}{2}f(x + ct) + \tfrac{1}{2}f(x - ct)$$

Next we apply the first of the boundary conditions (5.4):

$$0 = y(t, 0) = \tfrac{1}{2}f(ct) + \tfrac{1}{2}f(-ct)$$

and so

$$f(-ct) = -f(ct)$$

Therefore, f is an odd function. Last of all, we apply the second of the boundary conditions (5.4):

$$0 = y(t, L) = \tfrac{1}{2}f(L + ct) + \tfrac{1}{2}f(L - ct)$$

which implies that

$$f(L - ct) = -f(L + ct)$$
$$= f(-L - ct) \quad \text{since } f \text{ is odd}$$

Therefore, f has period $2L$. In summary, the unique solution of (5.1) satisfying the conditions (5.3) and (5.4) is

$$y = \tfrac{1}{2}f(x + ct) + \tfrac{1}{2}f(x - ct) \tag{5.5}$$

where the function f in condition (5.3) has its domain extended to the whole real line in such a way that f is odd and has period $2L$.

This solution has some deficiencies. First, we must assume that f is twice differentiable, since that is what we had to assume about F and since $f = 2F$. Thus, some initial positions of the string, such as the plucked shape pictured in Figure 5.3, might lead to motions not described by the solution (5.5). Second, the form of (5.5) is not useful in analyzing some of the properties of a vibrating string; for example, it does not help us describe the acoustical properties of a string, such as its fundamental frequency and its higher harmonics. And third, the method of solution is not sufficiently general for it to be applicable to problems involving partial differential equations other than those of the form (5.1).

A quite different approach to solving (5.1), one using infinite series, was begun in 1753 by Daniel Bernoulli. But it was not applied successfully until 1807 when Fourier used it to analyze a somewhat different differential equation, and it was not justified rigorously until 1829 when Dirichlet gave the first general theorem regarding the convergence of Fourier's series.

Even before 1753 it had been suggested that the equations of motion of a vibrating string ought to include functions of the form

$$y = \cos(\omega t)G(x)$$

Functions of this form, called "standing waves," combine a fixed shape of the string, $y = AG(x)$, and an amplitude $A = \cos(\omega t)$ that varies with time. Thus, the string retains its shape over time, although the amplitude of the deflection changes, and each point on the string vibrates periodically in much the same way as the prongs of a tuning fork or a weight suspended from a spring. These standing waves may be thought of as the motions which give rise to all the pure tones that the string is capable of producing. If we substitute $y = \cos(\omega t)G(x)$ into D'Alembert's equation (5.1), as Bernoulli did, we get

$$\cos(\omega t)G''(x) = \frac{-\omega^2}{c^2}\cos(\omega t)G(x)$$

Figure 5.3   A nondifferentiable plucked string

or

$$G''(x) + \frac{\omega^2}{c^2} G(x) = 0$$

Therefore, by Theorem D.6c,

$$G(x) = a_1 \cos\left(\frac{\omega x}{c}\right) + a_2 \sin\left(\frac{\omega x}{c}\right)$$

Next, we apply the boundary conditions $y(t,0) = y(t, L) = 0$ to the standing wave $y = \cos(\omega t)G(x)$. Since $\cos(\omega t)$ cannot be identically zero, we see that $G(0) = 0$ and $G(L) = 0$. Thus, by applying these restrictions to the above solution for G, we see that $a_1 = 0$ and $a_2 \sin(\omega L/c) = 0$. But if we are seeking only nonzero solutions G, we must have that $a_2 \neq 0$ and so

$$\frac{\omega L}{c} = \pi k \quad \text{for some positive integer } k$$

or

$$\omega = \frac{\pi k c}{L}$$

(Note that negative and positive values of k yield the same functions G, since for negative values $-k$, G has the form $a_2 \sin(\pi(-k)x/L) = -a_2 \sin(\pi kx/L)$.) Therefore, $G(x) = a_2 \sin(\pi kx/L)$ and so the only possible standing waves are the multiples of the solutions

$$y_k(t, x) = \cos\left(\frac{\pi k c t}{L}\right) \sin\left(\frac{\pi k x}{L}\right), \quad k = 1, 2, \dots$$

These solutions also satisfy the first of the two initial conditions (5.3).

Bernoulli then reasoned that all the solutions of D'Alembert's equation (5.1) satisfying the conditions (5.3) and (5.4) ought to be expressible as linear combinations of the standing waves; that is

$$y(t, x) = \sum_{k=1}^{\infty} b_k y_k(t, x) = \sum_{k=1}^{\infty} b_k \cos\left(\frac{\pi k c t}{L}\right) \sin\left(\frac{\pi k x}{L}\right) \tag{5.6}$$

Certainly these linear combinations ought to satisfy the differential equation (5.1), the boundary conditions (5.4), and the first of the two initial conditions (5.3), since each of the functions y_k does. Therefore, the coefficients b_k in (5.6) need only be chosen in such a way that $y(0, x) = f(x)$ is satisfied; that is

$$\sum_{k=1}^{\infty} b_k \sin\left(\frac{\pi k x}{L}\right) = f(x) \tag{5.7}$$

Bernoulli justified formula (5.6) on physical grounds. He suggested that each standing wave solution represents a pure tone in which the shape of the string is sinusoidal

$$y_k = A \sin\left(\frac{\pi k x}{L}\right)$$

and its amplitude

$$A = \cos\left(\frac{\pi k c t}{L}\right)$$

varies with a frequency $kc/2L$ that characterizes the tone. The standing wave y_1 has the fundamental frequency $c/2L$ of the string, while the other standing waves have frequencies that are integer multiples of the fundamental and thus represent the higher harmonics of the string. Therefore, Bernoulli reasoned, any motion of the string, by which he meant a sound of arbitrary complexity, ought to be decomposable into its components: the contribution of the fundamental tone and of each of the higher harmonics. He also suggested that it should be possible to solve (5.7) for the unknown coefficients b_k for an arbitrary initial position function f, even a nondifferentiable one as in Figure 5.3. Unfortunately, Bernoulli did not show how to solve (5.7) for the unknowns b_k, and he did not discuss the convergence of the series in (5.6) or (5.7).

Euler, the leading mathematician of the eighteenth century, took issue with Bernoulli's solution. He noted that $\sin(\pi k x/L)$ is odd and has period $2L$; hence, because of (5.7), f must have the same properties. This suggested further restrictions on f. Perhaps f, like $\sin(\pi k x/L)$, must have derivatives of all orders and be defined by a single formula; in the more modern language of Section 4.1, perhaps f must be analytic.

Not until 1807 did Fourier show how to define the coefficients b_k and then actually compute them for a variety of nonperiodic, nonanalytic, and even discontinuous functions. However, he proved convergence for only a few specific examples.

The series in (5.7) is actually a special form of the series treated by Fourier. The general form is

$$c + \sum_{k=1}^{\infty} \left[a_k \cos(\omega k x) + b_k \sin(\omega k x) \right]$$

These are called *trigonometric series*, and their partial sums are called *trigonometric polynomials*. Trigonometric polynomials were important for other reasons in eighteenth century mathematics: They were used to approximate periodic functions. For example, if the distance between two planets had been measured at several different times, then trigonometric polynomials were used to interpolate the distance function at intermediate times. In the eighteenth century, however, no one seems to have noted the similarities between trigonometric interpolation and Bernoulli's attempted solution of the vibrating string problem.

In Sections 5.1, 5.2, and 5.3 we develop some of the principal aspects of Fourier series, and in Section 5.4 we return to applications such as the vibrating string problem. In Sections 5.5, 5.6, and 5.7 we develop a more geometric approach to Fourier series, which enables us to prove some deeper theorems and develop a broader view of key parts of the theory. This theory is then applied to another class of problems in Section 5.8.

5.1 Definitions and Examples

Our first goal is to discover how to express a function f as the sum of a trigonometric series. To simplify this task, we make several assumptions that will later be removed or weakened. We assume that the domain of f is $[-\pi, \pi]$, that f can be written as the sum of a trigonometric series

$$f(x) = c + \sum_{k=1}^{\infty} \left[a_k \cos(kx) + b_k \sin(kx) \right] \tag{5.8}$$

for some unknown coefficients $c, a_1, a_2, \ldots, b_1, b_2, \ldots$, and that the series in (5.8) converges uniformly on $[-\pi, \pi]$. The method we use to find the unknown coefficients depends on the following crucial "orthogonality" properties of the cosine and sine functions.

Theorem 5.1 Let k and j be positive integers. Then

(a) $\displaystyle\int_{-\pi}^{\pi} \cos(kx)\cos(jx)\, dx = \begin{cases} 0 & \text{if } k \neq j \\ \pi & \text{if } k = j \end{cases}$

(b) $\displaystyle\int_{-\pi}^{\pi} \sin(kx)\sin(jx)\, dx = \begin{cases} 0 & \text{if } k \neq j \\ \pi & \text{if } k = j \end{cases}$

(c) $\displaystyle\int_{-\pi}^{\pi} \cos(kx)\sin(jx)\, dx = 0$

(d) $\displaystyle\int_{-\pi}^{\pi} \cos(kx)\, dx = \int_{-\pi}^{\pi} \sin(kx)\, dx = 0$

Proof The proof of (d) is easy; the other parts follow from the trigonometric identities

$$\cos x \cos y = \tfrac{1}{2}\left[\cos(x - y) + \cos(x + y)\right] \tag{5.9}$$
$$\sin x \sin y = \tfrac{1}{2}\left[\cos(x - y) - \cos(x + y)\right] \tag{5.10}$$
$$\sin x \cos y = \tfrac{1}{2}\left[\sin(x - y) + \sin(x + y)\right] \tag{5.11}$$

For example, when $k \neq j$

$$\int_{-\pi}^{\pi} \cos(kx)\cos(jx)\, dx = \frac{1}{2}\int_{-\pi}^{\pi} \left[\cos(k-j)x + \cos(k+j)x\right] dx$$

$$= \frac{1}{2}\left[\frac{\sin(k-j)x}{k-j} + \frac{\sin(k+j)x}{k+j}\right]\Bigg|_{-\pi}^{\pi}$$

$$= 0$$

and when $k = j$

$$\int_{-\pi}^{\pi} \cos(kx)\cos(jx)\, dx = \frac{1}{2}\int_{-\pi}^{\pi} \left[\cos(0) + \cos(2kx)\right] dx$$

$$= \frac{1}{2}\left[x + \frac{\sin(2kx)}{2k}\right]\Bigg|_{-\pi}^{\pi} = \pi$$

which proves part (a). The proofs of parts (b) and (c) are left to Exercise 5.

We now use Theorem 5.1 to find the unknown coefficients in (5.8). First we integrate both sides of that equation from $-\pi$ to π and then interchange the integration and summation signs; the interchange is permissible since we assumed that the convergence is uniform (Theorem 3.15b).

$$\int_{-\pi}^{\pi} f(x)\, dx = \int_{-\pi}^{\pi} c\, dx + \int_{-\pi}^{\pi} \sum_{k=1}^{\infty} \left[a_k \cos(kx) + b_k \sin(kx) \right] dx$$

$$= cx \Big|_{-\pi}^{\pi} + \sum_{k=1}^{\infty} \int_{-\pi}^{\pi} \left[a_k \cos(kx) + b_k \sin(kx) \right] dx$$

$$= 2\pi c + \sum_{k=1}^{\infty} \left[a_k \int_{-\pi}^{\pi} \cos(kx)\, dx + b_k \int_{-\pi}^{\pi} \sin(kx)\, dx \right]$$

$$= 2\pi c \qquad \text{(by Theorem 5.1d)}$$

Therefore

$$c = \frac{1}{2\pi} \int_{-\pi}^{\pi} f(x)\, dx \tag{5.12}$$

Next we multiply (5.8) by $\cos(kx)$, integrate both sides of the equation from $-\pi$ to π, and interchange the integration and summation signs; the convergence of the series is still uniform since $|\cos(kx)| \le 1$. Note the repeated use of Theorem 5.1.

$$\int_{-\pi}^{\pi} f(x) \cos(kx)\, dx = c \int_{-\pi}^{\pi} \cos(kx)\, dx$$

$$+ \int_{-\pi}^{\pi} \sum_{j=1}^{\infty} \left[a_j \cos(jx) + b_j \sin(jx) \right] \cos(kx)\, dx$$

$$= \sum_{j=1}^{\infty} \left[a_j \int_{-\pi}^{\pi} \cos(jx) \cos(kx)\, dx + b_j \int_{-\pi}^{\pi} \sin(jx) \cos(kx)\, dx \right]$$

$$= a_k \int_{-\pi}^{\pi} \cos(kx) \cos(kx)\, dx$$

$$= \pi a_k$$

Therefore

$$a_k = \frac{1}{\pi} \int_{-\pi}^{\pi} f(x) \cos(kx)\, dx \tag{5.13}$$

Similarly (Exercise 6)

$$b_k = \frac{1}{\pi} \int_{-\pi}^{\pi} f(x) \sin(kx)\, dx \tag{5.14}$$

Even when our three initial assumptions about f and the series $c + \sum_{k=1}^{\infty} (a_k \cos kx + b_k \sin kx)$ are not all true, we will find that this series, with coefficients c, a_k, and b_k defined by formulas (5.12), (5.13), and (5.14), is useful.

Definition 5.1   The *Fourier series* of a function f is the series

$$\frac{a_0}{2} + \sum_{k=1}^{\infty} \left[a_k \cos(kx) + b_k \sin(kx) \right] \tag{5.15}$$

where

$$a_k = \frac{1}{\pi} \int_{-\pi}^{\pi} f(x) \cos(kx)\, dx \quad \text{for } k = 0, 1, 2, \ldots \tag{5.16}$$

$$b_k = \frac{1}{\pi} \int_{-\pi}^{\pi} f(x) \sin(kx)\, dx \quad \text{for } k = 1, 2, \ldots \tag{5.17}$$

The numbers a_k, b_k are called the *Fourier coefficients* of f, and we write

$$f(x) \sim \frac{a_0}{2} + \sum_{k=1}^{\infty} \left[a_k \cos(kx) + b_k \sin(kx) \right]$$

(This formula means only that the expression on the right is the Fourier series of f, not that it necessarily equals f.)

Note that, by formula (5.16), $a_0/2 = (1/2\pi)\int_{-\pi}^{\pi} f(x)\, dx$, which is exactly the value of c in formula (5.12); thus, from now on, we use $a_0/2$ in place of c in the definition of Fourier series. Note also that we need to make some assumptions about f for Definition 5.1 to be legitimate; we must at least assume that f is defined on the interval $[-\pi, \pi]$ and that the integrals in (5.16) and (5.17) exist. From these assumptions, however, we cannot conclude that the series (5.15) converges or, if it does, that its sum is necessarily $f(x)$.

Unfortunately, the theory of the integral developed in Section A.4 limits our development of Fourier series. We proved the theorems of that section only for integrals of continuous functions. In this chapter, we want the integrals in (5.16) and (5.17) to make sense for certain discontinuous functions. The following shows how to define such integrals; Theorems A.21, A.24, and A.25 hold for this larger class of functions.

Definition 5.2 (See Figure 5.4)   A function f is *piecewise continuous* on the interval $[a, b]$ if $[a, b]$ can be partitioned into subintervals $[x_0, x_1], [x_1, x_2], \ldots, [x_{n-1}, x_n]$ (where $x_0 = a$ and $x_n = b$) in such a way that for each value of k, $k = 1, \ldots, n$,

1. f is continuous on the open subinterval (x_{k-1}, x_k); and
2. the one-sided limits $f(x_{k-1}^+)$ and $f(x_k^-)$ exist where

$$f(x_{k-1}^+) = \lim_{\substack{x \to x_{k-1} \\ x > x_{k-1}}} f(x), \qquad f(x_k^-) = \lim_{\substack{x \to x_k \\ x < x_k}} f(x)$$

Then the integral of f from a to b is defined by

Piecewise Continuous.

$$\int_a^b f(x)\, dx = \sum_{k=1}^{n} \int_{x_{k-1}}^{x_k} f(x)\, dx$$

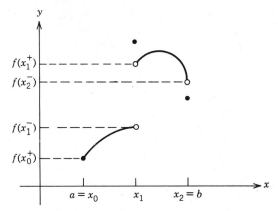

Figure 5.4   A typical piecewise continuous function

Our assumption that f is continuous on the interval (x_{k-1}, x_k) and has one-sided limits at the endpoints guarantees that the integral $\int_{x_{k-1}}^{x_k} f(x)\,dx$ exists. In fact, the value of this integral is the same as the integral of the continuous function which equals f on (x_{k-1}, x_k) and equals the limits $f(x_{k-1}^{+})$ and $f(x_k^{-})$, respectively, at the endpoints (see Exercise 10 in Section A.4).

EXAMPLE 5.1   Show that the following functions are piecewise continuous on $[-\pi, \pi]$, and find their integrals from $-\pi$ to π.

$$f(x) = \begin{cases} 2 & \text{if } \dfrac{\pi}{2} < x \\[2mm] 3x & \text{if } -\dfrac{\pi}{2} \le x \le \dfrac{\pi}{2} \\[2mm] -1 & \text{if } x < -\dfrac{\pi}{2} \end{cases} \qquad g(x) = \begin{cases} x & \text{if } x \ne -\pi, 0, \pi \\[2mm] 0 & \text{if } x = \pm\pi \\[2mm] 1 & \text{if } x = 0 \end{cases}$$

Note that f is continuous on the intervals $\left(-\pi, -\dfrac{\pi}{2}\right)$, $\left(-\dfrac{\pi}{2}, \dfrac{\pi}{2}\right)$, and $\left(\dfrac{\pi}{2}, \pi\right)$ and that

$$f(-\pi^{+}) = -1, \qquad f\left(-\dfrac{\pi}{2}^{-}\right) = -1, \qquad f\left(-\dfrac{\pi}{2}^{+}\right) = \dfrac{-3\pi}{2},$$

$$f\left(\dfrac{\pi}{2}^{-}\right) = \dfrac{3\pi}{2}, \qquad f\left(\dfrac{\pi}{2}^{+}\right) = 2, \qquad f(\pi^{-}) = 2$$

So

$$\int_{-\pi}^{\pi} f(x)\,dx = \int_{-\pi}^{-\pi/2}(-1)\,dx + \int_{-\pi/2}^{\pi/2} 3x\,dx + \int_{\pi/2}^{\pi} 2\,dx$$

$$= (-1)\left(\dfrac{\pi}{2}\right) + 0 + 2\left(\dfrac{\pi}{2}\right) = \dfrac{\pi}{2}$$

Note that g is continuous on the intervals $(-\pi, 0)$ and $(0, \pi)$ and that $g(-\pi^+) = -\pi$, $g(0^-) = 0$, $g(0^+) = 0$, $g(\pi^-) = \pi$. So

$$\int_{-\pi}^{\pi} g(x)\, dx = \int_{-\pi}^{0} x\, dx + \int_{0}^{\pi} x\, dx = \int_{-\pi}^{\pi} x\, dx = 0$$

The definition of the Fourier series of a function f makes sense whenever f is defined and piecewise continuous on $[-\pi, \pi]$; for, if f is piecewise continuous, so are $f(x)\cos(kx)$ and $f(x)\sin(kx)$, and thus the integrals in (5.16) and (5.17) exist. But it is not at all easy to see whether the Fourier series of f converges or, if it does, whether the sum of the series is f. In fact, there are many piecewise continuous functions f whose Fourier series fail to converge uniformly on $[-\pi, \pi]$ and even fail to converge pointwise to $f(x)$ for several values of x. For example, suppose f is discontinuous at some point of $[-\pi, \pi]$. Then the Fourier series of f cannot converge uniformly to f on $[-\pi, \pi]$; for, if it did, f would be a uniform sum of the continuous functions $a_0/2$ and $a_k \cos(kx) + b_k \sin(kx)$ and so f would also be continuous (Theorem 3.15a). For another example, suppose f does not have period 2π. Then the Fourier series of f cannot even converge pointwise to f on all of $[-\pi, \pi]$; for, if it did, f would be a sum of functions with period 2π and so would also have period 2π. Nevertheless, we will see that even a nonperiodic discontinuous function can be related to its Fourier series in a useful way.

Now we compute a few Fourier series. But first we recall a pair of facts from Section 2.5 (Exercise 28) that often simplify such computations:

$$\int_{-\pi}^{\pi} f(x)\, dx = 0 \qquad \text{if } f \text{ is odd}$$

$$\int_{-\pi}^{\pi} f(x)\, dx = 2\int_{0}^{\pi} f(x)\, dx \qquad \text{if } f \text{ is even}$$

In particular, if f is odd, so is $f(x)\cos(kx)$ and thus $a_k = \dfrac{1}{\pi}\displaystyle\int_{-\pi}^{\pi} f(x)\cos(kx)\, dx = 0$; and if f is even, then $f(x)\sin(kx)$ is odd and thus $b_k = \dfrac{1}{\pi}\displaystyle\int_{-\pi}^{\pi} f(x)\sin(kx)\, dx = 0$.

EXAMPLE 5.2   Find the Fourier series of f if $f(x) = x$.

Since f is odd, $a_k = 0$ for $k \geq 0$. Furthermore, for $k \geq 1$

$$b_k = \frac{1}{\pi}\int_{-\pi}^{\pi} x \sin(kx)\, dx = \frac{1}{\pi}\left[\frac{-x\cos(kx)}{k}\Big|_{-\pi}^{\pi} + \int_{-\pi}^{\pi} \frac{\cos(kx)}{k}\, dx\right]$$

← Integration by parts.

$$= \frac{-\cos(\pi k) - \cos(-\pi k)}{k} + \frac{\sin(kx)}{\pi k^2}\Big|_{-\pi}^{\pi}$$

$$= \frac{-2\cos(\pi k)}{k} = \frac{-2(-1)^k}{k} = \frac{2(-1)^{k+1}}{k}$$

Therefore

$$x \sim \sum_{k=1}^{\infty} \frac{2(-1)^{k+1}}{k} \sin(kx)$$

Note that if $x = \pm\pi$, then $\sin(kx) = 0$ and so the sum of the above Fourier series is 0. Hence, the Fourier series of $f(x) = x$ does not converge to f at $x = \pm\pi$. However, as we will learn later, this Fourier series does converge to f on the open interval $(-\pi, \pi)$ and the convergence is uniform on every interval of the form $[-c, c]$ whre $0 < c < \pi$. We can see these facts illustrated in Figure 5.5. The first few partial sums of the Fourier series approach $f(x) = x$ rapidly on most of $(-\pi, \pi)$, but the convergence is clearly far from uniform near $-\pi$ and π. Outside of the interval $[-\pi, \pi]$, the sum of the Fourier series is quite different from f. The series converges everywhere outside $[-\pi, \pi]$, but its sum has period 2π since all the terms of the sum have period 2π. The function f, on the other hand, is not periodic but grows linearly.

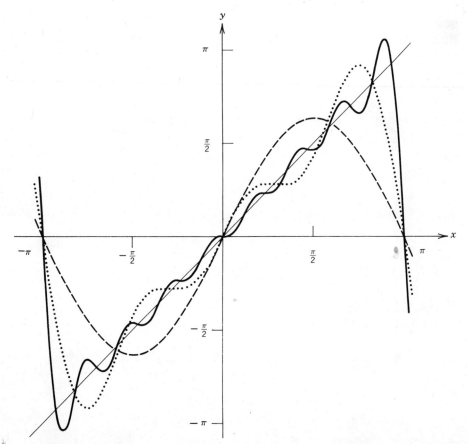

Figure 5.5 Partial sums of the Fourier series of $f(x) = x$. (_ _ _ _ _ _) $y = s_1(x)$; (. . . . . . .), $y = s_3(x)$; (_____), $y = s_8(x)$

EXAMPLE 5.3 Let $f(x) = 1$ if $0 \leq x \leq \pi$ and $f(x) = 0$ if $-\pi \leq x < 0$. Find the Fourier series of f.

$$a_k = \frac{1}{\pi} \int_{-\pi}^{\pi} f(x) \cos(kx)\, dx = \frac{1}{\pi} \int_0^{\pi} \cos(kx)\, dx$$

$$= \frac{\sin(kx)}{\pi k} \Big|_0^{\pi} = 0 \quad \text{when } k > 0$$

$$a_0 = \frac{1}{\pi} \int_0^{\pi} 1\, dx = 1$$

$$b_k = \frac{1}{\pi} \int_0^{\pi} \sin(kx)\, dx = \frac{-\cos(kx)}{\pi k} \Big|_0^{\pi}$$

$$= \frac{-\cos(\pi k) + 1}{\pi k} = \frac{1 - (-1)^k}{\pi k} = \begin{cases} \dfrac{2}{\pi k} & \text{if } k \text{ is odd} \\ 0 & \text{if } k \text{ is even} \end{cases}$$

Therefore

$$f(x) \sim \frac{1}{2} + \frac{2}{\pi} \sin(x) + \frac{2}{3\pi} \sin(3x) + \frac{2}{5\pi} \sin(5x) + \cdots$$

$$= \frac{1}{2} + \sum_{k=0}^{\infty} \frac{2 \sin(2k+1)x}{(2k+1)\pi}$$

Note that the sum of this Fourier series at $x = 0$ and $x = \pm\pi$ is $\frac{1}{2}$. Hence, the Fourier series of the function f in Example 5.3 does not converge to f at these three points. However, as we will learn later, this Fourier series does converge to f uniformly on every closed interval contained in $(-\pi, 0)$ or in $(0, \pi)$. Our discussion of the convergence of Fourier series begins in the next section.

In Table 5.1 we list a number of Fourier series for future reference. The verification of most of these formulas is left to the exercises.

Occasionally, special arguments can be used to find the sum of a particular Fourier series. The following is an adaptation of an argument given by Euler in 1754; it uses a few properties of complex numbers. Since $e^{ix} = \cos x + i \sin x$ and $(e^{ix})^k = e^{ikx}$, we can manipulate the geometric series $\sum_{k=0}^{\infty} (ae^{ix})^k$, where $|a| < 1$, as follows:

$$\sum_{k=0}^{\infty} (ae^{ix})^k = \sum_{k=0}^{\infty} a^k e^{ikx}$$

$$= \sum_{k=0}^{\infty} a^k [\cos(kx) + i \sin(kx)]$$

$$= \sum_{k=0}^{\infty} a^k \cos(kx) + i \sum_{k=0}^{\infty} a^k \sin(kx)$$

Table 5.1 Some Fourier Series

$f(x)$	Fourier series	Sum is $f(x)$ on
x	$\sum\limits_{k=1}^{\infty} \dfrac{2(-1)^{k+1}}{k}\sin(kx)$	$(-\pi,\pi)$
$\begin{cases}1 & \text{if } 0\le x\le\pi\\ 0 & \text{if } -\pi\le x<0\end{cases}$	$\dfrac{1}{2}+\sum\limits_{k=0}^{\infty}\dfrac{2\sin(2k+1)x}{(2k+1)\pi}$	$(-\pi,0)\cup(0,\pi)$
x^2	$\dfrac{\pi^2}{3}+\sum\limits_{k=1}^{\infty}\dfrac{4(-1)^k}{k^2}\cos(kx)$	$[-\pi,\pi]$
$\begin{cases}1 & \text{if } 0\le x\le\pi\\ -1 & \text{if } -\pi\le x<0\end{cases}$	$\sum\limits_{k=0}^{\infty}\dfrac{4\sin(2k+1)x}{(2k+1)\pi}$	$(-\pi,0)\cup(0,\pi)$
$\lvert x\rvert$	$\dfrac{\pi}{2}-\sum\limits_{k=0}^{\infty}\dfrac{4\cos(2k+1)x}{(2k+1)^2\pi}$	$[-\pi,\pi]$
$\begin{cases}x & \text{if } 0\le x\le\pi\\ 0 & \text{if } -\pi\le x<0\end{cases}$	$\dfrac{\pi}{4}-\sum\limits_{k=0}^{\infty}\dfrac{2\cos(2k+1)x}{(2k+1)^2\pi}-\sum\limits_{k=1}^{\infty}\dfrac{(-1)^k}{k}\sin(kx)$	$(-\pi,\pi)$
$\lvert\sin x\rvert$	$\dfrac{2}{\pi}-\sum\limits_{k=1}^{\infty}\dfrac{4\cos(2kx)}{(4k^2-1)\pi}$	$(-\infty,\infty)$
$\begin{cases}\cos x & \text{if } 0\le x\le\pi\\ -\cos x & \text{if } -\pi\le x<0\end{cases}$	$\sum\limits_{k=1}^{\infty}\dfrac{8k\sin(2kx)}{(4k^2-1)\pi}$	$(-\pi,0)\cup(0,\pi)$
e^x	$\dfrac{\sinh\pi}{\pi}\left(1+\sum\limits_{k=1}^{\infty}\left[\dfrac{2(-1)^k\cos(kx)}{k^2+1}+\dfrac{2(-1)^{k+1}\sin(kx)}{k^2+1}\right]\right)$	$(-\pi,\pi)$

But also

$$\sum_{k=0}^{\infty} \left(ae^{ix} \right)^k = \frac{1}{1 - ae^{ix}} = \frac{1 - ae^{-ix}}{\left(1 - ae^{ix}\right)\left(1 - ae^{-ix}\right)}$$

$$= \frac{1 - a\left(\cos x - i \sin x\right)}{1 - ae^{ix} - ae^{-ix} + a^2}$$

$$= \frac{1 - a \cos x + ia \sin x}{1 + a^2 - a\left(\cos x + i \sin x + \cos x - i \sin x\right)}$$

$$= \frac{1 - a \cos x}{1 + a^2 - 2a \cos x} + \frac{ia \sin x}{1 + a^2 - 2a \cos x}$$

Therefore, by equating real parts and equating imaginary parts of these two expressions for $\sum_{k=0}^{\infty}(ae^{ix})^k$, we get

$$\sum_{k=0}^{\infty} a^k \cos\left(kx\right) = \frac{1 - a \cos x}{1 + a^2 - 2a \cos x} \tag{5.19}$$

$$\sum_{k=1}^{\infty} a^k \sin\left(kx\right) = \frac{a \sin x}{1 + a^2 - 2a \cos x} \tag{5.20}$$

Note that $|a^k \cos kx| \le |a|^k$ for all x and that $\sum_{k=0}^{\infty}|a|^k$ converges; therefore, by Weierstrass' M-test, the series in (5.19) converges uniformly for all x, and so, by the argument given at the beginning of the section, this series must be the Fourier series of the function on the right side of the equation. The corresponding statement is true of the series in (5.20). Furthermore, the uniform convergence of these series permits us to integrate them term by term from 0 to x (Theorem 3.15b):

$$x + \sum_{k=1}^{\infty} \frac{a^k \sin\left(kx\right)}{k} = \int_0^x \frac{1 - a \cos t}{1 + a^2 - 2a \cos t}\, dt \tag{5.21}$$

$$-\sum_{k=1}^{\infty} \frac{a^k \cos\left(kx\right)}{k} + \sum_{k=1}^{\infty} \frac{a^k}{k} = \int_0^x \frac{a \sin t}{1 + a^2 - 2a \cos t}\, dt \tag{5.22}$$

These are again examples of Fourier series. Next, if we let the variable a approach -1 in (5.21), we confirm that the Fourier series of $f(x) = x$ converges to $f(x)$:

$$x + \sum_{k=1}^{\infty} \frac{(-1)^k \sin\left(kx\right)}{k} = \int_0^x \frac{1 + \cos t}{2 + 2 \cos t}\, dt$$

$$= \int_0^x \frac{1}{2}\, dt = \frac{x}{2}$$

Therefore

$$x = -2 \sum_{k=1}^{\infty} \frac{(-1)^k \sin\left(kx\right)}{k} = \sum_{k=1}^{\infty} \frac{2(-1)^{k+1} \sin\left(kx\right)}{k}$$

However, we were incautious in one step of this computation; we have not verified that the operation of taking the limit as a approaches -1 can be interchanged with the summation and integration operations. In fact, this interchange cannot be valid for all x since, as we saw in our discussion of Example 5.2, the Fourier series of $f(x) = x$ does not converge to $f(x)$ outside the interval $(-\pi, \pi)$. We will not go into these details, nor will we pursue this line of reasoning further.

Computations involving Fourier series can often be simplified by a systematic use of complex numbers. In particular, the Fourier series of f is sometimes defined in the following more symmetric form.

Definition 5.3   The *complex exponential Fourier series* of a function f is

$$\sum_{k=-\infty}^{\infty} c_k e^{ikx} \quad \text{where } c_k = \frac{1}{2\pi} \int_{-\pi}^{\pi} f(x) e^{-ikx} \, dx$$

The following equations show how Definition 5.3 is related to Definition 5.1 (see Exercise 20).

$$c_0 = \frac{a_0}{2} \tag{5.23}$$

and, for $k > 0$,

$$c_k = \frac{a_k - ib_k}{2}, \qquad c_{-k} = \frac{a_k + ib_k}{2} \tag{5.24}$$

$$c_k e^{ikx} + c_{-k} e^{-ikx} = a_k \cos(kx) + b_k \sin(kx) \tag{5.25}$$

Therefore

$$\sum_{k=-\infty}^{\infty} c_k e^{ikx} = \sum_{k=1}^{\infty} c_k e^{ikx} + c_0 + \sum_{k=-1}^{-\infty} c_k e^{ikx}$$

$$= c_0 + \sum_{k=1}^{\infty} \left(c_k e^{ikx} + c_{-k} e^{-ikx} \right)$$

$$= \frac{a_0}{2} + \sum_{k=1}^{\infty} \left[a_k \cos(kx) + b_k \sin(kx) \right]$$

EXERCISES

1. A spring is pulled 2 ft downward by a 15-lb weight. If the spring is then pulled 4 in. below this equilibrium position and released, find the period and frequency of its motion (see Example D.9).

2. If an 8-ft pendulum is held $\pi/12$ radians away from its equilibrium position and then released, find the period and frequency of its motion. [Use the equation (4.3).]

3. Verify that the functions (5.2) all satisfy d'Alembert's partial differential equation (5.1), if F and G are twice differentiable.

4. Suppose f has period p and is continuous at every point of the real line. Prove that, for every real number a,

$$\int_a^{a+p} f(x) \, dx = \int_0^p f(x) \, dx$$

5. Prove Theorems 5.1b and 5.1c.

6. Derive the formula

$$b_k = \frac{1}{\pi} \int_{-\pi}^{\pi} f(x) \sin(kx) \, dx$$

from the assumptions stated at the beginning of the section.

In Exercises 7–13, (a) find the Fourier series of f; (b) sketch the graph of the sum of the first two nonzero terms of the Fourier series of f on top of the graph of f. *Hint:* Graph each term separately and then add the terms graphically.

7. $f(x) = -1$ if $-\pi \le x < 0$; $f(x) = 1$ if $0 \le x \le \pi$

8. $f(x) = x^2$ **9.** $f(x) = |x|$ **10.** $f(x) = e^x$

11. $f(x) = 0$ if $-\pi \le x < 0$; $f(x) = x$ if $0 \le x \le \pi$

12. $f(x) = 0$ if $-\pi \le x < 0$; $f(x) = \cos x$ if $0 \le x \le \pi$

13. $f(x) = |\sin x|$

14. Show that, for all positive integers k and j,

(a) $\displaystyle\int_0^\pi \cos(kx)\cos(jx) \, dx = \begin{cases} 0 & \text{if } k \neq j \\ \dfrac{\pi}{2} & \text{if } k = j \end{cases}$

(b) $\displaystyle\int_0^\pi \sin(kx)\sin(jx) \, dx = \begin{cases} 0 & \text{if } k \neq j \\ \dfrac{\pi}{2} & \text{if } k = j \end{cases}$

(c) $\displaystyle\int_0^\pi \cos(kx) \, dx = 0$

15. If f has the Fourier coefficients a_k, b_k and g has the Fourier coefficients c_k, d_k, show that

(a) $f + g$ has the Fourier coefficients $a_k + c_k$, $b_k + d_k$;

(b) cf has the Fourier coefficients ca_k, cb_k.

16. Apply Exercise 15 to rows 1 and 3 of Table 5.1 to find the Fourier series of $3x - 5x^2$.

17. Derive the following consequences of the fact that

$$\frac{x}{2} = \sum_{k=1}^{\infty} \frac{(-1)^{k+1} \sin(kx)}{k} \qquad \text{whenever } -\pi < x < \pi$$

and the fact that the convergence is uniform on every interval of the form $[-t, t]$, where $0 < t < \pi$.

(a) $\displaystyle \frac{\pi}{4} = \sum_{k=1}^{\infty} \frac{(-1)^{k+1}}{2k-1}$

(b) $\displaystyle \frac{x^2}{4} = \sum_{k=1}^{\infty} \frac{(-1)^k \cos(kx)}{k^2} + \sum_{k=1}^{\infty} \frac{(-1)^{k+1}}{k^2}$ whenever $-\pi < x < \pi$.

(c) $\displaystyle \frac{\pi^2}{12} = \sum_{k=1}^{\infty} \frac{(-1)^{k+1}}{k^2}$ *Hint:* Use part (b) with $x = \dfrac{\pi}{2}$.

(d) Prove that the equation in part (b) holds for $x = \pm\pi$. *Hint:* Prove that the series of functions in part (b) converges uniformly on $[-\pi, \pi]$.

(e) $\displaystyle \frac{\pi^2}{6} = \sum_{k=1}^{\infty} \frac{1}{k^2}$ *Hint:* Use part (b) with $x = \pi$.

18.   Prove that the Fourier series in Exercise 8 converges uniformly on the whole real line

19.   Prove that the Fourier series in Exercise 9 converges uniformly on the whole real line.

20.   Derive (5.23), (5.24), and (5.25).

5.2 Convergence Theorems for Fourier Series

The properties of Fourier series are quite different from those of Taylor series. The Taylor series of f at x_0 usually converges quite rapidly at points near x_0 and less rapidly at points farther away; it converges uniformly on every closed interval with center x_0 and radius less than the radius of convergence of the series; and its sum, f, has derivatives of all orders. The sum of the Fourier series of f, on the other hand, can fail to be differentiable or even continuous; the convergence is quite often not uniform on $[-\pi, \pi]$ (and even when it is, the sum might only be continuous, not differentiable); and the convergence is most rapid on intervals where f is most smooth (that is, where f is many times differentiable).

In this section we make a start at proving some of these assertions about Fourier series; more of the theory of convergence is developed in Sections 5.5, 5.6, and 5.7. However, only a small portion of the theory can be included in an introductory text such as this [see Marsden (1974) for a more extensive treatment].

Theorem 5.2 (Bessel's inequality)   Suppose f is piecewise continuous on $[-\pi, \pi]$ and has Fourier coefficients a_k, b_k. Then

$$\frac{a_0^2}{2} + \sum_{k=1}^{\infty} \left[a_k^2 + b_k^2 \right] \leq \frac{1}{\pi} \int_{-\pi}^{\pi} \left[f(x) \right]^2 dx \qquad (5.26)$$

Proof   We obtain Bessel's inequality by a clever trick that makes use of the fact that a square is non-negative and the integral of a non-negative function is non-negative. Let

$$s_n(x) = \frac{a_0}{2} + \sum_{k=1}^{n} \left[a_k \cos(kx) + b_k \sin(kx) \right] \qquad (5.27)$$

which is the nth partial sum of the Fourier series of f. Then the trick is to note that

$$0 \leq \int_{-\pi}^{\pi} \left[f(x) - s_n(x) \right]^2 dx$$

$$= \int_{-\pi}^{\pi} \left(\left[f(x) \right]^2 - 2f(x)s_n(x) + \left[s_n(x) \right]^2 \right) dx$$

$$= \int_{-\pi}^{\pi} \left[f(x) \right]^2 dx - 2 \int_{-\pi}^{\pi} f(x)s_n(x) \, dx + \int_{-\pi}^{\pi} \left[s_n(x) \right]^2 dx \qquad (5.28)$$

But, by (5.27) and the definition of the Fourier coefficients,

$$\int_{-\pi}^{\pi} f(x)s_n(x)\, dx = \frac{a_0}{2}\int_{-\pi}^{\pi} f(x)\, dx + \sum_{k=1}^{n}\int_{-\pi}^{\pi} f(x)\left[a_k \cos(kx) + b_k \sin(kx)\right] dx$$

$$= \frac{a_0}{2}(\pi a_0) + \sum_{k=1}^{n}\left[\pi a_k^2 + \pi b_k^2\right]$$

$$= \pi\left(\frac{a_0^2}{2} + \sum_{k=1}^{n}\left[a_k^2 + b_k^2\right]\right)$$

And, by the orthogonality properties in Theorem 5.1,

$$\int_{-\pi}^{\pi}\left[s_n(x)\right]^2 dx$$

$$= \int_{-\pi}^{\pi}\left(\frac{a_0}{2} + \sum_{k=1}^{n}\left[a_k \cos(kx) + b_k \sin(kx)\right]\right)$$

$$\times\left(\frac{a_0}{2} + \sum_{j=1}^{n}\left[a_j \cos(jx) + b_j \sin(jx)\right]\right) dx$$

$$= \int_{-\pi}^{\pi}\left(\frac{a_0^2}{4} + \sum_{k=1}^{n}\left[a_k^2 \cos(kx)\cos(kx) + b_k^2 \sin(kx)\sin(kx)\right]\right) dx$$

$$= \pi\frac{a_0^2}{2} + \sum_{k=1}^{n}\left[\pi a_k^2 + \pi b_k^2\right]$$

$$= \pi\left(\frac{a_0^2}{2} + \sum_{k=1}^{n}\left[a_k^2 + b_k^2\right]\right)$$

Therefore, if we combine these two equations with the inequality (5.28), we get

$$0 \le \int_{-\pi}^{\pi}\left[f(x)\right]^2 dx - \pi\left(\frac{a_0^2}{2} + \sum_{k=1}^{n}\left[a_k^2 + b_k^2\right]\right)$$

which implies that

$$\frac{a_0^2}{2} + \sum_{k=1}^{n}\left[a_k^2 + b_k^2\right] \le \frac{1}{\pi}\int_{-\pi}^{\pi}\left[f(x)\right]^2 dx$$

But this inequality says that the nth partial sums of the positive term series $a_0^2/2 + \sum_{k=1}^{\infty}[a_k^2 + b_k^2]$ are bounded by the constant $\frac{1}{\pi}\int_{-\pi}^{\pi}[f(x)]^2 dx$. Therefore, the series converges (Corollary 3.7), and its sum is bounded by the same constant. This proves Bessel's inequality (5.26).

Our only use of Bessel's inequality in this section will be to prove the following preparatory result. In Section 5.7, on the other hand, Bessel's inequality will play a more crucial role in developing a geometric approach to Fourier series. In particular, we will see that the inequality is actually an equality.

Lemma 5.3 (Riemann–Lebesgue)   If f is piecewise continuous on $[-\pi, \pi]$, then

$$\lim_{k \to \infty} \int_{-\pi}^{\pi} f(x) \cos(kx)\, dx = 0, \qquad \lim_{k \to \infty} \int_{-\pi}^{\pi} f(x) \sin(kx)\, dx = 0$$

Proof   Bessel's inequality guarantees convergence of the series $\Sigma[a_k^2 + b_k^2]$, where a_k, b_k denote the Fourier coefficients of f. But this fact implies that the sequence of terms $a_k^2 + b_k^2$ has limit zero (Theorem 3.2), which implies the desired conclusion, according to the following chain of deductions:

$$0 \le a_k^2 \le a_k^2 + b_k^2 \to 0$$

which implies

$$a_k^2 \to 0$$

which implies

$$|a_k| = \sqrt{a_k^2} \to 0$$

which implies

$$\int_{-\pi}^{\pi} f(x) \cos(kx)\, dx = \pi a_k \to 0$$

The other limit is proved similarly.

Now we state our first major convergence theorem.

Theorem 5.4   Suppose f is defined on the entire real line, has period 2π, and is piecewise continuous on $[-\pi, \pi]$. Then the Fourier series of f converges pointwise to $f(x)$ for every point x at which f is differentiable.

The proof, which is essentially the one Dirichlet gave in 1829, will be divided into parts. The key computations involve the following "Dirichlet kernel" function:

$$D_n(x) = 1 + 2 \sum_{k=1}^{n} \cos(kx) \tag{5.29}$$

We denote the Fourier coefficients of f by a_k, b_k and the nth partial sum of its Fourier series by s_n. Thus

$$s_n(x) = \frac{a_0}{2} + \sum_{k=1}^{n} \left[a_k \cos(kx) + b_k \sin(kx) \right] \tag{5.30}$$

The Dirichlet kernel has the following properties.

Lemma 5.5

   (a) $\displaystyle \int_{-\pi}^{\pi} D_n(x)\, dx = 2\pi$

(b) $D_n(x) = \dfrac{\cos(x/2)\sin(nx)}{\sin(x/2)} + \cos(nx)$ for all $x \neq 2\pi k$, k an integer

(c) If f satisfies the hypotheses in the first sentence of Theorem 5.4, then

$$s_n(x) = \frac{1}{2\pi} \int_{-\pi}^{\pi} f(x-y) D_n(y)\, dy$$

Proof The proofs of properties (a) and (b) are not hard; they are left to Exercises 3 and 4. To prove property (c), we rewrite (5.30) as follows:

$$s_n(x) = \frac{1}{2\pi} \int_{-\pi}^{\pi} f(y)\, dy$$

$$+ \frac{1}{\pi} \sum_{k=1}^{n} \left[\int_{-\pi}^{\pi} f(y)\cos(ky)\, dy \cos(kx) + \int_{-\pi}^{\pi} f(y)\sin(ky)\, dy \sin(kx) \right]$$

$$= \frac{1}{2\pi} \int_{-\pi}^{\pi} f(y)\, dy + \frac{1}{\pi} \sum_{k=1}^{n} \int_{-\pi}^{\pi} f(y)[\cos(ky)\cos(kx) + \sin(ky)\sin(kx)]\, dy$$

$$= \frac{1}{2\pi} \int_{-\pi}^{\pi} f(y)\left[1 + 2\sum_{k=1}^{n} \cos(kx - ky) \right] dy$$

$$= \frac{1}{2\pi} \int_{-\pi}^{\pi} f(y) D_n(x-y)\, dy$$

$$= \frac{1}{2\pi} \int_{x+\pi}^{x-\pi} f(x-u) D_n(u)(-1)\, du \quad \text{where } u = x - y$$

$$= \frac{1}{2\pi} \int_{x-\pi}^{x+\pi} f(x-u) D_n(u)\, du$$

$$= \frac{1}{2\pi} \int_{-\pi}^{\pi} f(x-u) D_n(u)\, du$$

where the last step follows from Exercise 4 of Section 5.1, since both f and D_n have period 2π. This completes the proof.

Proof of Theorem 5.4 We must show that $s_n(x) \to f(x)$ for all x. But, by the three properties of the Dirichlet kernel in Lemma 5.5,

$$s_n(x) - f(x) = \frac{1}{2\pi} \int_{-\pi}^{\pi} f(x-y) D_n(y)\, dy - \frac{1}{2\pi} \int_{-\pi}^{\pi} D_n(y)\, dy\, f(x)$$

$$= \frac{1}{2\pi} \int_{-\pi}^{\pi} [f(x-y) - f(x)] D_n(y)\, dy$$

$$= \frac{1}{2\pi} \int_{-\pi}^{\pi} [f(x-y) - f(x)]\left[\frac{\cos(y/2)\sin(ny)}{\sin(y/2)} + \cos(ny) \right] dy$$

$$= \frac{1}{\pi} \int_{-\pi}^{\pi} g_x(y) \sin(ny)\, dy + \frac{1}{\pi} \int_{-\pi}^{\pi} h_x(y) \cos(ny)\, dy$$

where

$$g_x(y) = \frac{[f(x-y)-f(x)]\cos(y/2)}{2\sin(y/2)} \quad \text{for } y \neq 0$$

$$h_x(y) = \frac{f(x-y)-f(x)}{2}$$

Therefore, we may conclude from the Riemann–Lebesgue lemma that $s_n(x) - f(x) \to 0$, provided we know that the functions g_x and h_x are piecewise continuous on $[-\pi, \pi]$. The function h_x is piecewise continuous since f is; but to prove that g_x is piecewise continuous, we need to use our assumption that f is differentiable at x. We rewrite g_x to make use of that assumption:

$$g_x(y) = \frac{[f(x-y)-f(x)]}{-y}p(y)$$

where

$$p(y) = -\frac{(y/2)\cos(y/2)}{\sin(y/2)}$$

Thus

$$\lim_{y\to 0} g_x(y) = -f'(x)$$

since

$$\lim_{y\to 0} \frac{y/2}{\sin(y/2)} = 1 \quad \text{and} \quad \lim_{y\to 0} \frac{f(x-y)-f(x)}{-y} = f'(x)$$

Therefore g_x is continuous at $y = 0$, if we define $g_x(0)$ to be $-f'(x)$; now it follows that g_x is piecewise continuous on $[-\pi, \pi]$ since f is. This completes the proof.

Theorem 5.4 is certainly the most significant fact we have established about Fourier series. However, it is not sufficiently general for many applications. Sometimes we need to sum the Fourier series of nonperiodic functions or functions that are nondifferentiable or even discontinuous at a few points. Fortunately, the proof of Theorem 5.4 can be modified in a variety of ways to produce more widely applicable conclusions.

For example, in proving that g_x is piecewise continuous on $[-\pi, \pi]$, we showed that g_x is continuous at $y = 0$; but it would have been enough to prove that the two one-sided limits $g_x(0^+)$ and $g_x(0^-)$ exist. And these do exist if the following two "one-sided derivatives" of f exist:

$$\lim_{\substack{h\to 0 \\ h>0}} \frac{f(x+h)-f(x)}{h} \quad \text{and} \quad \lim_{\substack{h\to 0 \\ h<0}} \frac{f(x+h)-f(x)}{h} \tag{5.31}$$

Thus, even at points x at which f has two distinct one-sided derivatives (as in Figure 5.3), the Fourier series of f converges to $f(x)$.

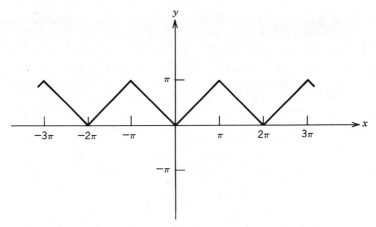

Figure 5.6 The 2π-periodic extension of $|x|$ outside $[-\pi, \pi]$

EXAMPLE 5.4 If $f(x) = |x|$, find the sum of the Fourier series of f.

Theorem 5.4 does not apply to the function f, since f does not have period 2π. However, the Fourier coefficients of f exist, and their values depend only on the values of $f(x)$ when $-\pi \le x \le \pi$. So f has the same Fourier coefficients as its "2π-periodic extension outside the interval $[-\pi, \pi]$," by which we mean the function F defined by (see Figure 5.6)

$$F(x) = f(x) \quad \text{if } -\pi \le x \le \pi$$

F has period 2π

This function does satisfy the hypotheses of Theorem 5.4; even though it is not differentiable at $x = \pi k$, it does have one-sided derivatives in the sense of formulas (5.31) at those points. Therefore, the Fourier series of F, which is identical to Fourier series of f, converges to $F(x)$ for all values of x.

The Fourier series of a function f can even converge for points x at which f is discontinuous; but then the sum is not likely to be $f(x)$. The following theorem, which is much more difficult to prove than Theorem 5.4, tells us what that sum is. [For a proof, see Marsden (1974).]

Theorem 5.6 Suppose f is defined on the entire real line, has period 2π, and is piecewise continuous and piecewise monotonic on the interval $[-\pi, \pi]$. Then the Fourier series of f converges pointwise to $[f(x^+) + f(x^-)]/2$ for every point x on the real line.

The assumption that f is piecewise monotonic on $[-\pi, \pi]$ just means that $[-\pi, \pi]$ can be partitioned into a finite number of intervals on each of which f is monotonic. The conclusion of Theorem 5.6 says that, if f is continuous at x, the Fourier series of f converges to $[f(x) + f(x)]/2 = f(x)$; but, if f is not continuous at x, the Fourier series converges to the average of the two one-sided limits of f at x. This conclusion must be interpreted with some care at the endpoints of the interval $[-\pi, \pi]$. It can happen that when we replace a function by its periodic extension outside the interval $[-\pi, \pi]$, we introduce discontinuities at $x = \pm\pi$ that the original function did not have. This did not happen in Example 5.4 because the absolute value function has identical values at $x = -\pi$ and $x = \pi$. But if f is a function with domain $[-\pi, \pi]$ and $f(-\pi) \neq f(\pi)$, we must define its periodic extension F somewhat differently:

$$F(x) = f(x) \quad \text{if } -\pi < x < \pi$$

F has period 2π

$$F(\pi) = F(-\pi) = \text{any given value}$$

We cannot define $F(\pm\pi)$ to be $f(\pm\pi)$, because then F would not have period 2π. However, as long as we give $F(\pi)$ and $F(-\pi)$ the same value, it does not matter what this value is, because the Fourier coefficients of f are integrals and an integral does not change its value when the value of the integrand is changed at finitely many points.

EXAMPLE 5.5 Let $f(x) = 1$ if $0 \leq x \leq \pi$ and $f(x) = 0$ if $-\pi \leq x < 0$. Draw the graph of the sum of the Fourier series of f.

Let F be a 2π-periodic extension of f outside $[-\pi, \pi]$. Then F satisfies the hypotheses of Theorem 5.6, and the Fourier coefficients of F equal the Fourier coefficients of f. Therefore, the sum of the Fourier series of f is $[F(x^+) + F(x^-)]/2$ for all x. Since f is continuous on the intervals $(-\pi, 0)$ and $(0, \pi)$, the sum of the Fourier series of f equals $f(x)$ for all x in these intervals. At $x = 0$ the sum is $\frac{1}{2}$, since $F(0^+) = f(0^+) = 1$ and $F(0^-) = f(0^-) = 0$. We must be more cautious, however, in finding the sum at an endpoint such as $x = \pi$. Note that $F(\pi^+) = F(-\pi^+)$ since F has period 2π; thus, $F(\pi^+) = f(-\pi^+) = 0$. Therefore, since $F(\pi^-) = f(\pi^-) = 1$, the sum at $x = \pi$ is also $\frac{1}{2}$. The rest of the graph of the sum of the Fourier series of f is determined by the fact that the sum of a Fourier series has period 2π (see Figure 5.7). Note especially the discontinuities of the sum at $x = \pm\pi$. These occur since $f(-\pi) \neq f(\pi)$ or, more precisely, since $f(-\pi^+) \neq f(\pi^-)$.

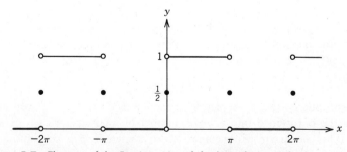

Figure 5.7 The sum of the Fourier series of f, where $f(x) = 1$ for $0 \leq x \leq \pi$ and $f(x) = 0$ for $-\pi \leq x < 0$

In Figure 5.8 we have again drawn the sum of the Fourier series of the function in Example 5.5, but now we have superimposed the graph of the 9th partial sum of the series. In addition to displaying the fairly rapid convergence of a typical Fourier series, the figure displays an interesting aspect of the behavior of the partial sums near a discontinuity of the sum. This is the "Gibbs phenomenon"—the fact that there is a sizable distance between the graph of the sum and the graph of each nth partial sum at certain points close to the discontinuity. It can be shown that, near any discontinuity x of the sum, the maximal distance between the sum $f(x)$ and its nth partial sum tends toward a certain nonzero limit as n goes to infinity, and this limiting distance is approximately equal to 9% of the jump $|f(x^+) - f(x^-)|$ [see Marsden (1974) for a partial proof].

Although we are omitting the proof of Theorem 5.6, we can still gain some insight into why $[f(x^+) + f(x^-)]/2$ is a reasonable choice for the sum of the Fourier series of f at x when f is discontinuous at x. Consider first the behavior of the Fourier series at $x = 0$. If we assume that f is an odd function, then $f(0^-) = -f(0^+)$ and so $[f(0^+) + f(0^-)]/2 = 0$. Furthermore, the Fourier coefficients a_k are zero for an odd function, and so at $x = 0$ the sum of the Fourier series of f is also 0. Next, we assume that f is even. But then $f(0^-) = f(0^+)$ and so $[f(0^+) + f(0^-)]/2 = f(0^+)$. Furthermore, if we force f to have the value $f(0^+)$ when $x = 0$, this does not cause the Fourier coefficients of f to change but does make f continuous at $x = 0$. Thus, we suspect that the sum of the Fourier series of f at $x = 0$ will be $f(0) = f(0^+) = [f(0^+) + f(0^-)]/2$. (*Warning:* A Fourier series can, on occasion, fail to converge at a point where the given function is continuous.) Even if f is neither even nor odd, f can be

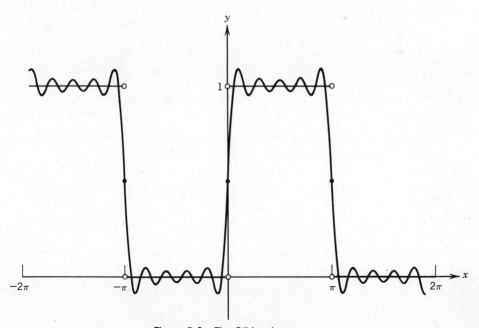

Figure 5.8   The Gibbs phenomenon

written as the sum of an even function g and an odd function h (Exercise 12). Thus, since

$$\frac{f(x^+) + f(x^-)}{2} = \frac{g(x^+) + g(x^-)}{2} + \frac{h(x^+) + h(x^-)}{2}$$

we would still expect the sum of the Fourier series of f at 0 to be $[f(0^+) + f(0^-)]/2$. Furthermore, these inferences about the Fourier series of f at $x = 0$ suggest similar inferences about the Fourier series of f at other points. For if we define a function g by $g(x) = f(x + x_0)$, then the Fourier series of f at $x = x_0$ is exactly the Fourier series of g at $x = 0$ (Exercise 13). These remarks suggest, therefore, that $[f(x^+) + f(x^-)]/2$ is always a reasonable candidate for the sum of the Fourier series of f at x.

EXERCISES

1. If f is differentiable and has period p, prove that f' also has period p.

2. If f is defined on an open interval containing a point x and if both one-sided derivatives (5.31) exist at x, prove that f is continuous at x.

3. Prove that $\displaystyle\int_{-\pi}^{\pi} D_n(x)\, dx = 2\pi$.

4. Prove Lemma 5.5b. *Hint:* Use the trigonometric identity (5.11) to show that

$$2 \sin\left(\frac{x}{2}\right) \cos(kx) = \sin\left(k + \tfrac{1}{2}\right)x - \sin\left(k - \tfrac{1}{2}\right)x$$

and use this to express $\sin(x/2)D_n(x)$ as a telescoping sum.

In Exercises 5–11, draw the graph of at least two periods of the sum of the Fourier series of f.

5. $f(x) = -1$ if $-\pi \le x < 0$; $f(x) = 1$ if $0 \le x \le \pi$

6. $f(x) = x^2$ 7. $f(x) = x$ 8. $f(x) = e^x$

9. $f(x) = 0$ if $-\pi \le x < 0$; $f(x) = x$ if $0 \le x \le \pi$

10. $f(x) = 0$ if $-\pi \le x < 0$; $f(x) = \cos x$ if $0 \le x \le \pi$

11. $f(x) = |\sin x|$

12. Suppose f is a function whose domain is $[-c, c]$. Show that there exists an even function g and an odd function h such that $f = g + h$. *Hint:* $f(x) + f(-x)$ is even.

13. Suppose f is piecewise continuous on $[-\pi, \pi]$ and has period 2π. Define another function g by $g(x) = f(x + x_0)$. Prove that $s_n(g, x) = s_n(f, x + x_0)$, where $s_n(f, x)$ denotes the value at x of the nth partial sum of the Fourier series of f.

In Exercises 14–21, find the sum of the given series by finding a Fourier series which, when x is replaced by an appropriate constant, equals the given series. Also, verify that Theorem 5.4 or 5.6 is applicable; that is, prove that your answer is indeed the sum. (Table 5.1 may be useful in locating an appropriate series.)

14. $\displaystyle\sum_{k=1}^{\infty} 1/k^2$ 15. $\displaystyle\sum_{k=1}^{\infty} (-1)^{k+1}/k^2$

16. $\displaystyle\sum_{k=1}^{\infty} 1/(4k^2 - 1)$ 17. $\displaystyle\sum_{k=1}^{\infty} (-1)^{k+1}/(4k^2 - 1)$

18. $\displaystyle\sum_{k=0}^{\infty} 1/(2k+1)^2$ **19.** $\displaystyle\sum_{k=0}^{\infty} (-1)^k/(2k+1)$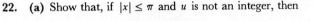

20. $\displaystyle\sum_{k=1}^{\infty} 1/(k^2+1)$ **21.** $\displaystyle\sum_{k=1}^{\infty} (-1)^{k+1}/(k^2+1)$

22. **(a)** Show that, if $|x| \le \pi$ and u is not an integer, then

$$\cos(ux) = \frac{\sin(\pi u)}{\pi u} + \frac{2u\sin(\pi u)}{\pi} \sum_{k=1}^{\infty} \frac{(-1)^k \cos(kx)}{u^2 - k^2}$$

(b) Deduce from (a) that, if u is not integer, then

$$\pi\cot(\pi u) - \frac{1}{u} = \sum_{k=1}^{\infty} \frac{2u}{u^2 - k^2}$$

(c) Deduce from (b) that if $0 \le u < 1$, then

$$\log\left(\frac{\sin(\pi u)}{\pi u}\right) = \sum_{k=1}^{\infty} \log\left(1 - \frac{u^2}{k^2}\right)$$

(d) Deduce from (c) that, if $0 \le u < 1$, then

$$\frac{\sin(\pi u)}{\pi u} = \lim_{n\to\infty} \prod_{k=1}^{n}\left(1 - \frac{u^2}{k^2}\right)$$

(e) Deduce Wallis' formula (Exercise 18 of Section 3.3) from (d).

23. Suppose f is defined and differentiable on the entire real line, f has period 2π, and f' is piecewise continuous on $[-\pi, \pi]$.

(a) Show that $a_k = -d_k/k$ and $b_k = c_k/k$ for $k \ge 1$, where a_k, b_k are the Fourier coefficients of f and c_k, d_k are the Fourier coefficients of f'. Also show that $c_0 = 0$. *Hint:* Integrate by parts.

(b) Prove that the Fourier series of f can be differentiated term by term at every point x at which $f''(x)$ exists; that is, prove that

$$f'(x) = \sum_{k=1}^{\infty} \left[-ka_k \sin(kx) + kb_k \cos(kx)\right]$$

5.3 Other Types of Fourier Series

The Fourier series we have studied so far are of limited use since they converge to functions with the specific period 2π. However, it is not difficult to adapt the theory of the two preceding sections to functions having any given period. Specifically, if f has period $2p$, we can define a related function g with period 2π by

$$g(t) = f\left(\frac{pt}{\pi}\right) \tag{5.32}$$

So if

$$g(t) \sim \frac{a_0}{2} + \sum_{k=1}^{\infty} \left[a_k \cos(kt) + b_k \sin(kt) \right]$$

it would seem reasonable to replace pt/π by x and thus write

$$f(x) \sim \frac{a_0}{2} + \sum_{k=1}^{\infty} \left[a_k \cos\left(\frac{\pi k x}{p} \right) + b_k \sin\left(\frac{\pi k x}{p} \right) \right] \tag{5.33}$$

Moreover

$$a_k = \frac{1}{\pi} \int_{-\pi}^{\pi} g(t) \cos(kt)\, dt$$

$$= \frac{1}{\pi} \int_{-\pi}^{\pi} f\left(\frac{pt}{\pi} \right) \cos(kt)\, dt$$

$$= \frac{1}{\pi} \int_{-p}^{p} f(x) \cos\left(\frac{\pi k x}{p} \right) \frac{\pi}{p}\, dx \quad \text{where } x = \frac{pt}{\pi}$$

$$= \frac{1}{p} \int_{-p}^{p} f(x) \cos\left(\frac{\pi k x}{p} \right) dx \tag{5.34}$$

Similarly

$$b_k = \frac{1}{p} \int_{-p}^{p} f(x) \sin\left(\frac{\pi k x}{p} \right) dx \tag{5.35}$$

Therefore, the series in formula (5.33) is called the *Fourier series of f with period* $2p$; the corresponding Fourier coefficients a_k, b_k are defined by (5.34) and (5.35).

EXAMPLE 5.6   Let $f(x) = 1$ if $0 \le x \le 2$ and $f(x) = 0$ if $-2 \le x < 0$. Find the Fourier series of f with period 4.

The fastest technique would be to transform the result of Example 5.3 using the above technique. Instead, we compute a_k and b_k directly from (5.34) and (5.35):

$$a_k = \frac{1}{2} \int_{-2}^{2} f(x) \cos\left(\frac{\pi k x}{2} \right) dx = \frac{1}{2} \int_{0}^{2} \cos\left(\frac{\pi k x}{2} \right) dx$$

$$= \frac{\sin(\pi k x/2)}{\pi k} \Big|_{0}^{2} = 0 \quad \text{when } k > 0$$

$$a_0 = \frac{1}{2} \int_{0}^{2} 1\, dx = 1$$

$$b_k = \frac{1}{2} \int_{0}^{2} \sin\left(\frac{\pi k x}{2} \right) dx = \frac{-\cos(\pi k x/2)}{\pi k} \Big|_{0}^{2}$$

$$= \frac{-\cos(\pi k) + 1}{\pi k} = \frac{1 - (-1)^k}{\pi k} = \begin{cases} \dfrac{2}{\pi k} & \text{if } k \text{ is odd} \\ 0 & \text{if } k \text{ is even} \end{cases}$$

Therefore, by (5.33),

$$f(x) \sim \frac{1}{2} + \frac{2}{\pi} \sin\left(\frac{\pi x}{2}\right) + \frac{2}{3\pi} \sin\left(\frac{3\pi x}{2}\right) + \frac{2}{5\pi} \sin\left(\frac{5\pi x}{2}\right) + \cdots$$

$$= \frac{1}{2} + \sum_{k=0}^{\infty} \frac{2\sin\left((2k+1)\pi x/2\right)}{(2k+1)\pi}$$

Furthermore, by an obvious modification of Theorem 5.6, this Fourier series converges, for all x, to $[F(x^+) + F(x^-)]/2$, where $F(x) = f(x)$ when $-2 < x < 2$ and F has period 4.

Other useful types of Fourier series are those in which all the cosine terms or all the sine terms are missing. As we saw in Section 5.1, these occur when f is an odd function or even function, respectively. But if we restrict the domain of f to the interval $[0, \pi]$ rather than $[-\pi, \pi]$, then we can always arrange for either the cosine terms or sine terms to be missing. To insure that there are no cosine terms, we form the "odd extension" of f to the domain $[-\pi, \pi]$:

Make f even or odd.

$$f(x) = -f(-x) \quad \text{when } -\pi \le x < 0 \tag{5.36}$$

So if a_k, b_k denote the Fourier coefficients of this extension of f, then $a_k = 0$ and

$$b_k = \frac{1}{\pi} \int_{-\pi}^{\pi} f(x) \sin(kx) \, dx$$

$$= \frac{2}{\pi} \int_0^{\pi} f(x) \sin(kx) \, dx$$

Similarly, we can form the "even extension" of f to the domain $[-\pi, \pi]$:

$$f(x) = f(-x) \quad \text{when } -\pi \le x < 0 \tag{5.37}$$

So if a_k, b_k denote the Fourier coefficients of this extension of f, then $b_k = 0$ and

$$a_k = \frac{1}{\pi} \int_{-\pi}^{\pi} f(x) \cos(kx) \, dx$$

$$= \frac{2}{\pi} \int_0^{\pi} f(x) \cos(kx) \, dx$$

These observations suggest the following definition.

Definition 5.4 If f is piecewise continuous on the interval $[0, \pi]$, then the *Fourier cosine series* of f is

$$\frac{a_0}{2} + \sum_{k=1}^{\infty} a_k \cos(kx) \tag{5.38}$$

where

$$a_k = \frac{2}{\pi} \int_0^{\pi} f(x) \cos(kx) \, dx \tag{5.39}$$

and the *Fourier sine series* of f is

$$\sum_{k=1}^{\infty} b_k \sin(kx) \tag{5.40}$$

where

$$b_k = \frac{2}{\pi} \int_0^{\pi} f(x) \sin(kx)\, dx \tag{5.41}$$

The reasoning preceding Definition 5.4 also shows that we can apply convergence theorems such as Theorem 5.6 to Fourier cosine series and Fourier sine series. However, extra care must be taken in finding the sum of a Fourier sine series at $x = 0$, since extending f to be an odd function can introduce a discontinuity at $x = 0$.

EXAMPLE 5.7   If $f(x) = \pi - x$ for $0 \le x \le \pi$, find the sum of both the Fourier cosine series of f and the Fourier sine series of f.

The Fourier cosine series of f has the same sum as the Fourier series of the even extension of f, whose graph is indicated by the solid lines in Figure 5.9. Since $f(-\pi) = f(\pi)$ and since f is continuous and piecewise monotonic on $[-\pi, \pi]$, the sum of the Fourier series of f equals f on $[-\pi, \pi]$. Two additional periods of the sum are indicated by the dashed lines in Figure 5.9.

The Fourier sine series of f has the same sum as the Fourier series of the odd extension of f, whose graph, except at $x = 0$, is indicated by the solid lines in Figure 5.10. Note that this function is discontinuous at $x = 0$; so the sum of the Fourier series at $x = 0$ is $[f(0^+) + f(0^-)]/2 = (\pi - \pi)/2 = 0$. Two additional periods of the sum are indicated by the dashed lines in Figure 5.10.

The theory of Fourier cosine series and Fourier sine series can also be adapted to functions with any given period. If f is piecewise continuous on $[0, p]$, its *Fourier cosine series with period $2p$* is

$$\frac{a_0}{2} + \sum_{k=1}^{\infty} a_k \cos\left(\frac{\pi k x}{p}\right) \tag{5.42}$$

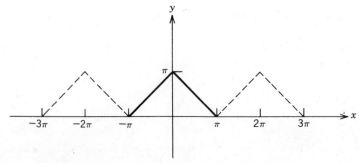

Figure 5.9   The sum of the Fourier cosine series of $f(x) = \pi - x$

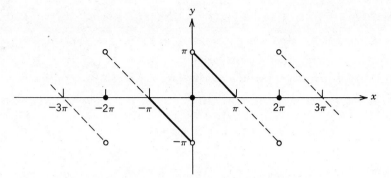

Figure 5.10   The sum of the Fourier sine series of $f(x) = \pi - x$

where

$$a_k = \frac{2}{p} \int_0^p f(x) \cos\left(\frac{\pi k x}{p}\right) dx \tag{5.43}$$

and its *Fourier sine series with period* $2p$ is

$$\sum_{k=1}^{\infty} b_k \sin\left(\frac{\pi k x}{p}\right) \tag{5.44}$$

where

$$b_k = \frac{2}{p} \int_0^p f(x) \sin\left(\frac{\pi k x}{p}\right) dx \tag{5.45}$$

EXAMPLE 5.8   Let $f(x) = x$ for $0 \leq x \leq 1$. Find the Fourier sine series of f with period 2.

$$b_k = 2 \int_0^1 x \sin(\pi k x)\, dx$$

$$= \frac{-2x \cos(\pi k x)}{\pi k}\bigg|_0^1 + \int_0^1 \frac{2 \cos(\pi k x)}{\pi k}\, dx$$

$$= \frac{-2 \cos(\pi k)}{\pi k} + \frac{2 \sin(\pi k x)}{\pi^2 k^2}\bigg|_0^1$$

$$= \frac{2(-1)^{k+1}}{\pi k}$$

So

$$x \sim \sum_{k=1}^{\infty} \frac{2(-1)^{k+1}}{\pi k} \sin(\pi k x)$$

EXERCISES

1. Let $f(x) = x$ if $0 \le x \le 1$ and $f(x) = 0$ if $-1 \le x < 0$.

    (a) Find the Fourier series of f with period 2.

    (b) Draw the graph of at least two periods of the sum of this Fourier series.

In Exercises 2–6, (a) find the Fourier cosine series of f; (b) find the Fourier sine series of f; (c) draw the graph of at least two periods of the sum of the Fourier cosine series; (d) draw the graph of at least two periods of the sum of the Fourier sine series.

2. $f(x) = \cos x$ 3. $f(x) = \sin x$ 4. $f(x) = 1$

5. $f(x) = 1$ if $0 \le x \le \pi/2$; $f(x) = 0$ if $\pi/2 < x \le \pi$

6. $f(x) = x^2$

7. Let $f(x) = x^2$.

    (a) Find the Fourier cosine series of f with period 2.

    (b) Find the Fourier sine series of f with period 2.

    (c) Draw the graph of at least two periods of the sum of the Fourier cosine series.

    (d) Draw the graph of at least two periods of the sum of the Fourier sine series.

8. (a) If f is continuous on $[0, \pi]$, prove that its even extension to the domain $[-\pi, \pi]$ [defined by (5.37)] is continuous at 0.

    (b) Give an example of a function f that is differentiable on $[0, \pi]$ but whose even extension to the domain $[-\pi, \pi]$ is not differentiable at 0.

5.4 Partial Differential Equations: An Application

In this section we study three partial differential equations that occur often in the mathematical analysis of problems in the physical sciences:

$$\frac{\partial^2 y}{\partial x^2} = \frac{1}{c^2} \frac{\partial^2 y}{\partial t^2} \quad \text{(the wave equation)} \tag{5.46}$$

$$\frac{\partial^2 u}{\partial x^2} = \frac{1}{c} \frac{\partial u}{\partial t} \quad \text{(the heat equation)} \tag{5.47}$$

$$\frac{\partial^2 u}{\partial x^2} + \frac{\partial^2 u}{\partial y^2} = 0 \quad \text{(Laplace's equation)} \tag{5.48}$$

The wave equation, which we began to analyze in the introduction to this chapter, occurs in far more contexts than just the vibrating string problem. It and its various generalizations are used to describe the motion of a vibrating membrane (such as a drum head), the motion of a vibrating elastic body (that is, a gas, liquid, or solid set in motion internally by mechanical forces such as sound waves), the height of water waves, and the magnitude of electric and magnetic fields (of, for example, light waves, radio waves, and X rays). The heat equation and its various generalizations are used to describe the time-dependent temperature in a physical object and the concentration of a chemical diffusing through a membrane. Laplace's equation and its various

generalizations are used to describe the steady-state (i.e., time-independent) temperature in a physical object, the gravitational potential in a region, the electrostatic
potential in a region, and the velocity potential of an incompressible fluid. In short, the
wave equation, the heat equation, and Laplace's equation are fundamental to the
mathematics of the physical sciences. We merely touch on this enormous field of study
by presenting a method of solution that uses Fourier series.

Each of the equations (5.46), (5.47), and (5.48) has infinitely many solutions.
However, when one of these equations arises in the analysis of a specific problem, it is
usually accompanied by initial and boundary conditions [such as (5.3) and (5.4)] that
narrow the set of solutions to the unique function that describes the solution of the
given problem. The method we use to find this unique solution is known as *separation of
variables*; it is an extension of Bernoulli's standing-wave method described in the
introduction to this chapter. First we find solutions that can be expressed as the
product of two functions of one variable (thus "separating" the two independent
variables); then we take all linear combinations of these solutions. Although there is
much more to the method of separation of variables than this, the details of the
method are more easily grasped by studying examples.

Our first example is a variant of the vibrating string problem discussed in the
introduction to this chapter.

EXAMPLE 5.9   Solve the wave equation

$$\frac{\partial^2 y}{\partial x^2} = \frac{1}{c^2}\frac{\partial^2 y}{\partial t^2} \quad \text{where } 0 < x < L \quad \text{and} \quad t \geq 0$$

with the boundary conditions

$$y(t,0) = 0, \qquad y(t,L) = 0 \quad \text{whenever } t \geq 0 \qquad (5.49)$$

and the initial conditions

$$y(0,x) = 0, \qquad \frac{\partial y}{\partial t}(0,x) = g(x) \quad \text{whenever } 0 \leq x \leq L \qquad (5.50)$$

This describes the displacement of a string whose ends are tied down and that is
started in motion not by an initial displacement but by a sudden initial velocity g
(by being struck, for example).

The first step in the method of separation of variables is to find all product
solutions $y = F(t)G(x)$ satisfying the homogeneous conditions $y(t,0) = 0$,
$y(t,L) = 0$, and $y(0,x) = 0$. When we substitute this product into the wave
equation, we get

$$F(t)G''(x) = \frac{1}{c^2}F''(t)G(x)$$

or

$$\frac{G''(x)}{G(x)} = \frac{F''(t)}{c^2 F(t)}$$

Thus, for each fixed value of t, $G''(x)/G(x)$ is a constant function of x; also, for each fixed value of x, $F''(t)/c^2F(t)$ is a constant function of t. But these two constants are the same; that is, there is a constant α such that

$$\frac{G''(x)}{G(x)} = \alpha \quad \text{and} \quad \frac{F''(t)}{c^2F(t)} = \alpha$$

and so

$$G''(x) - \alpha G(x) = 0 \quad \text{and} \quad F''(t) - \alpha c^2 F(t) = 0 \tag{5.51}$$

In addition to finding differential equations satisfied by the functions F and G, we can find initial or boundary conditions satisfied by them. Specifically, the conditions $y(t,0) = 0$, $y(t, L) = 0$, and $y(0, x) = 0$ imply, respectively, that

$$G(0) = 0, \qquad G(L) = 0, \qquad F(0) = 0 \tag{5.52}$$

(Note that we assumed that neither $G(x)$ nor $F(t)$ is identically zero; otherwise y would be identically zero and thus would not contribute to the nonhomogeneous initial condition $\frac{\partial y}{\partial t}(0, x) = g(x)$.)

To complete our search for product solutions $y = F(t)G(x)$, we must find the solutions of the differential equations (5.51) satisfying the conditions (5.52). We solve first for G since it has two conditions to satisfy. It can be shown (Exercise 1) that when $\alpha \geq 0$, the equation $G''(x) - \alpha G(x) = 0$ has no nonzero solution satisfying $G(0) = G(L) = 0$. So we assume that $\alpha = -\omega^2$ where $\omega > 0$. The general solution of $G''(x) + \omega^2 G(x) = 0$ is $G(x) = a_1 \cos(\omega x) + a_2 \sin(\omega x)$ (Theorem D.6c), and the conditions $G(0) = G(L) = 0$ imply that $a_1 = 0$ and $a_2 \sin(\omega L) = 0$. But since $G(x)$ is not identically zero, the coefficient a_2 cannot be zero and so

$$\omega = \frac{\pi k}{L}, \quad k \text{ a positive integer}$$

Thus, $G(x)$ is a scalar multiple of $\sin(\pi kx/L)$. Furthermore, since $-\alpha = \omega^2 = \pi^2 k^2/L^2$, we see that the differential equation (5.51) satisfied by F takes the form

$$F''(t) + \frac{\pi^2 k^2 c^2 F(t)}{L^2} = 0$$

The general solution of this equation is

$$F(t) = a_1 \cos\left(\frac{\pi kct}{L}\right) + a_2 \sin\left(\frac{\pi kct}{L}\right)$$

and the condition $F(0) = 0$ implies that $a_1 = 0$. Therefore, the only product functions $y = F(t)G(x)$ that satisfy the wave equation (5.46) and the three homogeneous conditions of the four conditions (5.49) and (5.50) are the multiplies of

$$y_k(t, x) = \sin\left(\frac{\pi kct}{L}\right) \sin\left(\frac{\pi kx}{L}\right), \quad k = 1, 2, \dots \tag{5.53}$$

Next, we form all the linear combinations of these product solutions

$$y(t, x) = \sum_{k=1}^{\infty} b_k y_k(t, x) = \sum_{k=1}^{\infty} b_k \sin\left(\frac{\pi kct}{L}\right) \sin\left(\frac{\pi kx}{L}\right) \qquad (5.54)$$

and try to choose the coefficients b_k so that the nonhomogeneous initial condition $\frac{\partial y}{\partial t}(0, x) = g(x)$ is satisfied. In the computations that follow, we defer all questions about whether the series (5.54) converges and whether it is legitimate to differentiate and integrate term by term. Thus

$$\frac{\partial y}{\partial t}(t, x) = \sum_{k=1}^{\infty} \frac{\pi kcb_k}{L} \cos\left(\frac{\pi kct}{L}\right) \sin\left(\frac{\pi kx}{L}\right) \qquad (5.55)$$

and so

$$g(x) = \frac{\partial y}{\partial t}(0, x) = \sum_{k=1}^{\infty} \frac{\pi kcb_k}{L} \sin\left(\frac{\pi kx}{L}\right)$$

Therefore, by formula (5.45) for the coefficients of the Fourier sine series of g with period $2L$,

$$\frac{\pi kcb_k}{L} = \frac{2}{L} \int_0^L g(x) \sin\left(\frac{\pi kx}{L}\right) dx$$

and so

$$b_k = \frac{2}{\pi kc} \int_0^L g(x) \sin\left(\frac{\pi kx}{L}\right) dx \qquad (5.56)$$

We claim that the unique solution of the wave equation (5.46) satisfying the conditions (5.49) and (5.50) is the infinite series in (5.54) with coefficients b_k defined by (5.56).

Next, we verify that, despite the shakiness of much of the reasoning in the preceding paragraph, the claimed solution is indeed a solution, provided g satisfies some reasonable conditions. We assume that the domain of g can be extended to the whole real line in such a way that g' exists and is piecewise continuous on $[-L, L]$, g is odd, and g has period $2L$. Then we define f to be the unique antiderivative of g satisfying $f(0) = 0$, and we define our solution y by

$$y(t, x) = \frac{1}{2c}[f(x + ct) - f(x - ct)] \qquad (5.57)$$

We show that this definition of y satisfies the wave equation (5.46) and the four conditions (5.49) and (5.50), and then we show that it satisfies (5.54) and (5.55), where b_k is defined by (5.56). In our proofs, we make use of the fact that f is even and has period $2L$ (Exercise 2). First, since f is even

$$y(t, 0) = \frac{1}{2c}[f(ct) - f(-ct)] = 0$$

Second, since f has period $2L$ and is even

$$y(t, L) = \frac{1}{2c}[f(L + ct) - f(L - ct)]$$

$$= \frac{1}{2c}[f(L + ct - 2L) - f(ct - L)] = 0$$

Third, since f is even, $y(0, x) = 0$. Fourth,

$$\frac{\partial y}{\partial t}(t, x) = \frac{c}{2c}[f'(x + ct) + f'(x - ct)]$$

$$= \tfrac{1}{2}[g(x + ct) + g(x - ct)] \tag{5.58}$$

and so

$$\frac{\partial y}{\partial t}(0, x) = \tfrac{1}{2}[g(x) + g(x)] = g(x)$$

Finally,

$$\frac{\partial^2 y}{\partial x^2} - \frac{1}{c^2}\frac{\partial^2 y}{\partial t^2} = \frac{1}{2c}[g'(x + ct) - g'(x - ct)]$$

$$- \frac{c^2}{2c^3}[g'(x + ct) - g'(x - ct)] = 0$$

Thus, we have verified that the function y defined by (5.57) satisfies the wave equation (5.46) and the conditions (5.49) and (5.50). Furthermore, according to Theorem 5.4, the Fourier sine series of g converges to g on the entire real line; that is,

$$g(x) = \sum_{k=1}^{\infty} d_k \sin\left(\frac{\pi k x}{L}\right) \quad \text{for all } x \tag{5.59}$$

where

$$d_k = \frac{2}{L}\int_0^L g(x)\sin\left(\frac{\pi k x}{L}\right) dx$$

Thus, $d_k = \pi k c b_k / L$ by (5.56); so, by the trigonometric identity (5.11),

$$\sum_{k=1}^{\infty} \frac{\pi k c b_k}{L} \cos\left(\frac{\pi k c t}{L}\right) \sin\left(\frac{\pi k x}{L}\right)$$

$$= \sum_{k=1}^{\infty} d_k \cos\left(\frac{\pi k c t}{L}\right) \sin\left(\frac{\pi k x}{L}\right)$$

$$= \frac{1}{2}\sum_{k=1}^{\infty} d_k\left[\sin\left(\frac{\pi k(x + ct)}{L}\right) + \sin\left(\frac{\pi k(x - ct)}{L}\right)\right]$$

$$= \tfrac{1}{2}[g(x + ct) + g(x - ct)] \quad \text{(by (5.59))}$$

$$= \frac{\partial y}{\partial t}(t, x) \quad \text{(by (5.58))}$$

which confirms (5.55). To verify that y satisfies (5.54), we argue similarly. That is, by Theorem 5.4,

$$f(x) = \sum_{k=1}^{\infty} c_k \cos\left(\frac{\pi k x}{L}\right) \quad \text{for all } x$$

where

$$c_k = \frac{2}{L} \int_0^L f(x) \cos\left(\frac{\pi k x}{L}\right) dx$$

$$= \frac{-2}{\pi k} \int_0^L f'(x) \sin\left(\frac{\pi k x}{L}\right) dx \quad \text{(integrating by parts)}$$

$$= -c b_k \quad \text{(by (5.56))}$$

Thus, by the trigonometric identity (5.10),

$$\sum_{k=1}^{\infty} b_k \sin\left(\frac{\pi k c t}{L}\right) \sin\left(\frac{\pi k x}{L}\right)$$

$$= \frac{-1}{c} \sum_{k=1}^{\infty} c_k \sin\left(\frac{\pi k c t}{L}\right) \sin\left(\frac{\pi k x}{L}\right)$$

$$= \frac{-1}{2c} \sum_{k=1}^{\infty} c_k \left[\cos\left(\frac{\pi k (x - ct)}{L}\right) - \cos\left(\frac{\pi k (x + ct)}{L}\right)\right]$$

$$= \frac{1}{2c} [f(x + ct) - f(x - ct)] = y(t, x)$$

This completes the verification of our solution of the wave equation (5.46) satisfying conditions (5.49) and (5.50). Examples involving other initial and boundary conditions occur in the exercises.

Before we leave this example, however, let us see what our solution tells us about the motion of the vibrating string. From (5.54) we infer that the motion is a superposition of standing waves $y_k = A_k \sin(\pi k x/L)$ whose time-dependent amplitudes $A_k = b_k \sin(\pi k c t/L)$ have frequency $ck/2L$. The standing wave y_1 gives rise to the fundamental frequency $c/2L$, and the other standing waves give rise to the higher harmonics whose frequencies $ck/2L$ are integral multiples of the fundamental. Thus, a complex sound is a combination of closely related pure tones.

Equation (5.57) gives quite different information about the motion of the string. It shows more clearly how the changing shape of the string depends on the initial conditions. In Figure 5.11 we have plotted y as a function of x for several values of t. We assumed that $2L/c$, the period of the fundamental frequency, is 4 and that the string was struck from below in its center so that ∂ is the zero function everywhere except along a small portion of the center of the string, where it is positive. We see that the sudden initial impact creates a bulge in the initially horizontal string, and that the bulge widens in both horizontal directions until $t = 1.0$. Then the bulge shrinks back until $t = 2.0$, when the string returns to its horizontal equilibrium position; that is, the string has the same shape at $t = 1.5$,

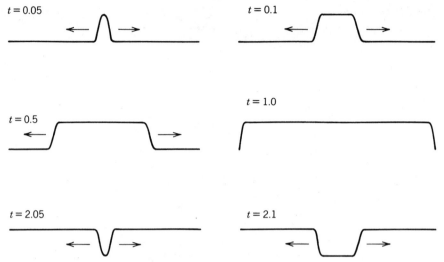

Figure 5.11 Solution of the wave equation with displacement plotted as a function of position along the string

1.9, and 1.95 as it does at $t = 0.5$, 0.1, and 0.05, but the directions of motion of the sides of the bulge are reversed. When $2 < t < 4$, the string bulges downward but otherwise repeats the same pattern. Thus, the period of this repeating pattern is also 4. More generally, this wave-like motion of the string has speed c since the time derivative of $x \pm ct$ is $\pm c$, and so the time it takes for the wave to return to its initial position and initial velocity is $2L/c$. Therefore, $2L/c$ is the period of the wave motion, and $c/2L$ is its frequency.

EXAMPLE 5.10 Solve the heat equation

$$\frac{\partial^2 u}{\partial x^2} = \frac{1}{c}\frac{\partial u}{\partial t} \quad \text{where } 0 < x < L \quad \text{and} \quad t > 0$$

with the boundary conditions

$$u(t,0) = 0, \qquad u(t,L) = 0 \quad \text{whenever } t > 0 \tag{5.60}$$

and the initial condition

$$u(0,x) = f(x) \quad \text{whenever } 0 < x < L \tag{5.61}$$

This describes the temperature u in a rod that extends from 0 to L along the x axis, has insulated surfaces except at its two ends, is kept at the temperature zero at its two ends, and has an initial temperature distribution f that varies only in the longitudinal direction x.

The first step is to find all product solutions $u = F(t)G(x)$ satisfying the homogeneous conditions (5.60). When we substitute this product into the heat equation, we get

$$F(t)G''(x) = \frac{1}{c}F'(t)G(x)$$

and so

$$\frac{G''(x)}{G(x)} = \frac{F'(t)}{cF(t)} = \alpha \quad (\alpha \text{ a constant})$$

$$G''(x) - \alpha G(x) = 0 \quad \text{and} \quad F'(t) - \alpha c F(t) = 0$$

In addition, the boundary conditions (5.60) imply that

$$G(0) = 0, \qquad G(L) = 0$$

Thus, since G satisfies the same differential equation and the same two boundary conditions as in the preceding example, G is a scalar multiple of $\sin(\pi k x/L)$ for some positive integer k, and $-\alpha = \pi^2 k^2/L^2$. Hence, F satisfies

$$F'(t) + \frac{\pi^2 k^2 c}{L^2} F(t) = 0$$

whose general solution is

$$F(t) = a \exp\left(\frac{-\pi^2 k^2 c t}{L^2} \right)$$

Therefore the only product functions $u = F(t)G(x)$ that satisfy the heat equation (5.47) and the homogeneous conditions (5.60) are the multiples of

$$u_k(t, x) = \exp\left(\frac{-\pi^2 k^2 c t}{L^2} \right) \sin\left(\frac{\pi k x}{L} \right), \qquad k = 1, 2, \ldots \qquad (5.62)$$

Next, we form all the linear combinations of these product solutions:

$$u(t, x) = \sum_{k=1}^{\infty} b_k u_k(t, x) = \sum_{k=1}^{\infty} b_k \exp\left(\frac{-\pi^2 k^2 c t}{L^2} \right) \sin\left(\frac{\pi k x}{L} \right) \qquad (5.63)$$

And we try to choose the coefficients b_k so that the initial condition $u(0, x) = f(x)$ is satisfied. Again we defer questions about convergence. Thus

$$f(x) = u(0, x) = \sum_{k=1}^{\infty} b_k \sin\left(\frac{\pi k x}{L} \right)$$

and so, by formula (5.45) for the coefficients of the Fourier sine series of f with period $2L$,

$$b_k = \frac{2}{L} \int_0^L f(x) \sin\left(\frac{\pi k x}{L} \right) dx \qquad (5.64)$$

We claim that the unique solution of the heat equation (5.47) satisfying the conditions (5.60) and (5.61) is the infinite series in (5.63) with coefficients b_k defined by (5.64).

Next, we verify that the claimed solution is indeed a solution, provided f is continuous and piecewise monotonic on $[0, L]$. First we show that, for each fixed positive value of t, the series in (5.63) converges uniformly for all x. To show this, we use the M-test:

$$|b_k| \le \frac{2}{L} \int_0^L \left| f(x) \sin\left(\frac{\pi kx}{L} \right) \right| dx \quad \text{(by (5.64))}$$

$$\le \frac{2}{L} \int_0^L |f(x)| \, dx$$

and so

$$\left| b_k \exp\left(\frac{-\pi^2 k^2 ct}{L^2} \right) \sin\left(\frac{\pi kx}{L} \right) \right| \le |b_k| \exp\left(\frac{-\pi^2 k^2 ct}{L^2} \right)$$

$$\le \frac{2}{L} \int_0^L |f(x)| \, dx \exp\left(\frac{-\pi^2 k^2 ct}{L^2} \right) = M_k$$

But, by the ratio test, ΣM_k converges. Therefore, for each fixed positive value of t, the series in (5.63) converges uniformly for all x, and so we may define $u(t, x)$ to be the sum of this series for $t > 0$ and $0 \le x \le L$. Then $u(t, 0) = 0$ and $u(t, L) = 0$, since each term in the sum equals 0 at $x = 0$ and $x = L$. Furthermore, we can define $u(0, x)$ to be the sum of the series in (5.63) with $t = 0$; in fact, it follows from Theorem 5.6 that this series converges and has the sum $[f(x^+) + f(x^-)]/2 = f(x)$ for $0 < x < L$. A deeper fact yet is that the temperature u is continuous at $t = 0$ for each fixed value of x, $0 < x < L$; that is,

$$\lim_{\substack{t \to 0 \\ t > 0}} u(t, x) = u(0, x) = f(x)$$

[For a proof of this, see Marsden (1974).] Finally, we verify that u satisfies the heat equation (5.47) for $0 < x < L$ and $t > 0$. But since each term of the series in (5.63) satisfies the heat equation on this domain, we need only show that $\dfrac{\partial^2 u}{\partial x^2}$ and $\dfrac{\partial u}{\partial t}$ exist and can be computed by differentiating the series term by term. The following variation of Theorem 3.15b is the tool we need (see Exercise 12 of Section 3.8):

Suppose the series $\Sigma_{k=1}^\infty f_k'$ converges uniformly on an interval I, and each function f_k' is continuous on I, and $\Sigma_{k=1}^\infty f_k(a)$ converges for some point a in I. Then $\Sigma_{k=1}^\infty f_k$ converges on I and

$$\frac{d}{dx} \left(\sum_{k=1}^\infty f_k(x) \right) = \sum_{k=1}^\infty f_k'(x) \quad \text{for all } x \text{ in } I$$

First we show that $\dfrac{\partial u}{\partial x}$ can be computed by differentiating the series (5.63) term by term. The series of differentiated terms is

$$\sum_{k=1}^\infty \frac{\pi k b_k}{L} \exp\left(\frac{-\pi^2 k^2 ct}{L^2} \right) \cos\left(\frac{\pi kx}{L} \right)$$

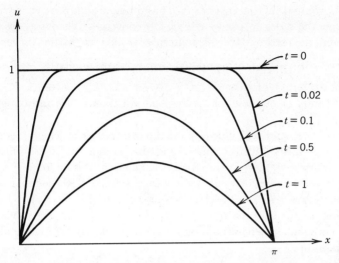

Figure 5.12   Solution of the heat equation with temperature plotted as a function of position

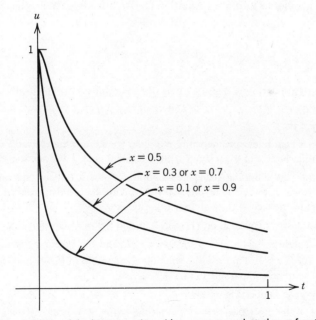

Figure 5.13   Solution of the heat equation with temperature plotted as a function of time

As above, we find that $|b_k \cos(\pi kx/L)|$ has a bound that is independent of k and x and that the series $\Sigma k \exp(-\pi^2 k^2 ct/L^2)$ converges. Therefore, by the M-test, the differentiated series converges uniformly for all x; so, by the above variation of Theorem 3.15b, $\dfrac{\partial u}{\partial x}$ can be computed term by term. Similarly, $\dfrac{\partial^2 u}{\partial x^2}$ and $\dfrac{\partial u}{\partial t}$ can be computed term by term since $\Sigma k^2 \exp(-\pi^2 k^2 ct/L^2)$ converges. This completes the verification of our solution of the heat equation (5.47) satisfying conditions (5.60) and (5.61).

Figures 5.12 and 5.13 illustrate the typical manner in which the temperature u of the rod varies with the time t and the position x. In the illustrations we assumed that $c = 1$, $L = \pi$, and $f(x) = 1$. Figure 5.12 is a plot of u as a function of x for several values of t; the temperature starts at 1 throughout the rod and decreases toward zero, but the change is most rapid near the ends, where the heat is escaping. Figure 5.13 is a plot of u as a function of t for several values of x; it illustrates the same phenomena but displays more clearly the exponential decay of the temperature.

This completes our discussion of the heat equation. Examples involving other initial and boundary conditions occur in the exercises. The method of separation of variables also applies to Laplace's equation (5.48), but we leave all the examples to the exercises.

EXERCISES

1. (a) Show that if $G''(x) = 0$ and $G(0) = G(L) = 0$, then G is identically zero.

   (b) Show that if $G''(x) - \omega^2 G(x) = 0$ where $\omega > 0$ and $G(0) = G(L) = 0$, then G is identically zero.

2. Suppose g is continuous on the entire real line, is odd, and has period $2L$. Let f be the unique antiderivative of g satisfying $f(0) = 0$. Show that f is even and has period $2L$.

3. Suppose that the vibrating string in Example 5.9 is 1.5 ft long and satisfies the wave equation in which the value of c is 900 ft/sec. Find the fundamental frequency and the frequencies of the second through fourth harmonics.

4. (a) Solve the wave equation on $\{(t, x)| t \geq 0 \text{ and } 0 < x < L\}$ with the initial conditions $y(0, x) = f(x)$ and $\dfrac{\partial y}{\partial t}(0, x) = 0$ for $0 < x < L$ and the boundary conditions $y(t, 0) = y(t, L) = 0$ for $t > 0$. Find a series solution comparable to (5.54) and (5.56) and also a functional solution comparable to (5.57).

   (b) Verify your solution rigorously.

5. In Exercise 4, let $L = 2$, let $c = 1$, and let the graph of f have the shape of the plucked string in Figure 5.3. Specifically, let $f(x) = [1 - |1 - x|]/2$.

   (a) Find an explicit formula for the coefficients in the series solution.

   (b) Sketch the graph of the functional solution for $t = 1$ and $t = 2$.

6. How can the solutions to Example 5.9 and Exercise 4 be combined to produce a solution of the wave equation with the initial conditions $y(0, x) = f(x)$ and $\dfrac{\partial y}{\partial t}(0, x) = g(x)$ for $0 < x < L$ and the boundary conditions $y(t, 0) = y(t, L) = 0$ for $t > 0$?

7. (a) Solve the wave equation on $\{(t, x) | t \geq 0 \text{ and } 0 < x < L\}$ with the initial conditions $y(0, x) = f(x)$ and $\frac{\partial y}{\partial t}(0, x) = 0$ for $0 < x < L$ and the boundary conditions $\frac{\partial y}{\partial x}(t, 0)$ $= \frac{\partial y}{\partial x}(t, L) = 0$ for $t > 0$. Find both a series solution comparable to (5.54) and (5.56) and a functional solution comparable to (5.57).

(b) Verify your solution rigorously.

8. (Temperature distribution in a rod with ends at different temperatures.)

(a) Solve the heat equation on $\{(t, x) | t > 0 \text{ and } 0 < x < L\}$ with the initial condition $u(0, x) = f(x)$ for $0 < x < L$ and the boundary conditions $u(t, 0) = 0$ and $u(t, L) = U$ (U a constant) for $t > 0$. *Hint:* Find functions r and g so that the function v defined by $v(t, x) = u(t, x) + r(x)$ satisfies the heat equation with the initial conditions $v(0, x) = g(x)$ for $0 < x < L$ and the boundary conditions $v(t, 0) = v(t, L) = 0$ for $t > 0$. Note that we solved for v in Example 5.10.

(b) Find an explicit formula for the coefficients in the series solution when $L = 1$, $c = 1$, and f is identically zero. Also find the steady-state temperature distribution: $\lim_{t \to \infty} u(t, x)$.

9. (Temperature distribution in a rod with both ends insulated.)

(a) Solve the heat equation on $\{(t, x) | t > 0 \text{ and } 0 < x < L\}$ with the initial condition $u(0, x) = f(x)$ for $0 < x < L$ and the boundary conditions $\frac{\partial u}{\partial x}(t, 0) = 0$ and $\frac{\partial u}{\partial x}(t, L) = 0$ for $t > 0$.

(b) Verify your solution rigorously.

10. Two rods of length 1 and made of the same material are placed end-to-end in perfect contact. Suppose one rod was at the uniform temperature $0°C$ and the other was at $100°C$ just before contact and that after contact the two exposed outer ends are kept at $0°C$.

(a) Find the temperature distribution in the two rods t seconds after contact, assuming that $c = 1$ in the heat equation. *Hint:* The solution to Example 5.10 can be applied here by assuming that the two rods form one rod of length 2 whose initial temperature distribution is a step function with a discontinuity at $x = 1$.

(b) Approximate the temperature at the interface 5 seconds after contact.

11. (a) Solve Laplace's equation on $\{(x, y) | 0 < x < L \text{ and } 0 < y < M\}$ with the boundary conditions $u(0, y) = u(L, y) = 0$ for $0 < y < M$ and $u(x, 0) = 0$, $u(x, M) = f(x)$ for $0 < x < L$.

(b) Verify your solution rigorously.

12. Show how to use your answer to Exercise 11 to solve Laplace's equation on the same rectangle but with the boundary conditions

(a) $u(0, y) = u(L, y) = 0$ for $0 < y < M$ and $u(x, 0) = g(x)$, $u(x, M) = 0$ for $0 < x < L$.

(b) $u(x, 0) = u(x, M) = 0$ for $0 < x < L$ and $u(0, y) = 0$, $u(L, y) = h(y)$ for $0 < y < M$.

(c) $u(x, 0) = u(x, M) = 0$ for $0 < x < L$ and $u(0, y) = k(y)$, $u(L, y) = 0$ for $0 < y < M$.

13. How can the solutions to Exercises 11, 12a, 12b, and 12c be combined to produce a solution of Laplace's equation with the boundary conditions

$$u(x, 0) = g(x), \qquad u(x, M) = f(x) \quad \text{for } 0 < x < L$$

and

$$u(0, y) = k(y), \qquad u(L, y) = h(y) \quad \text{for } 0 < y < M?$$

14. Solve Laplace's equation on the vertical strip $\{(x, y)|0 < x < L \text{ and } y > 0\}$ with the boundary conditions $u(0, y) = u(L, y) = 0$ for $y > 0$ and $u(x,0) = U$ (a constant) for $0 < x < L$ and with the added restriction that $|u(x, y)|$ be bounded.

5.5 Inner Products

In this section and the next two, we view Fourier series from a more geometric perspective. Our starting point is again Theorem 5.1, the theorem containing such orthogonality properties as

$$\int_{-\pi}^{\pi} \cos(kx) \cos(jx)\, dx = \begin{cases} 0 & \text{if } k \neq j \\ \pi & \text{if } k = j \end{cases}$$

Now, however, we view functions as vectors, and we interpret these orthogonality properties as saying that certain sequences of cosine and sine functions consist of mutually orthogonal vectors.

These ideas are best appreciated if one takes an abstract point of view that encompasses both ordinary geometric vectors in space and vector spaces of functions. Such a point of view also makes it possible for us to develop the ideas with sufficient generality to encompass a wide range of applications. The following definition provides an appropriately general setting for the ideas.

Definition 5.5 Suppose V is a vector space with real scalars. An *inner product* on V is an operation that associates to each pair of vectors **u** and **v** in V a scalar (i.e., a real number) $\mathbf{u} \cdot \mathbf{v}$ with the properties

(a) $\mathbf{u} \cdot \mathbf{v} = \mathbf{v} \cdot \mathbf{u}$
(b) $(\mathbf{u} + \mathbf{v}) \cdot \mathbf{w} = \mathbf{u} \cdot \mathbf{w} + \mathbf{v} \cdot \mathbf{w}$
(c) $(c\mathbf{u}) \cdot \mathbf{v} = c(\mathbf{u} \cdot \mathbf{v})$
(d) $\mathbf{u} \cdot \mathbf{u} \geq 0$ and $\mathbf{u} \cdot \mathbf{u} = 0$ iff $\mathbf{u} = \mathbf{0}$ (the zero vector)

for all vectors $\mathbf{u}, \mathbf{v}, \mathbf{w}$ in V and all real scalars c.

Some additional useful properties follow easily from these (Exercise 1):

$$\mathbf{u} \cdot (\mathbf{v} + \mathbf{w}) = \mathbf{u} \cdot \mathbf{v} = \mathbf{u} \cdot \mathbf{w} \tag{5.65}$$

$$\mathbf{u} \cdot (c\mathbf{v}) = c(\mathbf{u} \cdot \mathbf{v}) \tag{5.66}$$

The next two examples describe the two specific inner products that we will use most often.

EXAMPLE 5.11   The vector space of all n-tuples $\mathbf{u} = (u_1, \ldots, u_n)$ of real numbers $u_1, \ldots, u_n$ is called *Euclidean n-space* and is denoted by $\boldsymbol{R}^n$. The operations of vector addition and scalar multiplication on $\boldsymbol{R}^n$ are defined in the usual coordinate-wise fashion:

$$(u_1,\ldots,u_n) + (v_1,\ldots,v_n) = (u_1 + v_1,\ldots,u_n + v_n)$$

$$c(u_1,\ldots,u_n) = (cu_1,\ldots,cu_n)$$

The usual Euclidean inner product on R^n is defined by

$$(u_1,\ldots,u_n) \cdot (v_1,\ldots,v_n) = \sum_{k=1}^{n} u_k v_k \qquad (5.67)$$

We verify properties (b) and (d) of Definion 5.5 and leave the others to the exercises:

(b)
$$\begin{aligned}
(\mathbf{u} + \mathbf{v}) \cdot \mathbf{w} &= ((u_1,\ldots,u_n) + (v_1,\ldots,v_n)) \cdot (w_1,\ldots,w_n) \\
&= (u_1 + v_1,\ldots,u_n + v_n) \cdot (w_1,\ldots,w_n) \\
&= (u_1 + v_1)w_1 + \cdots + (u_n + v_n)w_n \\
&= (u_1 w_1 + \cdots + u_n w_n) + (v_1 w_1 + \cdots + v_n w_n) \\
&= \mathbf{u} \cdot \mathbf{w} + \mathbf{v} \cdot \mathbf{w}
\end{aligned}$$

(d) $\mathbf{u} \cdot \mathbf{u} = u_1^2 + \cdots + u_n^2$ is non-negative, since a square is always non-negative. Furthermore, if $\mathbf{u} \cdot \mathbf{u} = 0$, then every square in the above sum is 0, and so $(u_1,\ldots,u_n) = (0,\ldots,0) = \mathbf{0}$. Conversely, if $(u_1,\ldots,u_n) = (0,\ldots,0)$, then $\mathbf{u} \cdot \mathbf{u} = 0 + \cdots + 0 = 0$.

Equation (5.67) defines only one specific inner product operation on the vector space R^n; there are others (see Exercise 3).

EXAMPLE 5.12   The vector space of all real-valued piecewise continuous functions on $[a, b]$ is denoted by $PC[a, b]$. The operations of vector addition and scalar multiplication are the usual operations on functions:

$$f + g \text{ is defined by } (f + g)(x) = f(x) + g(x) \quad \text{for all } x \text{ in } [a, b]$$

$$cf \text{ is defined by } (cf)(x) = cf(x) \qquad\qquad \text{for all } x \text{ in } [a, b]$$

However, we alter the usual meaning of equality; we say that f and g are *equal as vectors* in $PC[a, b]$ if $f(x) = g(x)$ at all but finitely many values of x in $[a, b]$. If, instead, we were to stick with the usual meaning of equality between functions, we would have to accept distinct vectors that are zero distance apart (see Theorem 5.7a and the proof of property (d) in this example). The usual integral inner product on $PC[a, b]$ is defined by

$$f \cdot g = \int_a^b f(x)g(x)\, dx \qquad (5.68)$$

We verify properties (b) and (d) of Definition 5.5 and leave the others to the exercises:

(b)
$$(f + g) \cdot h = \int_a^b [f(x) + g(x)] h(x) \, dx$$

$$= \int_a^b [f(x)h(x) + g(x)h(x)] \, dx$$

$$= \int_a^b f(x)h(x) \, dx + \int_a^b g(x)h(x) \, dx$$

$$= f \cdot h + g \cdot h$$

(d)
$$f \cdot f = \int_a^b [f(x)]^2 \, dx \geq 0 \quad \text{since } [f(x)]^2 \geq 0.$$

Furthermore, if $f \cdot f = 0$, then

$$0 = \int_a^b [f(x)]^2 \, dx = \sum_{k=1}^n \int_{x_{k-1}}^{x_k} [f(x)]^2 \, dx$$

where f is continuous on each of the open intervals (x_{k-1}, x_k) (Definition 5.2). Therefore, by Lemma A.26, $f(x) = 0$ on each interval (x_{k-1}, x_k); hence, f equals the zero function at all but finitely many values of x in $[a, b]$, and so f and the zero function are equal as vectors. Conversely, if $f(x) = 0$ for all but finitely many values of x in $[a, b]$, then $f \cdot f = \int_a^b 0 \, dx = 0$.

Equation (5.68) defines only one specific inner product operation on the vector space $PC[a, b]$; there are others (see Exercise 5).

The resemblance between the integral inner product (5.68) and the Euclidean inner product (5.67) goes deeper than the fact that both satisfy the same algebraic properties. If we replace the integral in (5.68) by one of its Riemann sums, we see that $f \cdot g$ is approximately equal to

$$\sum_{k=1}^n f(z_k) g(z_k) \Delta x_k$$

which is quite similar to the Euclidean inner product (5.67). In a sense, therefore, integral inner products are limits of Euclidean inner products.

Still, it is the abstract properties (a)–(d) of Definition 5.5 that make inner products useful in geometry. From these properties alone, we can introduce several related geometric concepts and can analyze their properties.

Definition 5.6   Suppose V is a vector space with an inner product. Then, for all vectors $\mathbf{u}$ and $\mathbf{v}$ in V,

   (a) $\sqrt{\mathbf{u} \cdot \mathbf{u}}$ is called the *length* (or *norm*) of $\mathbf{u}$ and is denoted $\|\mathbf{u}\|$.
   (b) $\|\mathbf{u} - \mathbf{v}\|$ is called the *distance* between $\mathbf{u}$ and $\mathbf{v}$.
   (c) If $\mathbf{u} \cdot \mathbf{v} = 0$, $\mathbf{u}$ and $\mathbf{v}$ are said to be *orthogonal*.
   (d) If $\mathbf{u} \neq 0$, $\dfrac{\mathbf{v} \cdot \mathbf{u}}{\|\mathbf{u}\|^2} \mathbf{u}$ is called the *projection* of $\mathbf{v}$ *onto* $\mathbf{u}$.

EXAMPLE 5.13   If $\mathbf{u} = (1, -2, 0, 3)$ and $\mathbf{v} = (2, -3, 3, -2)$, their Euclidean inner product as vectors in $\mathbf{R}^4$ is

$$\mathbf{u} \cdot \mathbf{v} = 2 + 6 + 0 - 6 = 2$$

In particular, $\mathbf{u}$ and $\mathbf{v}$ are not orthogonal since $\mathbf{u} \cdot \mathbf{v} \neq 0$. Also, the length of $\mathbf{u}$ is

$$\|\mathbf{u}\| = \sqrt{\mathbf{u} \cdot \mathbf{u}} = \left[1^2 + (-2)^2 + 0^2 + 3^2 \right]^{1/2} = \sqrt{14}$$

and the projection of $\mathbf{v}$ onto $\mathbf{u}$ is

$$\tfrac{2}{14}(1, -2, 0, 3) = \left(\tfrac{1}{7}, -\tfrac{2}{7}, 0, \tfrac{3}{7} \right)$$

EXAMPLE 5.14   If $f(x) = \sin x$ and $g(x) = \cos x$, the integral inner product of f and g as vectors in $PC[0, \pi]$ is

$$f \cdot g = \int_0^\pi \sin x \cos x \, dx = \tfrac{1}{2} \sin^2 x \Big|_0^\pi = 0$$

Thus f and g are orthogonal, which means that the area of the region above the x axis and below the graph of $y = f(x)g(x)$ equals the area of the region below the x axis and above the graph of $y = f(x)g(x)$ (see Figure 5.14). Also

$$\|f\|^2 = f \cdot f = \int_0^\pi [\, f(x)\,]^2 \, dx = \int_0^\pi \sin^2 x \, dx$$

$$= \frac{1}{2} \int_0^\pi [1 - \cos(2x)] \, dx = \frac{1}{2}\left(x - \frac{1}{2}\sin(2x) \right)\Big|_0^\pi = \frac{\pi}{2}$$

and so the length of f as a vector in $PC[0, \pi]$ with the integral inner product is $\sqrt{\pi/2}$.

The formulas for the distance between two points in $\mathbf{R}^n$ or two functions in $PC[a, b]$ can be thought of as extensions of the usual formula

$$\sqrt{(x_1 - x_2)^2 + (y_1 - y_2)^2}$$

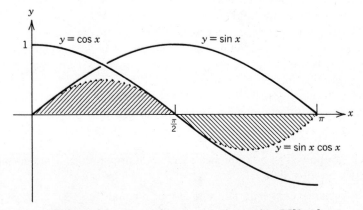

Figure 5.14 $\sin x$ and $\cos x$ are orthogonal in $PC[0, \pi]$

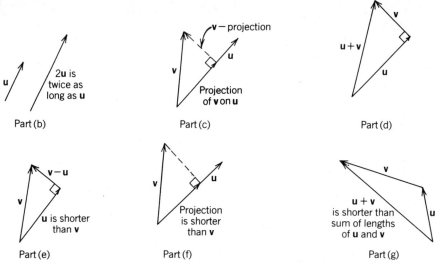

Figure 5.15   Illustrations of parts (b) through (g) of Theorem 5.7

for the distance between two points (x_1, y_1) and (x_2, y_2) in the plane. If $\mathbf{u} = (u_1, \ldots, u_n)$ and $\mathbf{v} = (v_1, \ldots, v_n)$ are points in R^n with the Euclidean inner product, then by Definition 5.6a and (5.67)

$$\|\mathbf{u} - \mathbf{v}\| = \left[\sum_{k=1}^{n} (u_k - v_k)^2 \right]^{1/2}$$

And if f and g are functions in $PC[a, b]$ with the integral inner product, then by Definition 5.6a and (5.68)

$$\|f - g\| = \left[\int_a^b [f(x) - g(x)]^2 \, dx \right]^{1/2}$$

The squares that appear in these distance formulas are quite distinctive; we will see them in several concepts and formulas, and the words we use to describe these concepts and formulas will often refer to the squares. For example, we will say that two vectors $\mathbf{u}$ and $\mathbf{v}$ in R^n are "close to one another in the least-squares sense" if the distance $\|\mathbf{u} - \mathbf{v}\|$ is small; and we will say that two functions f and g in $PC[a, b]$ are "close to one another in the least-squares sense" if the distance $\| f - g \|$ is small.

The next theorem describes some key properties associated with the geometric concepts introduced in Definition 5.6; the corresponding parts of Figure 5.15 illustrate properties (b) through (g) for geometric vectors.

Theorem 5.7   Suppose V is a vector space with an inner product. Then, for all vectors $\mathbf{u}$ and $\mathbf{v}$ in V and all scalars c,

(a) The distance between $\mathbf{u}$ and $\mathbf{v}$ is zero iff $\mathbf{u}$ equals $\mathbf{v}$.

(b) $\|c\mathbf{u}\| = |c| \, \|\mathbf{u}\|$

(c) If **u** is not the zero vector, then the difference of **v** and the projection of **v** onto **u** is orthogonal to **u**; that is,

$$\left(\mathbf{v} - \frac{\mathbf{v} \cdot \mathbf{u}}{\|\mathbf{u}\|^2} \mathbf{u}\right) \cdot \mathbf{u} = 0$$

(d) (Pythagorean theorem) If **u** and **v** are orthogonal then $\|\mathbf{u}\|^2 + \|\mathbf{v}\|^2 = \|\mathbf{u} + \mathbf{v}\|^2$.

(e) (Pythagorean inequality) If $\mathbf{v} - \mathbf{u}$ and **u** are orthogonal, then $\|\mathbf{v}\| \geq \|\mathbf{u}\|$ with equality iff $\mathbf{v} = \mathbf{u}$.

(f) (Cauchy–Schwarz inequality) $|\mathbf{u} \cdot \mathbf{v}| \leq \|\mathbf{u}\| \|\mathbf{v}\|$

(g) (Triangle inequality) $\|\mathbf{u} + \mathbf{v}\| \leq \|\mathbf{u}\| + \|\mathbf{v}\|$

Proof We will use properties (a)–(d) of Definition 5.5 and properties (5.65) and (5.66) without mentioning each use.

(a) By part (a) of Definition 5.6, $\|\mathbf{u} - \mathbf{v}\|^2 = (\mathbf{u} - \mathbf{v}) \cdot (\mathbf{u} - \mathbf{v})$. Therefore, $\|\mathbf{u} - \mathbf{v}\| = 0$ iff $(\mathbf{u} - \mathbf{v}) \cdot (\mathbf{u} - \mathbf{v}) = 0$, which holds iff $\mathbf{u} - \mathbf{v} = 0$, by part (d) of Definition 5.5. Furthermore, $\mathbf{u} - \mathbf{v} = 0$ iff $\mathbf{u} = \mathbf{v}$.

(b) $\|c\mathbf{u}\|^2 = (c\mathbf{u}) \cdot (c\mathbf{u}) = c(\mathbf{u} \cdot c\mathbf{u}) = c^2 \mathbf{u} \cdot \mathbf{u} = c^2 \|\mathbf{u}\|^2$. Property (b) follows by taking the square root of each side of the equation $\|c\mathbf{u}\|^2 = c^2 \|\mathbf{u}\|^2$.

(c) $\quad \left(\mathbf{v} - \dfrac{\mathbf{v} \cdot \mathbf{u}}{\|\mathbf{u}\|^2} \mathbf{u}\right) \cdot \mathbf{u} = \mathbf{v} \cdot \mathbf{u} - \dfrac{\mathbf{v} \cdot \mathbf{u}}{\|\mathbf{u}\|^2} \mathbf{u} \cdot \mathbf{u} = \mathbf{v} \cdot \mathbf{u} - \mathbf{v} \cdot \mathbf{u} = 0$

(d) $\quad \|\mathbf{u} + \mathbf{v}\|^2 = (\mathbf{u} + \mathbf{v}) \cdot (\mathbf{u} + \mathbf{v}) = (\mathbf{u} + \mathbf{v}) \cdot \mathbf{u} + (\mathbf{u} + \mathbf{v}) \cdot \mathbf{v}$
$$= \mathbf{u} \cdot \mathbf{u} + \mathbf{v} \cdot \mathbf{u} + \mathbf{u} \cdot \mathbf{v} + \mathbf{v} \cdot \mathbf{v}$$
$$= \|\mathbf{u}\|^2 + 2\mathbf{u} \cdot \mathbf{v} + \|\mathbf{v}\|^2 = \|\mathbf{u}\|^2 + \|\mathbf{v}\|^2$$

(e) By part (d), $\|\mathbf{v}\|^2 = \|\mathbf{u} + (\mathbf{v} - \mathbf{u})\|^2 = \|\mathbf{u}\|^2 + \|\mathbf{v} - \mathbf{u}\|^2 \geq \|\mathbf{u}\|^2$ with equality iff $\|\mathbf{v} - \mathbf{u}\|^2 = 0$, which holds iff $\mathbf{v} = \mathbf{u}$, by property (a).

(f) The Cauchy–Schwarz inequality is certainly true if $\mathbf{u} = \mathbf{0}$ (Exercise 13a). If **u** is not the zero vector, we use property (c) to deduce that

$$\left(\mathbf{v} - \frac{\mathbf{v} \cdot \mathbf{u}}{\|\mathbf{u}\|^2} \mathbf{u}\right) \cdot \left(\frac{\mathbf{v} \cdot \mathbf{u}}{\|\mathbf{u}\|^2} \mathbf{u}\right) = \frac{\mathbf{v} \cdot \mathbf{u}}{\|\mathbf{u}\|^2}\left(\mathbf{v} - \frac{\mathbf{v} \cdot \mathbf{u}}{\|\mathbf{u}\|^2} \mathbf{u}\right) \cdot \mathbf{u} = 0$$

Therefore, by properties (e) and (b),

$$\|\mathbf{v}\| \geq \left\|\frac{\mathbf{v} \cdot \mathbf{u}}{\|\mathbf{u}\|^2} \mathbf{u}\right\| = \frac{|\mathbf{v} \cdot \mathbf{u}|}{\|\mathbf{u}\|^2}\|\mathbf{u}\| = \frac{|\mathbf{v} \cdot \mathbf{u}|}{\|\mathbf{u}\|}$$

which implies property (f).

(g) $\quad \|\mathbf{u} + \mathbf{v}\|^2 = \|\mathbf{u}\|^2 + 2\mathbf{u} \cdot \mathbf{v} + \|\mathbf{v}\|^2$ (by the proof of property (d))
$$\leq \|\mathbf{u}\|^2 + 2\|\mathbf{u}\| \|\mathbf{v}\| + \|\mathbf{v}\|^2 \quad \text{(by property (f))}$$
$$= (\|\mathbf{u}\| + \|\mathbf{v}\|)^2$$

Then property (g) follows by taking the square root of each side.

Sometimes it is useful to write out the above results for specific inner products. For example, the Cauchy–Schwarz and triangle inequalities take the following forms for the Euclidean inner product on R^n and the integral inner product on $PC[a, b]$:

$$\left| \sum_{k=1}^{n} u_k v_k \right| \leq \left(\sum_{k=1}^{n} u_k^2 \right)^{1/2} \left(\sum_{k=1}^{n} v_k^2 \right)^{1/2} \tag{5.69}$$

$$\left(\sum_{k=1}^{n} (u_k + v_k)^2 \right)^{1/2} \leq \left(\sum_{k=1}^{n} u_k^2 \right)^{1/2} + \left(\sum_{k=1}^{n} v_k^2 \right)^{1/2} \tag{5.70}$$

$$\left| \int_a^b f(x) g(x) \, dx \right| \leq \left(\int_a^b [f(x)]^2 \, dx \right)^{1/2} \left(\int_a^b [g(x)]^2 \, dx \right)^{1/2} \tag{5.71}$$

$$\left(\int_a^b [f(x) + g(x)]^2 \, dx \right)^{1/2} \leq \left(\int_a^b [f(x)]^2 \, dx \right)^{1/2} + \left(\int_a^b [g(x)]^2 \, dx \right)^{1/2} \tag{5.72}$$

The concept of an inner product also makes sense for vector spaces with complex scalars, but some changes are necessary in this context.

Definition 5.7 Suppose V is a vector space with complex scalars. A *complex inner product* on V is an operation that associates to each pair of vectors $\mathbf{u}$ and $\mathbf{v}$ in V a scalar (i.e., a complex number) $(\mathbf{u}, \mathbf{v})$ with the properties

(a) $\overline{(\mathbf{u}, \mathbf{v})} = (\mathbf{v}, \mathbf{u})$ (where the bar denotes the complex conjugate)
(b) $(\mathbf{u} + \mathbf{v}, \mathbf{w}) = (\mathbf{u}, \mathbf{w}) + (\mathbf{v}, \mathbf{w})$
(c) $(c\mathbf{u}, \mathbf{v}) = c(\mathbf{u}, \mathbf{v})$
(d) $(\mathbf{u}, \mathbf{u})$ is real and non-negative, and $(\mathbf{u}, \mathbf{u}) = 0$ iff $\mathbf{u} = \mathbf{0}$

for all vectors $\mathbf{u}, \mathbf{v}, \mathbf{w}$ in V and all complex scalars c.

The analogs of properties (5.65) and (5.66) are (Exercise 21):

$$(\mathbf{u}, \mathbf{v} + \mathbf{w}) = (\mathbf{u}, \mathbf{v}) + (\mathbf{u}, \mathbf{w}) \tag{5.73}$$

$$(\mathbf{u}, c\mathbf{v}) = \bar{c}(\mathbf{u}, \mathbf{v}) \tag{5.74}$$

All four parts of Definition 5.6 make sense for complex inner products exactly as stated for real inner products, and all seven parts of Theorem 5.7 are true for complex inner products exactly as stated. However, in the proofs of parts (d) and (g), we must write $(\mathbf{v}, \mathbf{u}) + (\mathbf{u}, \mathbf{v})$ as the real part of $2(\mathbf{u}, \mathbf{v})$, not simply $2(\mathbf{u}, \mathbf{v})$ (Exercise 22). Also, in the proofs of parts (b) and (f), we must rewrite expressions of the form $(\mathbf{u}, c\mathbf{v})$ as $\bar{c}(\mathbf{u}, \mathbf{v})$, not $c(\mathbf{u}, \mathbf{v})$. With these changes, the proofs are correct for complex inner products.

The complex analogs of our two major examples of inner products are described next.

EXAMPLE 5.15 The vector space of all n-tuples $\mathbf{u} = (u_1, \ldots, u_n)$ of complex numbers $u_1, \ldots, u_n$ is called *complex n-space* and is denoted by C^n. The usual complex inner product on C^n is defined by

$$((u_1, \ldots, u_n), (v_1, \ldots, v_n)) = \sum_{k=1}^{n} u_k \bar{v}_k \tag{5.75}$$

EXAMPLE 5.16   The vector space of all complex-valued piecewise continuous functions on $[a, b]$ is denoted by $CPC[a, b]$. By saying that a complex-valued function f is piecewise continuous, we mean that its real and imaginary parts are piecewise continuous; that is, $f(x) = p(x) + iq(x)$ where p and q are piecewise continuous real-valued functions. The integral of a complex-valued function f is also defined by breaking up f into its real and imaginary parts:

$$\int_a^b f(x)\,dx = \int_a^b p(x)\,dx + i\int_a^b q(x)\,dx \tag{5.76}$$

The complex integral inner product on $CPC[a, b]$ is defined by

$$(f, g) = \int_a^b f(x)\overline{g(x)}\,dx \tag{5.77}$$

EXAMPLE 5.17   Define vectors f and g in $CPC[-\pi, \pi]$ by $f(x) = x$ and $g(x) = e^{ix}$. To compute their inner product, we use the fact that $\int e^{icx}\,dx = e^{icx}/ic$ for any nonzero real number c (Exercise 28). Thus

$$(f, g) = \int_{-\pi}^{\pi} xe^{-ix}\,dx = \frac{xe^{-ix}}{-i}\Big|_{-\pi}^{\pi} - \int_{-\pi}^{\pi} \frac{e^{-ix}}{-i}\,dx$$

$$= i\pi e^{-i\pi} + i\pi e^{i\pi} + e^{-ix}\Big|_{-\pi}^{\pi} = -2i\pi$$

since $e^{\pm i\pi} = -1$. Also

$$\|g\|^2 = \int_{-\pi}^{\pi} g(x)\overline{g(x)}\,dx = \int_{-\pi}^{\pi} e^{ix}e^{-ix}\,dx$$

$$= \int_{-\pi}^{\pi} 1\,dx = 2\pi$$

and so g has length $\sqrt{2\pi}$.

EXERCISES

1. Prove properties (5.65) and (5.66) for any inner product.

2. Verify properties (a) and (c) of the inner product (5.67).

3. **(a)** Let $A = [a_{ij}]$ be an $n \times n$ matrix of real numbers and define $\mathbf{u} \cdot \mathbf{v} = \sum_{i=1}^{n}\sum_{j=1}^{n} u_i a_{ij} v_j$ for all $\mathbf{u}$ and $\mathbf{v}$ in R^n. Show that this product satisfies properties (a) and (b) of Definition 5.5 and satisfies property (c) if A is symmetric (i.e., $a_{ij} = a_{ji}$ for all $i, j = 1, \ldots, n$).

    (b) Symmetric matrices A for which property (d) holds are said to be *positive definite*. Show that every 2×2 matrix of the form

$$A = \begin{bmatrix} a & b \\ b & d \end{bmatrix}$$

    where $a > 0$ and $ad - b^2 > 0$ is positive definite.

4. Verify properties (a) and (c) of the inner product (5.68).

5. Let w be a piecewise continuous function on $[a, b]$ and define

$$f \cdot g = \int_a^b f(x)g(x)w(x)\,dx \quad \text{for all } f \text{ and } g \text{ in } PC[a, b]$$

Show that this product satisfies properties (a)–(c) of Definition 5.5 and satisfies property (d) if $w(x) \geq 0$ for all x in $[a, b]$ and $w(x) = 0$ for only finitely many values of x in $[a, b]$.

6. If $\mathbf{u} = (2, -3, 1)$ and $\mathbf{v} = (4, 0, -1)$ find $\mathbf{u} \cdot \mathbf{v}$, $\|\mathbf{v}\|$, $\|\mathbf{u} - \mathbf{v}\|$, and the projection of $\mathbf{u}$ on $\mathbf{v}$. (Use the usual Euclidean inner product on $\mathbf{R}^3$.)

7. Find a nonzero vector in $\mathbf{R}^4$ that is orthogonal to both $(1, 2, 0, -3)$ and $(2, 0, 3, -1)$. (Use the usual Euclidean inner product on $\mathbf{R}^4$.)

8. If $f(x) = x$ and $g(x) = x^2$ and $0 \leq x \leq 1$, find $f \cdot g$, $\|g\|$, $\|f - g\|$, and the projection of f on g. (Use the usual integral inner product on $PC[0, 1]$.)

9. Find a nonzero function in $PC[-1, 1]$ that is orthogonal to the function g defined by $g(x) = x^2$. (Use the usual integral inner product on $PC[-1, 1]$.)

10. Verify inequalities (5.69) and (5.70) for n-tuples of real numbers.

11. Verify inequalities (5.71) and (5.72) for piecewise continuous functions on $[a, b]$.

12. If the infinite series Σa_k converges, where each a_k is a non-negative real number, prove that $\Sigma \sqrt{a_k}/k$ converges. *Hint:* Use the Cauchy–Schwarz inequality for $\mathbf{R}^n$.

Exercises 13–20 are to be carried out in the context of an abstract vector space V with an abstract inner product that satisfies Definition 5.5.

13. Prove that

    (a) the zero vector is orthogonal to every vector; and

    (b) the only vector orthogonal to every vector in V is the zero vector.

14. If $\mathbf{u}$ is a nonzero vector in V, for what values of c does $c\mathbf{u}$ have length 1?

15. If $|\mathbf{u} \cdot \mathbf{v}| = \|\mathbf{u}\| \|\mathbf{v}\|$ and $\mathbf{u}$ is not the zero vector in V, prove that $\mathbf{v}$ equals its projection on $\mathbf{u}$.

16. If $\mathbf{u}$ and $\mathbf{v}$ are nonzero orthogonal vectors in V of the same length, prove that $a\mathbf{u} + b\mathbf{v}$ and $c\mathbf{u} + d\mathbf{v}$ are orthogonal iff $ac + bd = 0$.

17. Prove that $\|\mathbf{u} + \mathbf{v}\|^2 + \|\mathbf{u} - \mathbf{v}\|^2 = 2\|\mathbf{u}\|^2 + 2\|\mathbf{v}\|^2$ for all vectors $\mathbf{u}$ and $\mathbf{v}$ in V.

18. Suppose $\mathbf{u}$ and $\mathbf{v}$ are vectors in V and $\mathbf{u}$ is nonzero. If, for some scalar c, $\mathbf{v} - c\mathbf{u}$ is orthogonal to $\mathbf{u}$, prove that $c\mathbf{u}$ is the projection of $\mathbf{v}$ on $\mathbf{u}$.

19. Prove that $|\|\mathbf{u}\| - \|\mathbf{v}\|| \leq \|\mathbf{u} - \mathbf{v}\|$ for all vectors $\mathbf{u}$ and $\mathbf{v}$ in V.

20. Suppose $\mathbf{u}$ and $\mathbf{v}$ are nonzero vectors in V.

    (a) Prove that $-1 \leq \dfrac{\mathbf{u} \cdot \mathbf{v}}{\|\mathbf{u}\| \|\mathbf{v}\|} \leq 1$.

    (b) Prove there exists a unique angle θ such that

    $$\cos \theta = \frac{\mathbf{u} \cdot \mathbf{v}}{\|\mathbf{u}\| \|\mathbf{v}\|} \quad \text{and} \quad 0 \leq \theta \leq \pi$$

    (θ is called the angle between $\mathbf{u}$ and $\mathbf{v}$.)

    (c) (Law of cosines) Prove that if θ is the angle between $\mathbf{u}$ and $\mathbf{v}$ then

    $$\|\mathbf{u} - \mathbf{v}\|^2 = \|\mathbf{u}\|^2 + \|\mathbf{v}\|^2 - 2\|\mathbf{u}\| \|\mathbf{v}\| \cos \theta$$

    (d) Prove that the length of the projection of $\mathbf{v}$ on $\mathbf{u}$ is $|\cos \theta| \|\mathbf{v}\|$, where θ is the angle between $\mathbf{u}$ and $\mathbf{v}$.

    (e) Find the angle between the vectors $(1, 2, 0, -3)$ and $(2, 0, 3, -1)$. (Use the usual Euclidean inner product on $\mathbf{R}^4$).

21. Prove properties (5.73) and (5.74) for any complex inner product.

22. Prove that, for any complex inner product, $(v, u) + (u, v) = 2 \operatorname{Re}(u, v)$, where $\operatorname{Re} z$ denotes the real part of the complex number z.

23. Prove that $\|u - v\|^2 = \|u\|^2 + \|v\|^2 - 2 \operatorname{Re}(u, v)$ for any complex inner product.

24. Verify the inner product property $\overline{(u, v)} = (v, u)$ for the product (5.75) on C^n.

25. Using the definition (5.76), prove that for any function f in $CPC[a, b]$

$$\overline{\int_a^b f(x)\, dx} = \int_a^b \overline{f(x)}\, dx$$

26. Verify the inner product property $\overline{(f, g)} = (g, f)$ for the product (5.77) on $CPC[a, b]$.

27. If $u = (i, 1 + i, 3)$ and $v = (-2i, 1 + i, 2 - i)$, find $u, v, \|u\|, \|u - v\|$, and the projection of v on u. (Use the usual complex inner product on C^3.)

28. Prove that, for any nonzero real number c,

$$\int e^{icx}\, dx = \frac{e^{icx}}{ic}$$

5.6 Orthogonal Sequences

We have not yet seen how Fourier series themselves fit into our discussion of the geometry of inner products. In this section we will see that representing a function by its Fourier series is akin to decomposing a vector in space into its coordinates along the three axes (Figure 5.16). The analog of a set of perpendicular coordinate axes that we will find useful is a sequence of orthogonal vectors in a vector space with an inner product.

Definition 5.8   Suppose V is a vector space with an inner product, and suppose $\{u_n\}$ is a finite or infinite sequence of vectors in V.

(a) $\{u_n\}$ is an *orthogonal* sequence if every vector u_n is nonzero and the vectors u_n are mutually orthogonal (that is, $u_n \cdot u_m = 0$ whenever $n \neq m$).

(b) $\{u_n\}$ is an *orthonormal* sequence if every vector u_n has length 1 and the vectors u_n are mutually orthogonal.

EXAMPLE 5.18   The standard basis

$$(1, 0, \ldots, 0), (0, 1, \ldots, 0), \ldots, (0, 0, \ldots, 1)$$

is an orthonormal sequence in R^n with the usual Euclidean inner product.

EXAMPLE 5.19 $\{(1, 1, -2), (2, 0, 1), (-1, 5, 2)\}$ is an orthogonal, but not orthonormal, sequence in R^3 with the usual Euclidean inner product.

For Fourier series, the next example is the key one.

EXAMPLE 5.20   Theorem 5.1 says that

$$1, \cos x, \sin x, \cos(2x), \sin(2x), \ldots, \cos(nx), \sin(nx), \ldots$$

is an orthogonal sequence in $PC[-\pi, \pi]$ with the usual integral inner product.

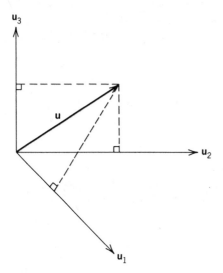

Figure 5.16 The vector **u** is the sum of its projections on u_1, u_2, and u_3

Furthermore, each vector in the sequence has length $\sqrt{\pi}$, except for the first, which has length $\sqrt{2\pi}$.

Our first theorem partly carries out the promise of the opening paragraph; its parts are all suggested by properties of coordinates. Part (a) says that if a vector **u** has coordinates at all, these coordinates are found by projecting **u** onto the coordinate axes. This interpretation is suggested by the fact that the vectors $(\mathbf{u} \cdot \mathbf{u}_k / \|\mathbf{u}_k\|^2)\mathbf{u}_k$ are the projections of **u** on the vectors $\mathbf{u}_k$ in the sense of Definition 5.6d. Equation (5.79) says that the length of **u** is then computed in essentially the same way as the length of a vector in the Euclidean space $\mathbf{R}^n$. In fact, if $\{\mathbf{u}_1, \ldots, \mathbf{u}_n\}$ is an orthonormal sequence, then the length of **u** is given precisely by the familiar square root of the sum of the squares formula, $(\sum_{k=1}^{n} |c_k|^2)^{1/2}$. Part (b) of Theorem 5.8 says that a projection of a vector never has greater length than the vector and that its length is the same as that of the vector only when the projection is the vector itself.

Theorem 5.8 Suppose V is a vector space with an inner product, and suppose $\{\mathbf{u}_1, \ldots, \mathbf{u}_n\}$ is a finite orthogonal sequence of vectors in V. Let **u** also be a vector in V.

   (a) If $\mathbf{u} = \sum_{k=1}^{n} c_k \mathbf{u}_k$ for some scalars $c_1, \ldots, c_n$, then

$$c_k = \frac{\mathbf{u} \cdot \mathbf{u}_k}{\|\mathbf{u}_k\|^2} \quad \text{for } k = 1, \ldots, n \tag{5.78}$$

and

$$\|\mathbf{u}\|^2 = \sum_{k=1}^{n} |c_k|^2 \|\mathbf{u}_k\|^2 \tag{5.79}$$

   (b) If $c_k = \mathbf{u} \cdot \mathbf{u}_k / \|\mathbf{u}_k\|^2$ for $k = 1, \ldots, n$, then

$$\left\| \sum_{k=1}^{n} c_k \mathbf{u}_k \right\| \leq \|\mathbf{u}\| \tag{5.80}$$

with equality iff $\sum_{k=1}^{n} c_k \mathbf{u}_k = \mathbf{u}$.

Proof

(a) For each value of k, $k = 1, \ldots, n$,

$$
\begin{aligned}
\mathbf{u} \cdot \mathbf{u}_k &= (c_1\mathbf{u}_1 + \cdots + c_n\mathbf{u}_n) \cdot \mathbf{u}_k \\
&= c_1\mathbf{u}_1 \cdot \mathbf{u}_k + \cdots + c_n\mathbf{u}_n \cdot \mathbf{u}_k \\
&= c_k\mathbf{u}_k \cdot \mathbf{u}_k \quad \text{since } \mathbf{u}_j \cdot \mathbf{u}_k = 0 \text{ when } j \neq k \\
&= c_k\|\mathbf{u}_k\|^2
\end{aligned}
$$

This implies formula (5.78) and also

$$
\|\mathbf{u}\|^2 = \mathbf{u} \cdot \mathbf{u} = \mathbf{u} \cdot \sum_{k=1}^{n} c_k\mathbf{u}_k = \sum_{k=1}^{n} c_k\mathbf{u} \cdot \mathbf{u}_k = \sum_{k=1}^{n} |c_k|^2\|\mathbf{u}_k\|^2
$$

which proves formula (5.79).

(b) According to the Pythagorean inequality (Theorem 5.7e), we need only show that the following inner product is zero:

$$
\begin{aligned}
\left(\mathbf{u} - \sum_{k=1}^{n} c_k\mathbf{u}_k\right) \cdot \left(\sum_{j=1}^{n} c_j\mathbf{u}_j\right) &= \sum_{j=1}^{n} c_j\left(\mathbf{u} - \sum_{k=1}^{n} c_k\mathbf{u}_k\right) \cdot \mathbf{u}_j \\
&= \sum_{j=1}^{n} c_j(\mathbf{u} \cdot \mathbf{u}_j - c_j\mathbf{u}_j \cdot \mathbf{u}_j) \\
&= \sum_{j=1}^{n} c_j(0) = 0
\end{aligned}
$$

Note that we used the fact that $\mathbf{u}_k \cdot \mathbf{u}_j = 0$ when $k \neq j$ and then used (5.78).

We want to generalize Theorem 5.8 to infinite orthogonal sequences $\{\mathbf{u}_n\}$, especially to the infinite orthogonal sequence in Example 5.20. But to do so, we must attach some meaning to infinite sums of the form $\sum_{k=1}^{\infty} c_k\mathbf{u}_k$. If the vectors $\mathbf{u}_k$ are familiar objects such as functions, we could use familiar concepts such as pointwise convergence or uniform convergence of series of functions. However, there is another meaning we can attach to the sum that, unlike the concepts of pointwise convergence and uniform convergence of series of functions, makes sense in any vector space with an inner product. The idea of the next definition is that we may think of two vectors as being close to one another if the distance between them (in the sense of Definition 5.6b) is small.

Definition 5.9 Suppose V is a vector space with an inner product, and suppose $\mathbf{u}, \mathbf{u}_1, \mathbf{u}_2, \ldots, \mathbf{u}_n, \ldots$ are vectors in V.

(a) We say that the sequence $\{\mathbf{u}_n\}$ *converges in norm* to $\mathbf{u}$ if $\lim_{n \to \infty} \|\mathbf{u}_n - \mathbf{u}\| = 0$. We then write

$$
\lim_{n \to \infty} \mathbf{u}_n = \mathbf{u} \text{ in norm} \quad (\text{or simply } \mathbf{u}_n \to \mathbf{u} \text{ in norm})
$$

(b) We say that the series $\sum_{k=1}^{\infty} \mathbf{u}_k$ *converges in norm* to the sum $\mathbf{u}$ if the sequence of partial sums $\{\sum_{k=1}^{n} \mathbf{u}_k\}$ converges in norm to $\mathbf{u}$. We then write

$$\sum_{k=1}^{\infty} \mathbf{u}_k = \mathbf{u} \text{ in norm} \quad \left(\text{or simply } \sum \mathbf{u}_k = \mathbf{u} \text{ in norm} \right)$$

Let us investigate the meaning of this new concept of convergence for our two standard examples of vector spaces with inner products. In R^n with the usual Euclidean inner product, convergence in norm is equivalent to convergence of each coordinate. That is, a sequence of vectors in R^n converges in norm iff the sequence of its first coordinates converges to the first coordinate of the limit and likewise for the other coordinates (Exercise 6.) In $PC[a, b]$ with the usual integral inner product, convergence in norm has a quite different meaning than either pointwise or uniform convergence of functions, although we will see that uniform convergence implies convergence in norm (Exercise 8). Suppose f_n and f are functions in $PC[a, b]$. Then (Exercise 7)

$$f_n \to f \text{ in norm} \quad \text{iff} \quad \int_a^b [f_n(x) - f(x)]^2 \, dx \to 0 \qquad (5.81)$$

$$\sum_{k=1}^{\infty} f_k = f \text{ in norm} \quad \text{iff} \quad \int_a^b \left[\sum_{k=1}^{n} f_k(x) - f(x) \right]^2 dx \to 0 \qquad (5.82)$$

We might describe (5.81) as saying, for example, that f_n converges to f in norm iff the area under the graph of $y = [f_n(x) - f(x)]^2$ converges to zero. Sometimes this type of convergence is referred to as "convergence in the least-squares sense."

EXAMPLE 5.21   Let $f_n(x) = x^n$, where $0 \le x \le 1$. Recall that f_n converges pointwise, but not uniformly, on $[0, 1]$ to the function f which equals 0 on $[0, 1)$ and equals 1 at $x = 1$. Find out if $f_n \to f$ in norm in the sense of the usual integral inner product on $PC[0, 1]$.

$$\| f_n - f \|^2 = \int_0^1 |f_n(x) - f(x)|^2 \, dx = \int_0^1 |x^n|^2 \, dx$$

$$= \int_0^1 x^{2n} \, dx = \frac{1}{2n + 1} \to 0$$

Therefore, $f_n \to f$ in norm.

EXAMPLE 5.22   Let $\{f_n\}$ be the sequence of functions in Example 1.32 and Figure 1.7. Thus, the graph of f_n consists of the two equal sides of the isosceles triangle with vertices at $(0, 0)$, $(1/2n, n)$, and $(1/n, 0)$ together with the segment $[1/n, 1]$ on the x axis. Recall that f_n converges pointwise, but not uniformly, to the zero function on $[0, 1]$. Find out if $f_n \to 0$ in norm in the sense of the usual integral inner product on $PC[0,1]$.

$$\|f_n - 0\|^2 = \int_0^1 |f_n(x) - 0|^2 \, dx = \int_0^{1/n} |f_n(x)|^2 \, dx$$

$$= 2\int_0^{1/2n} |2n^2 x|^2 \, dx = 8n^4 \int_0^{1/2n} x^2 \, dx$$

$$= \frac{8n^4}{3(2n)^3} = \frac{n}{3} \to \infty$$

Therefore, it is not true that $f_n \to f$ in norm. Figure 1.7 can help us see why. Even though squaring f_n changes the shape of the graph somewhat, we can still see that the areas do not converge to zero.

Next we derive an analog of Theorem 3.1 on the algebra of sums.

Theorem 5.9   Suppose V is a vector space with an inner product, and let $\mathbf{u}_k$, $\mathbf{v}_k$, $\mathbf{u}$, and $\mathbf{v}$ be vectors in V. If $\sum_{k=1}^{\infty} \mathbf{u}_k = \mathbf{u}$ in norm and $\sum_{k=1}^{\infty} \mathbf{v}_k = \mathbf{v}$ in norm, then

(a) $\sum_{k=1}^{\infty} (\mathbf{u}_k + \mathbf{v}_k) = \mathbf{u} + \mathbf{v}$ in norm
(b) $\sum_{k=1}^{\infty} (\mathbf{u}_k - \mathbf{v}_k) = \mathbf{u} - \mathbf{v}$ in norm
(c) $\sum_{k=1}^{\infty} (c\mathbf{u}_k) = c\mathbf{u}$ in norm   for all scalars c
(d) $\sum_{k=1}^{\infty} (\mathbf{w} \cdot \mathbf{u}_k) = \mathbf{w} \cdot \mathbf{u}$ for all vectors $\mathbf{w}$ in V

Proof   We leave the proofs of parts (a)–(c) to the exercises. To prove part (d), we must show that the sequence of partial sums $\sum_{k=1}^{n} (\mathbf{w} \cdot \mathbf{u}_k)$ converges to $\mathbf{w} \cdot \mathbf{u}$, where the meaning of convergence is that of Definition 3.1 since $\mathbf{w} \cdot \mathbf{u}_k$ and $\mathbf{w} \cdot \mathbf{u}$ are numbers. But

$$\left| \sum_{k=1}^{n} (\mathbf{w} \cdot \mathbf{u}_k) - \mathbf{w} \cdot \mathbf{u} \right| = \left| \mathbf{w} \cdot \left(\sum_{k=1}^{n} \mathbf{u}_k - \mathbf{u} \right) \right| \leq \|\mathbf{w}\| \left\| \sum_{k=1}^{n} \mathbf{u}_k - \mathbf{u} \right\|$$

by the Cauchy–Schwarz inequality. Therefore, since we are given that $\sum_{k=1}^{n} \mathbf{u}_k \to \mathbf{u}$ in norm and thus that $\|\sum_{k=1}^{n} \mathbf{u}_k - \mathbf{u}\| \to 0$, it follows that $\sum_{k=1}^{n} (\mathbf{w} \cdot \mathbf{u}_k) \to \mathbf{w} \cdot \mathbf{u}$.

Now we prove an infinite-dimensional analog of Theorem 5.8.

Theorem 5.10   Suppose V is a vector space with an inner product, and suppose $\{\mathbf{u}_n\}$ is an infinite orthogonal sequence of vectors in V. Let $\mathbf{u}$ also be a vector in V.

(a) If $\mathbf{u} = \sum_{k=1}^{\infty} c_k \mathbf{u}_k$ in norm for some sequence of scalars $\{c_k\}$, then

$$c_k = \frac{\mathbf{u} \cdot \mathbf{u}_k}{\|\mathbf{u}_k\|^2} \quad \text{for } k \geq 1 \tag{5.83}$$

(b) If $c_k = \mathbf{u} \cdot \mathbf{u}_k / \|\mathbf{u}_k\|^2$ for $k \geq 1$, then

$$\sum_{k=1}^{\infty} |c_k|^2 \|\mathbf{u}_k\|^2 \leq \|\mathbf{u}\|^2 \tag{5.84}$$

with equality iff $\sum_{k=1}^{\infty} c_k \mathbf{u}_k = \mathbf{u}$ in norm.

Proof

(a) By Theorem 5.9d and the fact that $\mathbf{u}_j \cdot \mathbf{u}_k = 0$ when $j \neq k$

$$\mathbf{u} \cdot \mathbf{u}_k = \left(\sum_{j=1}^{\infty} c_j \mathbf{u}_j \right) \cdot \mathbf{u}_k = \sum_{j=1}^{\infty} c_j (\mathbf{u}_j \cdot \mathbf{u}_k) = c_k (\mathbf{u}_k \cdot \mathbf{u}_k) = c_k \|\mathbf{u}_k\|^2$$

which implies formula (5.83).

(b) In the proof of Theorem 5.8b, we showed that $\mathbf{u} - \sum_{k=1}^{n} c_k \mathbf{u}_k$ is orthogonal to $\sum_{k=1}^{n} c_k \mathbf{u}_k$. Hence, by the Pythagorean theorem (Theorem 5.7d),

$$\left\| \mathbf{u} - \sum_{k=1}^{n} c_k \mathbf{u}_k \right\|^2 + \left\| \sum_{k=1}^{n} c_k \mathbf{u}_k \right\|^2 = \|\mathbf{u}\|^2$$

and so, by formula (5.79),

$$\left\| \mathbf{u} - \sum_{k=1}^{n} c_k \mathbf{u}_k \right\|^2 = \|\mathbf{u}\|^2 - \sum_{k=1}^{n} |c_k|^2 \|\mathbf{u}_k\|^2$$

Since the left side of this equation is non-negative, inequality (5.84) now follows. Furthermore, the equation shows that $\sum_{k=1}^{n} c_k \mathbf{u}_k \to \mathbf{u}$ in norm iff $\sum_{k=1}^{n} |c_k|^2 \|\mathbf{u}_k\|^2 \to \|\mathbf{u}\|^2$. This proves the last assertion in the theorem.

In the next section we exploit the new insights into Fourier series that this theorem offers.

Definitions 5.8 and 5.9 also make sense for vector spaces with complex scalars and complex inner products. Furthermore, Theorems 5.8, 5.9, and 5.10 are then still valid and their proofs require only minor modifications. The next example is the key one for Fourier series in this setting.

EXAMPLE 5.23   The sequence $\{ e^{inx} \}_{n=-\infty}^{\infty}$ is orthogonal in $CPC[-\pi, \pi]$ with the complex integral inner product, since, for $n \neq m$,

$$\int_{-\pi}^{\pi} e^{inx} \overline{e^{imx}} \, dx = \int_{-\pi}^{\pi} e^{inx} e^{-imx} \, dx = \int_{-\pi}^{\pi} e^{i(n-m)x} \, dx$$

$$= \frac{e^{i(n-m)x}}{i(n-m)} \bigg|_{-\pi}^{\pi} = \frac{(-1)^{n-m} - (-1)^{n-m}}{i(n-m)} = 0$$

Furthermore, every vector in the sequence has length $\sqrt{2\pi}$ since $\int_{-\pi}^{\pi} 1 \, dx = 2\pi$.

EXERCISES

1. Let $\mathbf{u}_1 = (1, 1, -1, -1)$, $\mathbf{u}_2 = (1, -1, 1, -1)$, $\mathbf{u}_3 = (1, 2, 2, 1)$.

    (a) Verify that $\{\mathbf{u}_1, \mathbf{u}_2, \mathbf{u}_3\}$ is an orthogonal sequence in R^4 in the sense of the usual Euclidean inner product.

    (b) Find scalars c_1, c_2, c_3 such that $\{ c_1 \mathbf{u}_1, c_2 \mathbf{u}_2, c_3 \mathbf{u}_3 \}$ is an orthonormal sequence.

(c) Use Theorem 5.8a to show that the vector $(1, 2, 3, 0)$ is not a linear combination of $\mathbf{u}_1, \mathbf{u}_2, \mathbf{u}_3$.

(d) Verify that the inequality (5.80) is strictly less than if $\mathbf{u} = (1, 2, 3, 0)$ and the orthogonal sequence is the given one, $\{\mathbf{u}_1, \mathbf{u}_2, \mathbf{u}_3\}$.

2. Find an orthogonal sequence $\{\mathbf{u}_1, \mathbf{u}_2, \mathbf{u}_3\}$ in R^3, given that the first two vectors are $\mathbf{u}_1 = (1, 2, -3)$ and $\mathbf{u}_2 = (-3, 3, 1)$. (Use the usual Euclidean inner product.)

3. Let $f_n(x) = \sin(n\pi x/L)$, where L is a fixed positive real number and n is a positive integer.

(a) Show that $\{f_n\}$ is an orthogonal sequence in $C[-L, L]$ in the sense of the usual integral inner product.

(b) Find scalars c_n such that $\{c_n f_n\}$ is an orthonormal sequence in $C[-L, L]$.

4. Show that $\{\cos(nx)\}_{n=0}^{\infty}$ is an orthogonal sequence in $C[0, \pi]$ in the sense of the usual integral inner product.

5. If n is a non-negative integer, Legendre's differential equation

$$(1 - x^2) y'' - 2xy' + n(n + 1) y = 0$$

has a polynomial solution of degree n (Exercise 11 in Section 4.1). The nth Legendre polynomial P_n is the unique such solution satisfying the initial condition $P_n(1) = 1$. Verify the following chain of deductions:

(a) $\dfrac{d}{dx}[(1 - x^2)P_n'] + n(n + 1)P_n = 0$

(b) $\dfrac{d}{dx}[(1 - x^2)(P_n'P_m - P_m'P_n)] + [m(m + 1) - n(n + 1)]P_n P_m = 0$

(c) $\{P_n\}$ is an orthogonal sequence in $C[-1, 1]$ in the sense of the usual integral product.

6. Let $\{\mathbf{u}_n\}$ be a sequence of vectors in R^k and let $(u_{n1}, \ldots, u_{nk}) = \mathbf{u}_n$. Also let $(u_1, \ldots, u_k) = \mathbf{u}$ be a vector in R^k. Show that $\{\mathbf{u}_n\}$ converges in norm to $\mathbf{u}$ (in the sense of the usual Euclidean inner product) iff $\lim_{n \to \infty} u_{nj} = u_j$ for $j = 1, \ldots, k$.

7. Let $\{f_n\}$ be a sequence of functions in $PC[a, b]$ and let f be a function in $PC[a, b]$. Verify statements (5.81) and (5.82).

8. Let $\{f_n\}$ be a sequence of functions in $PC[a, b]$ and let f be a function in $PC[a, b]$.

(a) If $f_n \to f$ uniformly on $[a, b]$, prove that $f_n \to f$ in norm in the sense of the usual integral inner product on $PC[a, b]$.

(b) If $\sum_{k=1}^{\infty} f_k = f$ uniformly on $[a, b]$, prove that $\sum_{k=1}^{\infty} f_k = f$ in norm in the sense of the usual integral inner product on $PC[a, b]$.

Exercises 9–14 are to be carried out in the context of an abstract vector space V with an abstract inner product.

9. If $\{\mathbf{u}, \mathbf{v}\}$ is an orthonormal sequence, show that $\|\mathbf{u} - \mathbf{v}\| = \sqrt{2}$.

10. Prove Theorems 5.9a, 5.9b, and 5.9c.

11. (a) If $\mathbf{u}$ is orthogonal to each of the vectors $\mathbf{u}_1, \ldots, \mathbf{u}_n$, show that $\mathbf{u}$ is orthogonal to every one of their linear combinations $\sum_{k=1}^{n} c_k \mathbf{u}_k$.

(b) If $\mathbf{u}$ is orthogonal to each of the vectors in the infinite sequence $\{\mathbf{u}_k\}$ and if the series $\sum_{k=1}^{\infty} c_k \mathbf{u}_k$ converges in norm for some sequence of the scalars $\{c_k\}$, show that $\mathbf{u}$ is orthogonal to this sum.

12.   Let $\{\mathbf{u}_k\}$ be an infinite orthonormal sequence.

   (a) If $\sum_{k=1}^{\infty} c_k \mathbf{u}_k = \mathbf{u}$ in norm for some sequence of scalars $\{c_k\}$, show that

$$\|\mathbf{u}\|^2 = \sum_{k=1}^{\infty} |c_k|^2$$

   (b) If $\sum_{k=1}^{\infty} c_k \mathbf{u}_k = \mathbf{u}$ in norm and $\sum_{k=1}^{\infty} d_k \mathbf{u}_k = \mathbf{v}$ in norm for some sequences of scalars $\{c_k\}$ and $\{d_k\}$, show that

$$\mathbf{u} \cdot \mathbf{v} = \sum_{k=1}^{\infty} c_k d_k$$

13.   If $\mathbf{u}_n \to \mathbf{u}$ in norm and $\mathbf{v}_n \to \mathbf{v}$ in norm, show that $\mathbf{u}_n \cdot \mathbf{v}_n \to \mathbf{u} \cdot \mathbf{v}$.

14.   Suppose $\{\mathbf{u}_1, \ldots, \mathbf{u}_n\}$ is an orthogonal sequence; let $c_k = \mathbf{u} \cdot \mathbf{u}_k / \|\mathbf{u}_k\|^2$ for $k = 1, \ldots, n$. Show that

$$\left\| \mathbf{u} - \sum_{k=1}^{n} c_k \mathbf{u}_k \right\| \le \left\| \mathbf{u} - \sum_{k=1}^{n} d_k \mathbf{u}_k \right\|$$

   for all choices of scalars $d_1, \ldots, d_k$, with equality iff $c_k = d_k$ for $k = 1, \ldots, n$. *Hint:* Use the Pythagorean inequality (Theorem 5.7e).

In Exercises 15–18, find out whether $f_n \to 0$ in norm in the sense of the usual integral inner product on $PC[0, 1]$.

15. $f_n(x) = nx^n$, $0 \le x \le 1$ 16. $f_n(x) = nx^n(1 - x)$, $0 \le x \le 1$

17. $f_n(x) = \sqrt{n}$ if $0 \le x \le 1/n$; $f_n(x) = 0$ if $1/n < x \le 1$

18. $f_n(x) = \sqrt[3]{n}$ if $0 \le x \le 1/n$; $f_n(x) = 0$ if $1/n < x \le 1$

5.7 Fourier Series Revisited

As we observed in the preceding section, Theorem 5.1 says that

$$1, \cos x, \sin x, \cos (2x), \sin (2x), \ldots, \cos (nx), \sin (nx), \ldots \qquad (5.85)$$

is an orthogonal sequence in $PC[-\pi, \pi]$ with the usual integral inner product. Thus, the problem of expressing a function f as the sum of its Fourier series

$$\frac{a_0}{2} + \sum_{k=1}^{\infty} \left[a_k \cos (kx) + b_k \sin (kx) \right]$$

may be restated as the problem of expressing f in the form $\sum_{k=1}^{\infty} c_k \mathbf{u}_k$, where $\mathbf{u}_k$ is the kth term of the orthogonal sequence (5.85). In this section, we apply the concepts and conclusions of Sections 5.5 and 5.6 to the theory of Fourier series, and so we restrict our attention to the vector space $PC[-\pi, \pi]$ with the usual integral inner product and to the orthogonal sequence (5.85). Thus, we now regard f and $\mathbf{u}_k$ as vectors in $PC[-\pi, \pi]$, and we interpret $\int_{-\pi}^{\pi} f(x) \cos (kx)\, dx$ as the inner product of f and $\mathbf{u}_{2k}$ and $\int_{-\pi}^{\pi} f(x) \sin (kx)\, dx$ as the inner product of f and $\mathbf{u}_{2k+1}$. Furthermore, since each vector in the sequence (5.85), except for the first one, has length $\sqrt{\pi}$, we see that the

Fourier coefficients

$$a_k = \frac{1}{\pi} \int_{-\pi}^{\pi} f(x) \cos(kx) \, dx$$

$$b_k = \frac{1}{\pi} \int_{-\pi}^{\pi} f(x) \sin(kx) \, dx$$

(except for a_0) have the form $c_k = f \cdot \mathbf{u}_k / \|\mathbf{u}_k\|^2$ of formula (5.83). Specifically, $a_k = c_{2k}$ and $b_k = c_{2k+1}$ for $k \geq 1$, and $a_0/2 = c_1$ since $\mathbf{u}_1$ has length $\sqrt{2\pi}$. This shows that the Fourier series of f has the form $\sum_{k=1}^{\infty} c_k \mathbf{u}_k$ where the terms $c_k \mathbf{u}_k$ are the projections of f on the vectors (5.85). Thus, Theorem 5.10 can be restated as follows.

Theorem 5.11 Let f be any piecewise continuous function on $[-\pi, \pi]$.

    (a) If $a_0'/2 + \sum_{k=1}^{\infty} [a_k' \cos(kx) + b_k' \sin(kx)]$ converges in norm to f for some scalars a_k' and b_k', then a_k' and b_k' are the Fourier coefficients of f.

    (b) (Bessel's inequality) Let a_k, b_k denote the Fourier coefficients of f. Then

$$\frac{|a_0|^2}{2} + \sum_{k=1}^{\infty} \left(|a_k|^2 + |b_k|^2 \right) \leq \frac{1}{\pi} \int_{-\pi}^{\pi} |f(x)|^2 \, dx \tag{5.86}$$

with equality iff the Fourier series of f converges in norm to f.

Even more is true. As we show in Theorem 5.13, the Fourier series of any piecewise continuous function f converges in norm to f; thus, Bessel's inequality (5.86) is actually an equality. The key to the proof is to find a sufficiently large number of functions f whose Fourier series converge uniformly to f.

Theorem 5.12 Suppose f is continuous at every point of the closed interval $[-\pi, \pi]$ and satisfies $f(-\pi) = f(\pi)$. Suppose also that f is differentiable at all but a finite number of points in $[-\pi, \pi]$, that the one-sided derivatives (5.31) exist at the remaining points, and that f' is piecewise continuous on $[-\pi, \pi]$. Then the Fourier series of f converges uniformly to f on $[-\pi, \pi]$.

Proof We already know from Theorem 5.4 that the Fourier series of f converges pointwise to f on $[-\pi, \pi]$. To help us prove uniform convergence of the Fourier series, we find a relationship between the Fourier coefficients a_k, b_k of f and the Fourier coefficients c_k, d_k of f':

$$\pi a_k = \int_{-\pi}^{\pi} f(x) \cos(kx) \, dx$$

$$= f(x) \frac{\sin(kx)}{k} \Big|_{-\pi}^{\pi} - \int_{-\pi}^{\pi} f'(x) \frac{\sin(kx)}{k} \, dx$$

$$= \frac{-\pi d_k}{k}$$

The integration by parts that was used in the middle step above is valid even though f' is only piecewise continuous; to verify it, just integrate by parts on each of the intervals on which f' is continuous and add up the integrals. Thus $a_k = -d_k/k$ and, similarly, $b_k = c_k/k$. Therefore

$$\left| a_k \cos(kx) \right| \le |a_k| = \frac{|d_k|}{k}$$

and

$$\left| b_k \sin(kx) \right| \le |b_k| = \frac{|c_k|}{k}$$

So if we can show that the series $\Sigma |d_k|/k$ and $\Sigma |c_k|/k$ converge, it will follow from the M-test (Theorem 3.16) that $\Sigma a_k \cos(kx)$ and $\Sigma b_k \sin(kx)$ converge uniformly on $(-\infty, \infty)$. We first use the Cauchy–Schwarz inequality (5.69):

$$\left| \sum_{k=n+1}^{m} \frac{|d_k|}{k} \right| \le \left(\sum_{k=n+1}^{m} |d_k|^2 \right)^{1/2} \left(\sum_{k=n+1}^{m} \frac{1}{k^2} \right)^{1/2}$$

But $\Sigma 1/k^2$ is a convergent p-series with $p = 2$, and $\Sigma |d_k|^2$ converges by Bessel's inequality (5.86) applied to f'. So, by the Cauchy criterion of Corollary 3.9, for any $\varepsilon > 0$ there exists a positive integer N such that

$$\sum_{k=n+1}^{m} |d_k|^2 < \varepsilon \quad \text{and} \quad \sum_{k=n+1}^{m} \frac{1}{k^2} < \varepsilon$$

whenever $n \ge N$ and $m > n$. Therefore, for these same values of n and m,

$$\left| \sum_{k=n+1}^{m} \frac{|d_k|}{k} \right| \le \sqrt{\varepsilon}\sqrt{\varepsilon} = \varepsilon$$

which proves that $\Sigma |d_k|/k$ converges. Similarly, $\Sigma |c_k|/k$ converges. This completes the proof.

Theorem 5.13 If f is piecewise continuous on $[-\pi, \pi]$, then the Fourier series of f converges in norm to f. Thus, Parseval's equation holds:

$$\frac{|a_0|^2}{2} + \sum_{k=1}^{\infty} \left[|a_k|^2 + |b_k|^2 \right] = \frac{1}{\pi} \int_{-\pi}^{\pi} |f(x)|^2 \, dx \tag{5.87}$$

where a_k, b_k denote the Fourier coefficients of f.

Proof For any positive real number ε, we will show that there exists a function g satisfying the hypotheses of Theorem 5.12 such that $\| f - g \| < \varepsilon$. But before we do, we show that this conclusion implies the truth of the theorem. By Theorem 5.12, the Fourier series of g converges uniformly on $[-\pi, \pi]$ to g and therefore, by Exercise 8 of Section 5.6, converges in norm to g. To show that this implies the theorem, we express the Fourier series of f and g in the more compact and geometric notation of Section 5.6

$$\sum_{k=1}^{\infty} \frac{f \cdot \mathbf{u}_k}{\|\mathbf{u}_k\|^2} \mathbf{u}_k \quad \text{and} \quad \sum_{k=1}^{\infty} \frac{g \cdot \mathbf{u}_k}{\|\mathbf{u}_k\|^2} \mathbf{u}_k$$

rather than the cumbersome notation of Definition 5.1. Since the Fourier series of g converges in norm to g, there exists N such that

$$\left\| g - \sum_{k=1}^{n} \frac{g \cdot \mathbf{u}_k}{\|\mathbf{u}_k\|^2} \mathbf{u}_k \right\| < \varepsilon \quad \text{for all } n \geq N$$

Furthermore, by Bessel's inequality (5.80) applied to the vector $g - f$,

$$\left\| \sum_{k=1}^{n} \frac{g \cdot \mathbf{u}_k}{\|\mathbf{u}_k\|^2} \mathbf{u}_k - \sum_{k=1}^{n} \frac{f \cdot \mathbf{u}_k}{\|\mathbf{u}_k\|^2} \mathbf{u}_k \right\| = \left\| \sum_{k=1}^{n} \frac{(g - f) \cdot \mathbf{u}_k}{\|\mathbf{u}_k\|^2} \mathbf{u}_k \right\| \leq \|g - f\| < \varepsilon$$

Therefore, by the triangle inequality (Theorem 5.7g),

$$\left\| f - \sum_{k=1}^{n} \frac{f \cdot \mathbf{u}_k}{\|\mathbf{u}_k\|^2} \mathbf{u}_k \right\|$$

$$= \left\| (f - g) + \left(g - \sum_{k=1}^{n} \frac{g \cdot \mathbf{u}_k}{\|\mathbf{u}_k\|^2} \mathbf{u}_k \right) + \left(\sum_{k=1}^{n} \frac{g \cdot \mathbf{u}_k}{\|\mathbf{u}_k\|^2} \mathbf{u}_k - \sum_{k=1}^{n} \frac{f \cdot \mathbf{u}_k}{\|\mathbf{u}_k\|^2} \mathbf{u}_k \right) \right\|$$

$$\leq \|f - g\| + \left\| g - \sum_{k=1}^{n} \frac{g \cdot \mathbf{u}_k}{\|\mathbf{u}_k\|^2} \mathbf{u}_k \right\| + \left\| \sum_{k=1}^{n} \frac{g \cdot \mathbf{u}_k}{\|\mathbf{u}_k\|^2} \mathbf{u}_k - \sum_{k=1}^{n} \frac{f \cdot \mathbf{u}_k}{\|\mathbf{u}_k\|^2} \mathbf{u}_k \right\|$$

$$< \varepsilon + \varepsilon + \varepsilon = 3\varepsilon$$

whenever $n \geq N$. Since ε was an arbitrary positive number, it follows that the Fourier series of f converges in norm to f.

Next, we verify the claim in the opening sentence of the proof. To simplify the notations in the proof, we assume that f has only one discontinuity in the interior of the interval $[-\pi, \pi]$. That is, we assume there exists a point c in the open interval $(-\pi, \pi)$ such that f is continuous on $(-\pi, c)$ and on (c, π) and the limits $f(-\pi^+), f(c^-), f(c^+), f(\pi^-)$ exist. (If f were to have more discontinuities, but still just finitely many, a quite similar proof would work.) Given these above assumptions about f, we construct g as follows (see Figure 5.17). Let δ and γ be (small) positive numbers. On the interval $[-\pi, -\pi + \delta]$, define g so that its graph is the line joining the points $(-\pi, f(\pi^-))$ and $(-\pi + \delta, f(-\pi + \delta))$. On the interval $[-\pi + \delta, c - \delta]$, define g to be a continuous piecewise linear function such that $|f(x) - g(x)| < \gamma$

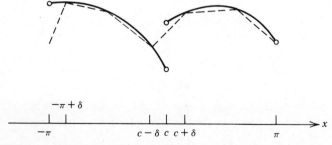

Figure 5.17 A piecewise continuous function $y = f(x)$ (_____) approximated by a continuous, periodic, piecewise linear function $y = g(x)$ (------)

whenever $-\pi + \delta \le x \le c - \delta$ and also that $g(-\pi + \delta) = f(-\pi + \delta)$ and $g(c - \delta)$ $= f(c - \delta)$. (g exists by Exercise 8 of Section A.2.) On the interval $[c - \delta, c + \delta]$, define g so that its graph is the line joining the points $(c - \delta, f(c - \delta))$ and $(c + \delta, f(c + \delta))$. And on the interval $[c + \delta, \pi]$, define g to be a continuous piecewise linear function such that $|f(x) - g(x)| < \gamma$ whenever $c + \delta \le x < \pi$ and also that $g(c + \delta) = f(c + \delta)$ and $g(\pi) = f(\pi^-)$. Therefore g satisfies the hypotheses of Theorem 5.12. To see that $\|g - f\| < \varepsilon$ when δ and γ are sufficiently small, we let M denote an upper bound for $\{|f(x) - g(x)| \mid -\pi \le x \le \pi\}$ and note that

$$\int_{-\pi}^{-\pi + \delta} |f(x) - g(x)|^2 \, dx \le \int_{-\pi}^{-\pi + \delta} M^2 \, dx = M^2 \delta$$

$$\int_{-\pi + \delta}^{c - \delta} |f(x) - g(x)|^2 \, dx \le \int_{-\pi + \delta}^{c - \delta} \gamma^2 \, dx \le (c + \pi)\gamma^2$$

$$\int_{c - \delta}^{c + \delta} |f(x) - g(x)|^2 \, dx \le \int_{c - \delta}^{c + \delta} M^2 \, dx = 2M^2 \delta$$

$$\int_{c + \delta}^{\pi} |f(x) - g(x)|^2 \, dx \le \int_{c + \delta}^{\pi} \gamma^2 \, dx \le (\pi - c)\gamma^2$$

Therefore, since $\int_{-\pi}^{\pi} |f(x) - g(x)|^2 \, dx$ is the sum of the four above integrals in the left column,

$$\|f - g\|^2 = \int_{-\pi}^{\pi} |f(x) - g(x)|^2 \, dx \le 3M^2 \delta + 2\pi\gamma^2$$

Hence, $\|f - g\| < \varepsilon$ when δ and γ are sufficiently small.

Finally, the last sentence of the theorem follows from Theorem 5.11b.

EXAMPLE 5.24 Apply Parseval's equation to the function f defined by $f(x) = x$. Show that $\sum_{k=1}^{\infty} 1/k^2 = \pi^2/6$.

According to Table 5.1, $a_k = 0$ for all $k \ge 0$ and $b_k = 2(-1)^{k+1}/k$. Therefore, by Parseval's equation (5.87),

$$\sum_{k=1}^{\infty} \frac{4}{k^2} = \frac{1}{\pi} \int_{-\pi}^{\pi} x^2 \, dx = \frac{x^3}{3\pi} \Big|_{-\pi}^{\pi} = \frac{2}{3}\pi^2$$

which implies the desired result.

EXERCISES

1. If f is piecewise continuous on $[-\pi, \pi]$ and is orthogonal to every function in the sequence (5.85) (in the sense of the integral inner product on $PC[-\pi, \pi]$), prove that $f = 0$ at all but a finite number of points of $[-\pi, \pi]$.

2. Let a_k, b_k denote the Fourier coefficients of f and c_k, d_k denote the Fourier coefficients of g, where f and g are assumed to be piecewise continuous on $[-\pi, \pi]$. Prove that

$$\frac{a_0 c_0}{2} + \sum_{k=1}^{\infty} (a_k c_k + b_k d_k) = \frac{1}{\pi} \int_{-\pi}^{\pi} f(x) g(x) \, dx$$

3. **(a)** If f is piecewise continuous on $[0, \pi]$, prove that the Fourier sine series of f converges in norm to f, where the norm is derived from the integral inner product on $PC[0, \pi]$. *Hint:* Form the odd extension of f to the domain $[-\pi, \pi]$.

    (b) If f is piecewise continuous on $[0, \pi]$, prove that the Fourier cosine series of f converges in norm to f, where the norm is derived from the integral inner product on $PC[0, \pi]$. *Hint:* Form the even extension of f to the domain $[-\pi, \pi]$.

4. (Uniform approximation of continuous functions) Suppose f is continuous at every point of $[-\pi, \pi]$ and that $f(-\pi) = f(\pi)$. For any positive real number ε, prove there exists a trigonometric polynomial p

$$ p(x) = c_0 + \sum_{k=1}^{n} \left[a_k \cos(kx) + b_k \sin(kx) \right] $$

such that $|f(x) - p(x)| < \varepsilon$ for all x in $[-\pi, \pi]$. (*Warning:* It can happen that no partial sum of the Fourier series of f satisfies this condition.) *Hint:* Use Exercise 8 in Section A.2 to approximate f by a piecewise linear function.

In Exercises 5–8, use Parseval's equation (5.87) to find the sum of the given series. (Table 5.1 may be useful in locating an appropriate Fourier series.)

5. $\displaystyle\sum_{k=0}^{\infty} 1/(2k + 1)^2$

6. $\displaystyle\sum_{k=1}^{\infty} 1/k^4$

7. $\displaystyle\sum_{k=0}^{\infty} 1/(2k + 1)^4$

8. $\displaystyle\sum_{k=1}^{\infty} 1/(k^2 + 1)^2$

5.8 Systems of Linear Equations

In this section we apply the ideas of Sections 5.5 and 5.6 to the problem of solving linear systems of equations:

$$
\begin{aligned}
a_{11}u_1 + \cdots + a_{1n}u_n &= v_1 \\
\vdots \qquad\qquad \vdots \quad &\;\; \vdots \\
a_{m1}u_1 + \cdots + a_{mn}u_n &= v_m
\end{aligned}
\tag{5.88}
$$

The coefficients a_{ij} and the right-hand terms v_i are given, and we want a method of finding a solution $(u_1, \ldots, u_n)$. Furthermore, since modern applications involving linear systems often require the solution of linear systems with hundreds or even thousands of equations and unknowns, we want a method of solution that can be implemented effectively in a computer program; that is, we want an algorithm that leads to relatively little round-off error and is relatively fast.

A linear system, of course, need not have any solution. In many applications, however, a so-called least-squares solution, which might not be a solution at all, is quite satisfactory. So we amend our goal further; we now seek a method for finding solutions and least-squares solutions of the linear system (5.88), and we want the method to be one that can be implemented effectively in a computer program.

Least-squares solutions and their properties are most easily understood by means of the geometric concepts of Sections 5.5 and 5.6 applied to the vector space R^m with the usual Euclidean inner product. For example, we can rewrite the linear system (5.88) as

$$\sum_{k=1}^{n} u_k \mathbf{a}_k = \mathbf{v} \tag{5.89}$$

where $\mathbf{v} = (v_1, \ldots, v_m)$ and $\mathbf{a}_k = (a_{1k}, \ldots, a_{mk})$ are vectors in $\mathbf{R}^m$. To explain what we mean by a least-squares solution of the system (5.89), we use the Euclidean inner product and the associated distance concept for points in $\mathbf{R}^m$.

Definition 5.10 The n-tuple $(u_1, \ldots, u_n)$ is a *least-squares solution* of the linear system (5.89) if

$$\left\| \sum_{k=1}^{n} u_k \mathbf{a}_k - \mathbf{v} \right\| \leq \left\| \sum_{k=1}^{n} w_k \mathbf{a}_k - \mathbf{v} \right\| \tag{5.90}$$

for all choices of the scalars $w_1, \ldots, w_n$.

In particular, if the linear system has a solution $(u_1, \ldots, u_n)$, this solution is also a least-squares solution since the left side of the inequality (5.90) is then zero. And if the linear system has no solution, we can think of a least-squares solution as one for which the "error" $\left\| \sum_{k=1}^{n} u_k \mathbf{a}_k - \mathbf{v} \right\|$ is not zero but is the least it could be. To see where the phrase "least-squares" comes from, we use Definition 5.6b and (5.67) to rewrite this error term as

$$\left\| \sum_{k=1}^{n} u_k \mathbf{a}_k - \mathbf{v} \right\| = \left(\sum_{i=1}^{n} \left(\sum_{k=1}^{n} u_k a_{ik} - v_i \right)^2 \right)^{1/2}$$

Thus, the error is least if the sum of squares $\sum_{i=1}^{n}(\sum_{k=1}^{n} u_k a_{ik} - v_i)^2$ is least. However, this quantity of detail obscures the geometry, and so we usually prefer the compact vector notation.

We can reason geometrically to see how we might find such a least-squares solution. If we were to project $\mathbf{v}$ onto the subspace of $\mathbf{R}^m$ spanned by the vectors $\mathbf{a}_1, \ldots, \mathbf{a}_n$, then this projection would be some linear combination $\sum_{k=1}^{n} u_k \mathbf{a}_k$. And, as Figure 5.18 suggests, no linear combination of $\mathbf{a}_1, \ldots, \mathbf{a}_n$ is likely to be closer to $\mathbf{v}$ than this one. Furthermore, the difference vector $\sum_{k=1}^{n} u_k \mathbf{a}_k - \mathbf{v}$ ought to be orthogonal to each of $\mathbf{a}_1, \ldots, \mathbf{a}_n$. Thus, for $j = 1, \ldots, n$

$$0 = \left(\sum_{k=1}^{n} u_k \mathbf{a}_k - \mathbf{v} \right) \cdot \mathbf{a}_j$$

$$= \left(\sum_{k=1}^{n} u_k \mathbf{a}_k \right) \cdot \mathbf{a}_j - \mathbf{v} \cdot \mathbf{a}_j$$

$$= \sum_{k=1}^{n} u_k (\mathbf{a}_k \cdot \mathbf{a}_j) - \mathbf{v} \cdot \mathbf{a}_j$$

This suggests, but does not prove, that to find a least-squares solution of the linear

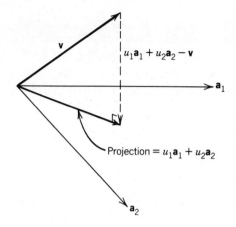

Figure 5.18   The difference of **v** and its projection onto the plane spanned by $\mathbf{a}_1$ and $\mathbf{a}_2$ is orthogonal to $\mathbf{a}_1$ and $\mathbf{a}_2$

system (5.89) we need only solve the system of so-called *normal equations*:

$$\sum_{k=1}^{n} u_k(\mathbf{a}_k \cdot \mathbf{a}_j) = \mathbf{v} \cdot \mathbf{a}_j \quad \text{for } j = 1, \ldots, n \tag{5.91}$$

We will prove that this equivalence is indeed valid, but first we show that the normal equations always have a solution.

Theorem 5.14   The system (5.91) of normal equations always has a solution. This solution is unique iff the vectors $\mathbf{a}_1, \ldots, \mathbf{a}_n$ are linearly independent.

Proof   First let us suppose that $\mathbf{a}_1, \ldots, \mathbf{a}_n$ are linearly independent. We show that if M is the $n \times n$ matrix whose (k, j) entry is $\mathbf{a}_k \cdot \mathbf{a}_j$, then M is nonsingular. If

$$\sum_{j=1}^{n} (\mathbf{a}_k \cdot \mathbf{a}_j)u_j = 0 \quad \text{for } k = 1, \ldots, n$$

then

$$\left\| \sum_{j=1}^{n} u_j\mathbf{a}_j \right\|^2 = \left(\sum_{k=1}^{n} u_k\mathbf{a}_k \right) \cdot \left(\sum_{j=1}^{n} u_j\mathbf{a}_j \right)$$

$$= \sum_{k=1}^{n} u_k \left(\mathbf{a}_k \cdot \sum_{j=1}^{n} u_j\mathbf{a}_j \right)$$

$$= \sum_{k=1}^{n} u_k \left(\sum_{j=1}^{n} u_j(\mathbf{a}_k \cdot \mathbf{a}_j) \right)$$

$$= \sum_{k=1}^{n} u_k(0) = 0$$

Therefore, $\sum_{j=1}^{n} u_j\mathbf{a}_j = 0$, and so $u_1 = \cdots = u_n = 0$ since $\mathbf{a}_1, \ldots, \mathbf{a}_n$ are linearly independent. This proves that M is nonsingular and thus that the normal equations (5.91) have a unique solution.

Next let us suppose that $\mathbf{a}_1, \ldots, \mathbf{a}_n$ are linearly dependent. Then we can select a proper subset of $\{\mathbf{a}_1, \ldots, \mathbf{a}_n\}$ consisting of vectors that are linearly independent and yet span the same subspace as $a_1, \ldots, a_n$. Furthermore, by rearranging the subscripts, we may assume that, for some $p < n$, $a_1, \ldots, a_p$ are linearly independent and that $a_{p+1}, \ldots, a_n$ are linear combinations of $a_1, \ldots, a_p$. Then, by the preceding paragraph, we know that

$$\sum_{k=1}^{p} u_k(a_k \cdot a_j) = v \cdot a_j \quad \text{for } j = 1, \ldots, p$$

has a solution $(u_1, \ldots, u_p)$. If we next choose $u_{p+1} = \cdots = u_n = 0$, it follows that

$$\sum_{k=1}^{n} u_k(a_k \cdot a_j) = v \cdot a_j \quad \text{for } j = 1, \ldots, p$$

And, since $a_{p+1}, \ldots, a_n$ are linear combinations of $a_1, \ldots, a_p$, these last equations also hold for $j = p + 1, \ldots, n$. Finally, we show that the solution $(u_1, \ldots, u_n)$ is not unique. Since $a_1, \ldots, a_n$ are linearly dependent, there exist scalars $w_1, \ldots, w_n$, not all zero, such that $\sum_{k=1}^{n} w_k a_k = 0$. Therefore

$$\sum_{k=1}^{n} (u_k + w_k)(a_k \cdot a_j) = \sum_{k=1}^{n} u_k(a_k \cdot a_j) + \left(\sum_{k=1}^{n} w_k a_k \right) \cdot a_j$$

$$= v \cdot a_j + 0 \cdot a_j = v \cdot a_j$$

and so $(u_1 + w_1, \ldots, u_n + w_n)$ is another solution of the normal equations.

Theorem 5.15   The n-tuple $(u_1, \ldots, u_n)$ is a least-squares solution of the linear system (5.89) iff $(u_1, \ldots, u_n)$ is a solution of the system of normal equations (5.91).

Proof   First suppose that $(u_1, \ldots, u_n)$ is a solution of the system of normal equations. We use the Pythagorean inequality (Theorem 5.7e) to verify the inequality (5.90); thus, we need only check that the following inner product is zero:

$$\left(\sum_{k=1}^{n} u_k a_k - v \right) \cdot \left[\left(\sum_{k=1}^{n} w_k a_k - v \right) - \left(\sum_{k=1}^{n} u_k a_k - v \right) \right]$$

$$= \left(\sum_{k=1}^{n} u_k a_k - v \right) \cdot \left(\sum_{j=1}^{n} w_j a_j - \sum_{j=1}^{n} u_j a_j \right)$$

$$= \left(\sum_{k=1}^{n} u_k a_k - v \right) \cdot \sum_{j=1}^{n} (w_j - u_j) a_j$$

$$= \sum_{j=1}^{n} (w_j - u_j) \left(\sum_{k=1}^{n} u_k a_k - v \right) \cdot a_j$$

$$= \sum_{j=1}^{n} (w_j - u_j) \left(\sum_{k=1}^{n} u_k(a_k \cdot a_j) - v \cdot a_j \right)$$

$$= \sum_{j=1}^{n} (w_j - u_j)(0) = 0 \quad \text{(by (5.91))}$$

Conversely, suppose that $(u_1, \ldots, u_n)$ is a least-squares solution of the linear system (5.89). We know from Theorem 5.14 that the normal equations have some solution $(w_1, \ldots, w_n)$, and we know from the preceding paragraph that $(w_1, \ldots, w_n)$ is also a least-squares solution of (5.89). Thus, $\sum_{k=1}^{n} w_k \mathbf{a}_k - \mathbf{v}$ and $\sum_{k=1}^{n} u_k \mathbf{a}_k - \mathbf{v}$ have the same length, and therefore, by the Pythagorean inequality, they are the same vector. It then follows that, for $j = 1, \ldots, n$

$$\sum_{k=1}^{n} u_k(\mathbf{a}_k \cdot \mathbf{a}_j) - \mathbf{v} \cdot \mathbf{a}_j = \left(\sum_{k=1}^{n} u_k \mathbf{a}_k - \mathbf{v} \right) \cdot \mathbf{a}_j$$

$$= \left(\sum_{k=1}^{n} w_k \mathbf{a}_k - \mathbf{v} \right) \cdot \mathbf{a}_j$$

$$= \sum_{k=1}^{n} w_k(\mathbf{a}_k \cdot \mathbf{a}_j) - \mathbf{v} \cdot \mathbf{a}_j = 0$$

Thus, $(u_1, \ldots, u_n)$ is also a solution of the normal equations.

The above concepts and results can be expressed most succinctly in matrix notation. Let

$$A = \begin{bmatrix} a_{11} & \cdots & a_{1n} \\ \vdots & & \vdots \\ a_{m1} & \cdots & a_{mn} \end{bmatrix} \qquad U = \begin{bmatrix} u_1 \\ \vdots \\ u_n \end{bmatrix} \qquad V = \begin{bmatrix} v_1 \\ \vdots \\ v_m \end{bmatrix}$$

In other words, let A be the matrix whose kth column consists of the coordinates of $\mathbf{a}_k$, and U and V be the one-column matrices whose entries are the coordinates of $(u_1, \ldots, u_n)$ and $\mathbf{v}$, respectively. Then the linear system (5.89) and its related system of normal equations (5.91) take the respective forms

$$AU = V \qquad A^{\mathrm{T}}AU = A^{\mathrm{T}}V$$

where A^{T} denotes the transpose of A (see Exercise 5).

EXAMPLE 5.25 Find the least-squares solution of

$$u_1 = 1$$
$$u_2 = 2$$
$$u_1 - u_2 = 0$$

Then $A^{\mathrm{T}}AU = A^{\mathrm{T}}V$ becomes

$$\begin{bmatrix} 1 & 0 & 1 \\ 0 & 1 & -1 \end{bmatrix} \begin{bmatrix} 1 & 0 \\ 0 & 1 \\ 1 & -1 \end{bmatrix} \begin{bmatrix} u_1 \\ u_2 \end{bmatrix} = \begin{bmatrix} 1 & 0 & 1 \\ 0 & 1 & -1 \end{bmatrix} \begin{bmatrix} 1 \\ 2 \\ 0 \end{bmatrix}$$

or

$$\begin{bmatrix} 2 & -1 \\ -1 & 2 \end{bmatrix} \begin{bmatrix} u_1 \\ u_2 \end{bmatrix} = \begin{bmatrix} 1 \\ 2 \end{bmatrix}$$

and so $(u_1, u_2) = (\frac{4}{3}, \frac{5}{3})$.

The normal equations are much easier to solve when the columns $\mathbf{a}_1, \ldots, \mathbf{a}_n$ of A form an orthogonal sequence. Then equations (5.91) can be rewritten as

$$\mathbf{v} \cdot \mathbf{a}_j = \sum_{k=1}^{n} u_k(\mathbf{a}_k \cdot \mathbf{a}_j) = u_j(\mathbf{a}_j \cdot \mathbf{a}_j) = u_j\|\mathbf{a}_j\|^2$$

and so

$$u_k = \frac{\mathbf{v} \cdot \mathbf{a}_k}{\|\mathbf{a}_k\|^2} \quad \text{for } k = 1, \ldots, n \tag{5.92}$$

Since $(\mathbf{v} \cdot \mathbf{a}_k/\|\mathbf{a}_k\|^2)\mathbf{a}_k$ is the projection of $\mathbf{v}$ on $\mathbf{a}_k$, this result can be thought of as saying that the projection $\sum_{k=1}^{n} u_k\mathbf{a}_k$ is the sum of the projections of $\mathbf{v}$ on each of the vectors $\mathbf{a}_1, \ldots, \mathbf{a}_n$ (which is not true when $\mathbf{a}_1, \ldots, \mathbf{a}_n$ are not orthogonal).

EXAMPLE 5.26   Find the least-squares solution of

$$u_1 + u_2 = 1$$
$$u_2 = 2$$
$$u_1 - u_2 = -1$$

Since $\mathbf{a}_1 = (1, 0, 1)$ and $\mathbf{a}_2 = (1, 1, -1)$ are orthogonal,

$$u_1 = \frac{\mathbf{v} \cdot \mathbf{a}_1}{\|\mathbf{a}_1\|^2} = \frac{(1, 2, -1) \cdot (1, 0, 1)}{\|(1, 0, 1)\|^2} = 0$$

$$u_2 = \frac{\mathbf{v} \cdot \mathbf{a}_2}{\|\mathbf{a}_2\|^2} = \frac{(1, 2, -1) \cdot (1, 1, -1)}{\|(1, 1, -1)\|^2} = \frac{4}{3}$$

This method is so appealing that, even when the columns $\mathbf{a}_1, \ldots, \mathbf{a}_n$ are not orthogonal, we try to replace them by vectors that are orthogonal. Specifically, we seek vectors $\mathbf{b}_1, \ldots, \mathbf{b}_n$ that span the same subspace of R^m as $\mathbf{a}_1, \ldots, \mathbf{a}_n$ but are orthogonal. One method for doing this is called the *Gram–Schmidt process*, and the idea behind it again involves projections. We assume that $\mathbf{a}_1, \ldots, \mathbf{a}_n$ are linearly independent.

Gram–Schmidt process: As Figure 5.19 suggests, we choose $\mathbf{b}_1$ to be $\mathbf{a}_1$, and we choose $\mathbf{b}_2$ to be the difference between $\mathbf{a}_2$ and the projection of $\mathbf{a}_2$ on $\mathbf{b}_1$. Thus

$$\mathbf{b}_1 = \mathbf{a}_1 \quad \text{and} \quad \mathbf{b}_2 = \mathbf{a}_2 - \frac{\mathbf{a}_2 \cdot \mathbf{b}_1}{\|\mathbf{b}_1\|^2} \mathbf{b}_1$$

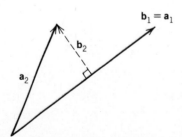

Figure 5.19 The first two vectors produced by the Gram–Schmidt process

and by Theorem 5.7c, $\mathbf{b}_1 \cdot \mathbf{b}_2 = 0$. We construct $\mathbf{b}_3, \ldots, \mathbf{b}_n$ in like fashion; that is, we subtract from $\mathbf{a}_k$ its projections onto $\mathbf{b}_1, \ldots, \mathbf{b}_{k-1}$.

The next theorem summarizes this process; its proof is straightforward (Exercise 8).

Theorem 5.16 Suppose $\mathbf{a}_1, \ldots, \mathbf{a}_n$ are linearly independent vectors in $\boldsymbol{R}^m$, and $\mathbf{b}_1, \ldots, \mathbf{b}_n$ are defined by

$$\mathbf{b}_1 = \mathbf{a}_1$$

and, for $k = 2, \ldots, n$

$$\mathbf{b}_k = \mathbf{a}_k - \frac{\mathbf{a}_k \cdot \mathbf{b}_1}{\|\mathbf{b}_1\|^2}\mathbf{b}_1 - \cdots - \frac{\mathbf{a}_k \cdot \mathbf{b}_{k-1}}{\|\mathbf{b}_{k-1}\|^2}\mathbf{b}_{k-1}$$

Then $\mathbf{b}_1, \ldots, \mathbf{b}_n$ is an orthogonal sequence that spans the same subspace of $\boldsymbol{R}^m$ as $\mathbf{a}_1, \ldots, \mathbf{a}_n$.

EXAMPLE 5.27 Use the Gram–Schmidt process to construct an orthonormal sequence from

$$\mathbf{a}_1 = (1, 0, -1), \qquad \mathbf{a}_2 = (1, -1, 3), \qquad \mathbf{a}_3 = (0, -1, 1)$$

Then

$$\mathbf{b}_1 = \mathbf{a}_1 = (1, 0, -1) \quad \text{and so } \|\mathbf{b}_1\|^2 = 2$$

$$\mathbf{b}_2 = \mathbf{a}_2 - \frac{\mathbf{a}_2 \cdot \mathbf{b}_1}{\|\mathbf{b}_1\|^2}\mathbf{b}_1 = \mathbf{a}_2 + \mathbf{b}_1 = (2, -1, 2)$$

and so

$$\|\mathbf{b}_2\|^2 = 9$$

$$\mathbf{b}_3 = \mathbf{a}_3 - \frac{\mathbf{a}_3 \cdot \mathbf{b}_1}{\|\mathbf{b}_1\|^2}\mathbf{b}_1 - \frac{\mathbf{a}_3 \cdot \mathbf{b}_2}{\|\mathbf{b}_2\|^2}\mathbf{b}_2 = \mathbf{a}_3 + \tfrac{1}{2}\mathbf{b}_1 - \tfrac{3}{9}\mathbf{b}_2$$

$$= (0, -1, 1) + (\tfrac{1}{2}, 0, -\tfrac{1}{2}) - (\tfrac{2}{3}, -\tfrac{1}{3}, \tfrac{2}{3}) = (-\tfrac{1}{6}, -\tfrac{2}{3}, -\tfrac{1}{6})$$

and so

$$\|\mathbf{b}_3\|^2 = \frac{1 + 16 + 1}{36} = \frac{1}{2}$$

Finally, we convert the orthogonal sequence $\mathbf{b}_1, \mathbf{b}_2, \mathbf{b}_3$ into an orthonormal sequence $\mathbf{q}_1, \mathbf{q}_2, \mathbf{q}_3$ by dividing each vector $\mathbf{b}_k$ by its length:

$$\mathbf{q}_1 = \frac{1}{\sqrt{2}}\mathbf{b}_1 = \left(\frac{1}{\sqrt{2}}, 0, -\frac{1}{\sqrt{2}}\right), \qquad \mathbf{q}_2 = \frac{1}{3}\mathbf{b}_2 = \left(\frac{2}{3}, -\frac{1}{3}, \frac{2}{3}\right),$$

$$\mathbf{q}_3 = \sqrt{2}\,\mathbf{b}_3 = \left(-\frac{\sqrt{2}}{6}, -\frac{2\sqrt{2}}{3}, -\frac{\sqrt{2}}{6}\right)$$

There is also a matrix formulation of the Gram–Schmidt process. We illustrate it by rewriting the conclusion of the preceding example in matrix form:

$$\mathbf{a}_1 = \mathbf{b}_1 = \sqrt{2}\ \mathbf{q}_1 = \sqrt{2}\ \mathbf{q}_1 + 0\mathbf{q}_2 + 0\mathbf{q}_3$$

$$\mathbf{a}_2 = -\mathbf{b}_1 + \mathbf{b}_2 = -\sqrt{2}\ \mathbf{q}_1 + 3\mathbf{q}_2 + 0\mathbf{q}_3$$

$$\mathbf{a}_3 = -\frac{1}{2}\mathbf{b}_1 + \frac{1}{3}\mathbf{b}_2 + \mathbf{b}_3 = -\frac{1}{\sqrt{2}}\ \mathbf{q}_1 + \mathbf{q}_2 + \frac{1}{\sqrt{2}}\mathbf{q}_3$$

Therefore, if A and Q denote the matrices whose columns are $\mathbf{a}_1, \mathbf{a}_2, \mathbf{a}_3$ and $\mathbf{q}_1, \mathbf{q}_2, \mathbf{q}_3$, respectively, then

$$A = QR \quad \text{where } R = \begin{bmatrix} \sqrt{2} & -\sqrt{2} & -\dfrac{1}{\sqrt{2}} \\ 0 & 3 & 1 \\ 0 & 0 & \dfrac{1}{\sqrt{2}} \end{bmatrix} \tag{5.93}$$

The next theorem is just this matrix formulation of the Gram–Schmidt process.

Theorem 5.17 (*QR* factorization)   If the columns of a matrix A are linearly independent, then A can be factored as $A = QR$, where Q is a matrix whose columns form an orthonormal sequence and R is an upper triangular, invertible matrix.

The QR factorization of A leads to a method for finding a least-squares solution of $AU = V$ that is closer to the method of Example 5.26 than to that of Example 5.25. To see this resemblance, we rewrite the normal equations as follows:

$$A^{\mathrm{T}}AU = A^{\mathrm{T}}V$$

or, since $A = QR$,

$$(QR)^{\mathrm{T}}(QR)U = (QR)^{\mathrm{T}}V$$

or, since $(BC)^{\mathrm{T}} = C^{\mathrm{T}}B^{\mathrm{T}}$,

$$R^{\mathrm{T}}Q^{\mathrm{T}}QRU = R^{\mathrm{T}}Q^{\mathrm{T}}V$$

or, since $Q^{\mathrm{T}}Q =$ identity (Exercise 5),

$$R^{\mathrm{T}}RU = R^{\mathrm{T}}Q^{\mathrm{T}}V$$

or, since R, and thus R^{T}, is invertible,

$$RU = Q^{\mathrm{T}}V \tag{5.94}$$

Even if the columns of A are not linearly independent, an analogous formula can be found.

EXAMPLE 5.28   Use formula (5.94) to find the least-squares solution of

$$u_1 + u_2 = 3$$

$$-u_2 = -2$$

$$-u_1 + 3u_2 = -1$$

Then $\mathbf{a}_1 = (1, 0, -1)$ and $\mathbf{a}_2 = (1, -1, 3)$, which are the first two vectors of Example 5.27. Therefore, (5.94) becomes

$$\begin{bmatrix} \sqrt{2} & -\sqrt{2} \\ 0 & 3 \end{bmatrix}\begin{bmatrix} u_1 \\ u_2 \end{bmatrix} = \begin{bmatrix} \dfrac{1}{\sqrt{2}} & 0 & -\dfrac{1}{\sqrt{2}} \\ 2/3 & -1/3 & 2/3 \end{bmatrix}\begin{bmatrix} 3 \\ -2 \\ -1 \end{bmatrix} = \begin{bmatrix} 2\sqrt{2} \\ 2 \end{bmatrix}$$

and so, by back substitution,

$$3u_2 = 2 \qquad \text{which implies } u_2 = \tfrac{2}{3}$$
$$\sqrt{2}\,u_1 - \sqrt{2}\,u_2 = 2\sqrt{2} \qquad \text{which implies } u_1 = \tfrac{2}{3} + 2 = \tfrac{8}{3}$$

The method used in Example 5.28 resembles the method in Example 5.26 in that the product $Q^{\mathrm{T}}V$ was computed in both. The matrix R, of course, was the identity matrix in Example 5.26 but not in Example 5.28. Still, the system (5.94) is always easy to solve; since R is upper triangular, we can find one unknown at a time by solving the last equation first, then the next-to-last, and so on (which is called "back substitution"). It cannot be claimed, however, that the method is faster than that of Example 5.25; although solving the system (5.94) is fast, finding the matrices Q and R is not. The real advantage of the QR-factorization is that, when it is implemented properly, it does a very good job of controlling the accumulation of round-off error.

Any judgment about the speed or reliability of a numerical algorithm, however, requires a precise statement and careful analysis and testing of the algorithm. Since

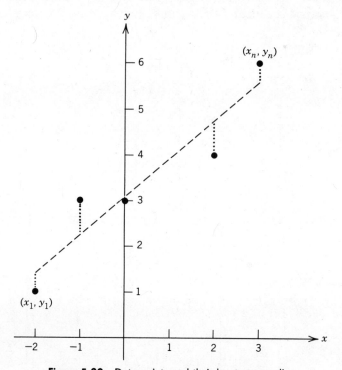

Figure 5.20 Data points and their least-squares line

even a precise statement of a good algorithm for performing the QR-factorization would take a lengthy development, we will carry this discussion no further, except to give one hint. The Gram–Schmidt process is actually not the best method for performing the QR-factorization; there are other ways to accomplish it that lead to less round-off error. One of the best is to construct Q^T as a product of matrices called "Householder transformations," which have the form $2UU^T - I$ where U is a one-column matrix whose entries are the coordinates of a vector of length 1 and I is the identity matrix [see the discussions in Stewart (1973) and the excellent algorithms in Wilkinson and Reinsch (1971)].

Least-squares solutions of linear systems occur most commonly in problems that involve fitting equations to real data. For example, suppose we have a collection of at least two data points $(x_1, y_1), \ldots, (x_n, y_n)$, where all the x coordinates are distinct and the points appear to be close to a straight line $y = ax + b$ (Figure 5.20). If the points were all exactly on the line $y = ax + b$, we would have the following system of n equations and 2 unknowns:

$$
\begin{aligned}
ax_1 + b &= y_1 \\
&\vdots \\
ax_n + b &= y_n
\end{aligned}
\tag{5.95}
$$

But since real data are unlikely to be so well behaved, we do not expect the system (5.95) to have a solution (a, b) when $n > 2$. As we showed in Theorems 5.14 and 5.15, however, the system does have a least-squares solution (a, b), and this solution satisfies the following normal equations:

$$
\begin{bmatrix} x_1 & \cdots & x_n \\ 1 & \cdots & 1 \end{bmatrix}
\begin{bmatrix} x_1 & 1 \\ \vdots & \vdots \\ x_n & 1 \end{bmatrix}
\begin{bmatrix} a \\ b \end{bmatrix}
=
\begin{bmatrix} x_1 & \cdots & x_n \\ 1 & \cdots & 1 \end{bmatrix}
\begin{bmatrix} y_1 \\ \vdots \\ y_n \end{bmatrix}
$$

or

$$
\begin{bmatrix} \sum_{k=1}^{n} x_k^2 & \sum_{k=1}^{n} x_k \\ \sum_{k=1}^{n} x_k & n \end{bmatrix}
\begin{bmatrix} a \\ b \end{bmatrix}
=
\begin{bmatrix} \sum_{k=1}^{n} x_k y_k \\ \sum_{k=1}^{n} y_k \end{bmatrix}
$$

or

$$
\begin{bmatrix} a \\ b \end{bmatrix}
=
\frac{1}{n \sum_{k=1}^{n} x_k^2 - \left(\sum_{k=1}^{n} x_k \right)^2}
\begin{bmatrix} n & -\sum_{k=1}^{n} x_k \\ -\sum_{k=1}^{n} x_k & \sum_{k=1}^{n} x_k^2 \end{bmatrix}
\begin{bmatrix} \sum_{k=1}^{n} x_k y_k \\ \sum_{k=1}^{n} y_k \end{bmatrix}
$$

$$
=
\frac{1}{n \sum_{k=1}^{n} x_k^2 - \left(\sum_{k=1}^{n} x_k \right)^2}
\begin{bmatrix} n \sum_{k=1}^{n} x_k y_k & - \left(\sum_{k=1}^{n} x_k \right)\left(\sum_{k=1}^{n} y_k \right) \\ \left(\sum_{k=1}^{n} x_k^2 \right)\left(\sum_{k=1}^{n} y_k \right) & - \left(\sum_{k=1}^{n} x_k \right)\left(\sum_{k=1}^{n} x_k y_k \right) \end{bmatrix}
$$

These are the usual formulas for finding the coefficients of the least-squares line for a set of data points. Geometrically, this is the line for which the sum of the squares of the vertical distances marked in Figures 5.20 is least.

EXAMPLE 5.29 Find the least-squares line for the data points $(-2, 1)$, $(-1, 3)$, $(0, 3)$, $(2, 4)$, $(3, 6)$.

Although we could use the formulas just derived, it is quite easy to use the normal equations themselves:

$$\begin{bmatrix} -2 & -1 & 0 & 2 & 3 \\ 1 & 1 & 1 & 1 & 1 \end{bmatrix} \begin{bmatrix} -2 & 1 \\ -1 & 1 \\ 0 & 1 \\ 2 & 1 \\ 3 & 1 \end{bmatrix} \begin{bmatrix} a \\ b \end{bmatrix} = \begin{bmatrix} -2 & -1 & 0 & 2 & 3 \\ 1 & 1 & 1 & 1 & 1 \end{bmatrix} \begin{bmatrix} 1 \\ 3 \\ 3 \\ 4 \\ 6 \end{bmatrix}$$

or

$$\begin{bmatrix} 18 & 2 \\ 2 & 5 \end{bmatrix} \begin{bmatrix} a \\ b \end{bmatrix} = \begin{bmatrix} 21 \\ 17 \end{bmatrix}$$

or

$$\begin{bmatrix} a \\ b \end{bmatrix} = \frac{1}{90 - 4} \begin{bmatrix} 5 & -2 \\ -2 & 18 \end{bmatrix} \begin{bmatrix} 21 \\ 17 \end{bmatrix} = \begin{bmatrix} 71/86 \\ 264/86 \end{bmatrix} = \begin{bmatrix} 0.83 \\ 3.07 \end{bmatrix}$$

The line $y = 0.83x + 3.07$ and the five given data points are graphed in Figure 5.20.

Similar methods can be used for fitting data to curves other than straight lines; two such examples appear in Exercises 14 and 15. Also, if our data points have more than two coordinates, we can fit them to surfaces in higher dimensional spaces. For example, suppose our data points have the form $(x_1, y_1, z_1), \ldots, (x_n, y_n, z_n)$, and suppose we have reason to believe that there exist coefficients c_1, c_2, c_3 such that

$$c_1 x_k + c_2 y_k + c_3 = z_k \quad \text{for } k = 1, \ldots, n \tag{5.96}$$

is nearly true. (We are supposing that the data points nearly lie on a plane in space.) Then we can find a least-squares solution (c_1, c_2, c_3) of this system by solving the normal equations where A has the columns $x_1, \ldots, x_n$; $y_1, \ldots, y_n$; $1, \ldots, 1$ and where V is the column $z_1, \ldots, z_n$. The numbers c_1, c_2, c_3 are then called the "multiple-regression" coefficients of the system (5.96). Multiple-regression coefficients are defined similarly for data points with any number of coordinates.

EXAMPLE 5.30 Table 5.2 contains made up data for eight test plots of Iowa corn; it suggests that the yield depends on the amounts of fertilizer and moisture (rainfall plus stored subsoil moisture) during the growing season. Find the multiple regression coefficients for yield as a linear function of fertilizer and rainfall.

We must find the least-squares solution of the equation $AU = V$, where A is the 8×3 matrix whose first two columns are the first two columns of Table 5.2 and third column is all ones, and where V is the 8×1 matrix whose entries consist of the third column of Table 5.2. Since multiple regressions are tedious to do by

Table 5.2 Corn Yield

Fertilizer (lb/acre)	Moisture (in.)	Yield (bushels)
80	24	86
100	16	72
120	21	103
140	23	114
160	12	91
180	20	137
200	18	132
220	25	170

hand, a computer package was used to solve this equation. The solution U has the coordinates

$$c_1 = 0.57 \qquad c_2 = 3.37 \qquad c_3 = -39.36$$

Therefore, the following relationship is approximately true, where z represents bushels of corn, x represents pounds of fertilizer, and y represents inches of rainfall:

$$z = 0.57x + 3.37y - 39.36$$

EXERCISES

1. Find the least-squares solution of the system

$$u_1 + u_2 + u_3 = 0$$
$$u_1 + u_2 - u_3 = 1$$
$$u_1 - u_2 + u_3 = 2$$
$$u_1 - u_2 - u_3 = 1$$

2. Use the Gram–Schmidt process to construct an orthonormal sequence from $(1, -1, 1)$, $(0, 1, 0)$, $(0, 0, 1)$.

3. (a) Use the Gram–Schmidt process to construct an orthonormal sequence from $(1, 2, 2)$, $(1, 3, 1)$.

   (b) Use part (a) to find the QR-factorization of

$$A = \begin{bmatrix} 1 & 1 \\ 2 & 3 \\ 2 & 1 \end{bmatrix}$$

   (c) Use part (b) to find the least-squares solution of

$$u_1 + u_2 = 1$$
$$2u_1 + 3u_2 = 1$$
$$2u_1 + u_2 = 1$$

   (d) Confirm your answer to (c) by checking that it satisfies the normal equations.

4. Show that the least-squares solution of the linear system

$$u_1 = v_1, \ldots, u_1 = v_n \quad (n \text{ equations, 1 unknown})$$

is $(v_1 + \cdots + v_n)/n$ (the *average* of the numbers $v_1, \ldots, v_n$).

5. Show that the (k, j) entry of $A^{\mathrm{T}}A$ is $\mathbf{a}_k \cdot \mathbf{a}_j$, where $\mathbf{a}_k$ is the vector whose coordinates make up the kth column of A.

6. Use the Gram–Schmidt process to construct an orthonormal sequence from $(2, i, 1 + i), (-i, 2 + i, 1)$. (Use the complex inner product on C^3.)

7. Define functions f, g, and h in $PC[-1, 1]$ by $f(x) = 1$, $g(x) = x$, and $h(x) = x^2$. Use the Gram–Schmidt process to construct an orthonormal sequence from f, g, h. (Use the usual integral inner product on $PC[-1, 1]$.)

8. Use the principle of induction to prove that the sequence of vectors $\mathbf{b}_1, \ldots, \mathbf{b}_n$ defined in Theorem 5.16 is an orthogonal sequence.

9. If Q is a square matrix, prove that each of the following four conditions implies the other three. (A square matrix satisfying any one of the four conditions, and thus the others, is said to be *orthogonal*.)

    (a) The columns of Q form an orthonormal sequence.

    (b) $Q^{\mathrm{T}}Q = I$ (I is the identity matrix).

    (c) $QQ^{\mathrm{T}} = I$.

    (d) The rows of Q form an orthonormal sequence.

10. Verify that every matrix of the form

$$\begin{bmatrix} \cos\theta & -\sin\theta \\ \sin\theta & \cos\theta \end{bmatrix}$$

is orthogonal. (Such a matrix is called a *rotation* matrix since the corresponding linear transformation rotates every vector in the plane through the angle θ.)

11. Verify that every matrix of the form

$$\begin{bmatrix} \cos(2\theta) & \sin(2\theta) \\ \sin(2\theta) & -\cos(2\theta) \end{bmatrix}$$

is orthogonal. (Such a matrix is called a *reflection* matrix since the corresponding linear transformation reflects every vector in the plane across the line $y = (\tan\theta)x$.)

12. Let $Q_{\mathbf{u}} = 2UU^{\mathrm{T}} - I$, where U is a one-column matrix whose entries are the coordinates of a vector $\mathbf{u}$ in R^n of length 1.

    (a) Show that $Q_{\mathbf{u}}$ is orthogonal.

    (b) Suppose that $\mathbf{u} = (\mathbf{v} + \mathbf{w})/\|\mathbf{v} + \mathbf{w}\|$, where $\|\mathbf{v}\| = \|\mathbf{w}\|$ and $\mathbf{w} \neq -\mathbf{v}$. Show that $Q_{\mathbf{u}}V = W$, where V and W are the one-column matrices whose entries are the coordinates of $\mathbf{v}$ and $\mathbf{w}$, respectively.

13. Find the least-squares line for the data points $(-2, 6)$, $(-1, 5)$, $(1, 4)$, $(2, 4)$, $(4, 3)$, $(5, 2)$. Sketch the graph of the line and the six points.

14. (Semilog plot)   Data points can be fitted to an exponential curve $y = ae^{bx}$ as follows. By applying the logarithm function to both sides of $y = ae^{bx}$, we get $\log y = \log a + bx$; so if $\log y$ is plotted against x, we get a straight line. Therefore, data points (x_k, y_k) that lie close to an exponential curve can be transformed into the points $(x_k, \log y_k)$, and then the

Table 5.3 Population of the United States

Year	Population (in millions)
1900	76
1910	92
1920	106
1930	123
1940	132
1950	151
1960	179
1970	203

least-squares line can be found for these. Find the least-squares exponential curve for the data points $(0, 3)$, $(1, 2)$, $(3, 0.9)$, $(4, 0.4)$.

15. (log–log plot) Data points can be fitted to a general power function $y = ax^b$ as follows. By applying the logarithm function to both sides of $y = ax^b$, we get $\log y = \log a + b \log x$; so if $\log y$ is plotted against $\log x$, we get a straight line. Therefore, data points (x_k, y_k) that lie close to the graph of a power function can be transformed into the points $(\log x_k, \log y_k)$, and then the least-squares line can be found for these. Find the least-squares power function for the data points $(1, 0.3)$, $(2, 1)$, $(3, 1.6)$, $(5, 3.3)$.

16. (a) Find the least-squares line for the data points in Table 5.3. (The computations will be simpler if the time scale is changed so that the years are given the values $0, 1, \ldots, 7$.)

(b) Find the least-squares exponential curve for the same data.

(c) On a single coordinate system, plot the data points and sketch the graphs of both curves. Does one curve fit the data better than the other?

17. Table 5.4 suggests that the weekly revenue of Company X depends on both the amount of money the company spends on television advertising and on the amount it spends on newspaper advertising. Find the multiple-regression coefficients for revenue as a linear

Table 5.4 Weekly Revenue of Company X

Ad costs (in thousands)		Revenue (in thousands)
Television	Newspaper	
2.0	4.1	67.5
2.4	3.5	68.0
2.7	2.6	67.5
3.0	4.6	66.0
3.3	1.8	68.0
3.6	2.3	70.0
3.8	4.3	72.5
4.2	3.1	72.0
4.7	1.5	71.0
5.1	1.2	72.0

Table 5.5 Grade-Point Average (GPA) for a Small Student Sample

SAT scores		
Math	**Verbal**	**GPA**
500	480	2.03
510	500	2.11
530	530	2.23
560	570	2.46
580	570	2.53
590	450	2.39
620	700	2.86
640	600	2.93
650	620	2.96
680	520	3.11
700	630	3.27

function of the two types of advertisement costs. (It is advisable to use a computer package to solve the linear system.)

18. Table 5.5 suggests that the grade-point averages of the 11 students depends on both their math SAT scores and their verbal SAT scores. Find the multiple-regression coefficients for GPA as a linear function of the two types of SAT scores. (It is advisable to use a computer package to solve the linear system.)

Continuous functions and mean value theorems

A.1 Continuous Functions and the Intermediate Value Theorem

Although we begin this appendix with the familiar ε–δ definition of a continuous function, we immediately reformulate the definition by using sequences. This enables us to use the results of Chapter 1 to give fairly quick proofs of some theorems about continuous functions. The functions in this appendix are always assumed to take real numbers to real numbers.

Definition A.1 Suppose x_0 is a point (i.e., number) in the domain of a function f. f is *continuous at* x_0 if for every positive real number ε there exists a positive real number δ such that

$$|f(x) - f(x_0)| < \varepsilon$$

whenever x is a point in the domain of f satisfying

$$|x - x_0| < \delta$$

Roughly speaking, the next theorem says that a function f is continuous at a point x_0 iff f takes points near x_0 to points near $f(x_0)$.

Theorem A.1 Suppose x_0 is a point in the domain of a function f. f is continuous at x_0 iff $f(t_n) \to f(x_0)$ whenever $\{t_n\}$ is a sequence of points in the domain of f satisfying $t_n \to x_0$.

Proof First, suppose f is continuous at x_0. If $\{t_n\}$ is a sequence of points in the domain of f such that $t_n \to x_0$, we must show that $f(t_n) \to f(x_0)$. But, for every $\varepsilon > 0$, there exists $\delta > 0$ such that

$$|f(x) - f(x_0)| < \varepsilon \quad \text{whenever } |x - x_0| < \delta$$

Furthermore, since $t_n \to x_0$, there exists N such that

$$|t_n - x_0| < \delta \quad \text{for all } n \geq N$$

Therefore, putting these two inequalities together, we see that

$$|f(t_n) - f(x_0)| < \varepsilon \quad \text{for all } n \geq N$$

Hence, $f(t_n) \to f(x_0)$.

To prove the converse, we suppose that f is not continuous at x_0. Thus, there exists $\varepsilon > 0$ such that, for every $\delta > 0$, there exists a point x in the domain of f such that

$$|x - x_0| < \delta \quad \text{but} \quad |f(x) - f(x_0)| \geq \varepsilon$$

We use this result for the following values of δ: $1, 1/2, \ldots, 1/n, \ldots$. That is, for every positive integer n, there exists a point t_n in the domain of f such that

$$|t_n - x_0| < \frac{1}{n} \quad \text{but} \quad |f(t_n) - f(x_0)| \geq \varepsilon$$

Therefore, $t_n \to x_0$ but $\{ f(t_n) \}$ does not converge to $f(x_0)$; this completes the proof of the theorem.

The next three theorems are now easy consequences of Theorems A.1 and 1.4; we leave most of their proofs to the exercises.

Theorem A.2   The functions f and g defined by

(a) $f(x) = x$ for all real numbers x
(b) $g(x) = y_0$ for all real numbers x (y_0 a fixed real number)

are continuous at every point of the real line.

Theorem A.3   If f and g are continuous at x_0, then (a) $f + g$, (b) $f - g$, (c) fg, and (d) f/g are continuous at x_0. (In (d) we must also assume that $g(x_0) \neq 0$.)

Proof of (a)   Suppose $\{ t_n \}$ is a sequence of points in the domain of both f and g such that $t_n \to x_0$. Then, by Theorem A.1,

$$f(t_n) \to f(x_0) \quad \text{and} \quad g(t_n) \to g(x_0)$$

and so, by Theorem 1.4b,

$$f(t_n) + g(t_n) \to f(x_0) + g(x_0)$$

Therefore $f + g$ is continuous at x_0 (Theorem A.1).

Theorem A.4   If f is continuous at x_0 and g is continuous at $f(x_0)$, then their composite $g \circ f$, defined by $(g \circ f)(x) = g(f(x))$, is continuous at x_0.

We next use Theorem A.1 to prove a fundamental property of continuous functions called the *intermediate value theorem*. First we prove a special case that is used frequently in discussions of approximate solutions of equations $f(x) = 0$. In fact, the proof of Theorem A.5 uses the bisection method of Section 2.1.

Theorem A.5   Suppose f is continuous at every point in the closed interval $[a, b]$. If $f(a)$ and $f(b)$ have opposite signs, then $f(z) = 0$ for some point z in the open interval (a, b).

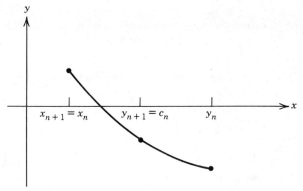

Figure A.1   The definition of x_{n+1} and y_{n+1} in the proof of Theorem A.5 when $f(c_n) < 0$

Proof   We assume that $f(a) > 0$ and $f(b) < 0$; the other case is proved similarly. We define, recursively, two sequences $\{x_n\}$ and $\{y_n\}$ so that x_n and y_n are on opposite sides of the desired number z and so that the distance between x_n and y_n is half the distance between their previous values. Specifically, let $x_1 = a$ and $y_1 = b$; and for each $n \geq 1$, let $c_n = (x_n + y_n)/2$ (the midpoint of x_n and y_n) and let (see Figure A.1)

$$x_{n+1} = x_n \quad \text{and} \quad y_{n+1} = c_n \quad \text{if } f(c_n) < 0$$

$$x_{n+1} = c_n \quad \text{and} \quad y_{n+1} = y_n \quad \text{if } f(c_n) \geq 0$$

Then $\{x_n\}$ is a monotonically increasing bounded sequence and $\{y_n\}$ is a monotonically decreasing bounded sequence. Therefore, by Theorem 3.6, $x_n \to x$ and $y_n \to y$, for some numbers x and y; so $f(x_n) \to f(x)$ and $f(y_n) \to f(y)$ (Theorem A.1). But we selected x_n and y_n so that $f(x_n) \geq 0$ and $f(y_n) \leq 0$; hence, $f(x) \geq 0$ and $f(y) \leq 0$ (Exercises 17 and 18 of Section 1.2). We also selected x_n and y_n so that $|y_n - x_n| = (b - a)/2^{n-1}$; hence, $y_n - x_n \to 0$, and so $y = x$. Therefore, if we let z denote the common value of x and y, we can write $0 \leq f(x) = f(z) = f(y) \leq 0$ and so conclude that $f(z) = 0$.

Corollary A.6 (Intermediate value theorem)   Suppose f is continuous at every point in the closed interval $[a, b]$. If y_0 is any point strictly between $f(a)$ and $f(b)$, then $f(x_0) = y_0$ for some point x_0 in the open interval (a, b).

Proof   Define a function h by $h(x) = f(x) - y_0$ for all x in $[a, b]$. Then $h(a)$ and $h(b)$ have opposite signs, and h is continuous (Theorems A.2b and A.3b). Therefore, by Theorem A.5, $h(x_0) = 0$ for some x_0 in (a, b), and so $f(x_0) = y_0$.

EXERCISES

1. Prove Theorem A.2.
2. Prove Theorems A.3b, A.3c, and A.3d.
3. Prove Theorem A.4.

4. Let $f(x) = x \cos(3x^2 - 5)$. Use Theorems A.2 through A.4 to prove that f is continuous at every point of the real line. (You may assume that the cosine function is continuous.)

5. Let $f(x) = 2x^2 + 3$ if $x \geq 1$ and $f(x) = 3 - 2x$ if $x < 1$. Show that f is discontinuous (i.e., not continuous) at $x = 1$.

6. Prove that all polynomials are continuous at every point of the real line. (Use induction on the degree of the polynomial.)

7. If $f(x) = x^3 + 3x + 1$, show that $f(x) = 0$ for at least one value of x.

8. If $f(x) = x^4 - 3x^2 + x - 2$, show that $f(x) = 0$ for at least two values of x.

9. Suppose f is continuous at every point of the closed interval $[a, b]$. Show that f has a "continuous extension" to the entire real line; that is, show there exists a function g that is defined and continuous at every point of the real line and equals f at every point of $[a, b]$.

10. Give an example of a function that is defined and continuous at every point of some interval but does not have a continuous extension to the entire real line (see Exercise 9).

11. Suppose f is continuous at every point of the closed interval $[0, 1]$ and $0 \leq f(x) \leq 1$ whenever $0 \leq x \leq 1$. Prove that $f(x) = x$ for at least one value of x in $[0, 1]$. *Hint:* Show $f(x) - x = 0$ for some x in $[0, 1]$.

A.2 Two Fundamental Theorems on Continuous Functions

Two of the deepest properties of a continuous function whose domain is a closed finite interval are that the function has a maximum and a minimum value and that the number δ in Definition A.1 can be chosen independently of the point x_0. To help us prove these properties, we first prove a more basic result (due to the nineteenth century mathematicians Bolzano and Weierstrass) about sequences of points in a closed finite interval.

Definition A.2   The sequence $\{u_n\}$ is a *subsequence* of the sequence $\{x_n\}$ if there exist subscripts $n_1 < n_2 < n_3 < \cdots$ such that

$$u_1 = x_{n_1}, u_2 = x_{n_2}, u_3 = x_{n_3}, \ldots$$

For example, if $x_n = 1/n$ then $1, 1/4, 1/7, \ldots, 1/(3n - 2), \ldots$ is one of its many subsequences. We leave the proof of the first small theorem to the exercises.

Theorem A.7   If $x_n \to x$ then every subsequence of $\{x_n\}$ also converges to the limit x.

Theorem A.8 (Bolzano–Weierstrass theorem)   If $\{x_n\}$ is a sequence of points in the closed interval $[a, b]$, then $\{x_n\}$ has a subsequence that converges to a limit in $[a, b]$.

Proof   We define, recursively, two sequences $\{y_n\}$ and $\{z_n\}$ so that the interval $[y_n, z_n]$ contains infinitely many terms of the sequence $\{x_n\}$ and so that the distance between y_n and z_n is half the distance between their previous values. Specifically, let $y_1 = a$ and $z_1 = b$; and for each $n \geq 1$, let $c_n = (y_n + z_n)/2$ (the midpoint of y_n and

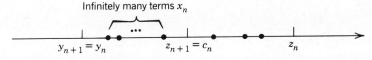

Infinitely many terms x_n

Figure A.2   The definition of y_{n+1} and z_{n+1} in the proof of Theorem A.8 when $[y_n, c_n]$ contains infinitely many terms of $\{x_n\}$

z_n) and let (see Figure A.2)

$$y_{n+1} = y_n \quad \text{and} \quad z_{n+1} = c_n \quad \text{if } [\, y_n, c_n] \text{ contains infinitely many terms of } \{x_n\}$$

$$y_{n+1} = c_n \quad \text{and} \quad z_{n+1} = z_n \quad \text{otherwise}$$

(Here, when we refer to "infinitely many terms" of x_n, we mean that terms with different subscripts are to be regarded as different terms even if they have the same value.) Next, we define a subsequence $\{u_n\}$ of $\{x_n\}$ by letting $u_1 = x_1$ and, for $n \geq 2$, letting u_n be any term of $\{x_n\}$ in $[\, y_n, z_n]$ other than the terms $u_1, u_2, \ldots, u_{n-1}$. Note that this is possible since $[\, y_n, z_n]$ contains infinitely many terms of $\{x_n\}$. Furthermore, if $m > n$ then u_m is also in $[\, y_n, z_n]$ (since $[\, y_m, z_m] \subset [\, y_n, z_n]$), and so

$$|u_m - u_n| \leq \text{length of } [\, y_n, z_n] = (b - a)/2^{n-1}$$

Since $(b - a)/2^{n-1} \to 0$, it follows from Exercise 15 of Section 3.3 that the sequence $\{u_n\}$ converges to a limit x. Finally, $a \leq x \leq b$, according to Exercises 17 and 18 of Section 1.2.

Now we prove the two promised properties of continuous functions.

Theorem A.9   If f is continuous at every point of the closed interval $[a, b]$, then f has a maximum and a minimum value. (That is, the set of points $\{ f(x) | x \in [a, b] \}$ has a largest and a smallest member.)

Proof   Here is a likely candidate for the maximum value of f: $M = \text{lub} \{ f(x) | x \in [a, b] \}$. (Recall from Definition 3.3 that $\text{lub} \, A$ denotes the "least upper bound" of the set A, by which we mean the smallest of the upper bounds of the set A. Recall also that the Axiom of Completeness of Section 3.3 assures us that the least upper bound of A exists whenever A is nonempty and has an upper bound. If $\{ f(x) | x \in [a, b] \}$ has no upper bound, we let $M = \infty$.) It remains only to show that $M = f(x)$ for some x in the interval $[a, b]$. But, by Exercise 10 of Section 3.3, we may select a sequence $\{x_n\}$ in $[a, b]$ such that $f(x_n) \to M$; this is even true if $M = \infty$. Then, by Theorem A.8, $\{x_n\}$ has a subsequence $\{u_n\}$ that converges to a limit x in $[a, b]$. And since f is continuous at x, it follows from Theorem A.1 that $f(u_n) \to f(x)$. On the other hand, $f(u_n) \to M$ (Theorem A.7), and so $M = f(x)$. Therefore, M is the maximum value of f. Similarly, it can be shown that f has a minimum value.

Theorem A.10   If f is continuous at every point of the closed interval $[a, b]$, then f has the following "uniform continuity" property on $[a, b]$: For every positive real number ε there exists a positive real number δ such that

$$|f(x_1) - f(x_2)| < \varepsilon$$

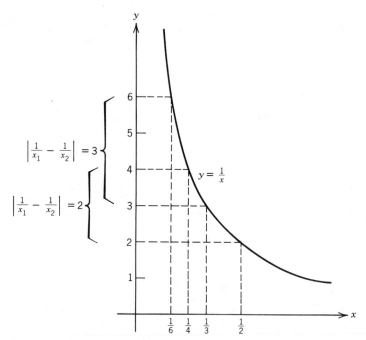

Figure A.3   If $x_1 = \delta/2$ and $x_2 = \delta$ then $|1/x_1 - 1/x_2| \geq 1$. ($\delta = \frac{1}{2}$ and $\delta = \frac{1}{3}$ are illustrated.)

whenever x_1 and x_2 are any two points in $[a, b]$ satisfying

$$|x_1 - x_2| < \delta$$

(Here δ depends only on ε, not on x_1 or x_2.)

Before we prove this theorem, we discuss an example that illustrates the difference between continuity and uniform continuity. The function defined by $f(x) = 1/x$ is certainly continuous at every point of the interval $(0, 1]$ (Theorems A.2 and A.3d). But we cannot conclude from Theorem A.10 that f has the uniform continuity property on $(0, 1]$, since that interval is not closed. In fact, Figure A.3 shows that the conclusion of the theorem is false even for $\varepsilon = 1$; that is, there is no positive real number δ such that

$$\left| \frac{1}{x_1} - \frac{1}{x_2} \right| < 1 \quad \text{whenever } |x_1 - x_2| < \delta \text{ and } x_1, x_2 \text{ are in } (0, 1]$$

A pair of x values for which this condition fails is $x_1 = \delta/2$, $x_2 = \delta$ (assuming $\delta \leq 1$).

Proof of Theorem A.10   Suppose f does not have the uniform continuity property on $[a, b]$. Then there exists $\varepsilon > 0$ such that for every $\delta > 0$ there are points x and y in $[a, b]$ satisfying

$$|x - y| < \delta \quad \text{and} \quad |f(x) - f(y)| \geq \varepsilon$$

We use this result for the following values of δ: $1, 1/2, \ldots, 1/n, \ldots$. Thus, for every positive integer n, there exist points x_n and y_n in $[a, b]$ such that

$$|x_n - y_n| < \frac{1}{n} \quad \text{and} \quad |f(x_n) - f(y_n)| \geq \varepsilon \tag{A.1}$$

By Theorem A.8, we can choose a subsequence $\{u_n\}$ of $\{x_n\}$ which converges to a limit x in $[a, b]$. Then we let $\{v_n\}$ be the subsequence of $\{y_n\}$ corresponding to $\{u_n\}$; that is, if $u_n = x_{m_n}$, let $v_n = y_{m_n}$. But, by the first of the inequalities (A.1), $x_n - y_n \to 0$; hence, $u_n - v_n \to 0$ (Theorem A.7). Since $u_n \to x$ it follows that $v_n \to x$; so, by Theorem A.1, $f(u_n) \to f(x)$ and $f(v_n) \to f(x)$. Furthermore, by the second of the inequalities (A.1), $|f(u_n) - f(v_n)| \geq \varepsilon$; therefore, by letting $n \to \infty$ we see that $0 = |f(x) - f(x)| \geq \varepsilon > 0$. This contradiction proves that f has the uniform continuity property on $[a, b]$.

EXERCISES

1. Prove Theorem A.7.

2. Suppose $\{I_n\}$ is a sequence of nonempty closed bounded intervals such that $I_{n+1} \subset I_n$ for all $n \geq 1$. Prove that there exists a point x that belongs to every I_n. *Hint:* Pick one point (any point) in each I_n and apply Theorem A.8 to this sequence.

3. Find a sequence $\{I_n\}$ of nonempty intervals such that $I_{n+1} \subset I_n$ for all $n \geq 1$ and yet no point belongs to every I_n.

4. For each of the following, give an example of a continuous function f defined on the interval $(0, 1]$ and that satisfies the given condition.

   (a) f has a maximum value and a minimum value, but f is not constant.

   (b) f has a maximum value, but no minimum value.

   (c) f has neither a maximum value nor a minimum value.

5. If the domain of f is a closed bounded interval and f is continuous at every point of this domain, prove that the range of f is a closed bounded interval. (The "range" of f is $\{f(x) | x \text{ is in the domain of } f\}$.)

6. Let $f(x) = x^2$. Show that f satisfies the uniform continuity property of Theorem A.10 on every bounded interval but not the interval $[0, \infty)$.

7. Give an example of a function that is defined and continuous at every point of the interval $(0, 1]$ and is also bounded on this interval but does not satisfy the uniform continuity property of Theorem A.10 on $(0, 1]$.

8. Suppose f is defined and continuous at every point of a closed interval $[a, b]$, and let ε be a given positive number. Prove that there exists a function g that is also defined and continuous on $[a, b]$, that is piecewise linear on $[a, b]$, and that satisfies $|f(x) - g(x)| < \varepsilon$ for all x in $[a, b]$. (g is "piecewise linear" on $[a, b]$ iff $[a, b]$ can be expressed as the union of finitely many closed intervals on each of which the graph of g is a straight line.) Also show that g can be chosen so that $g(a) = f(a)$ and $g(b) = f(b)$.

A.3 Derivatives and Mean Value Theorems

The mean value theorem is our principal tool for finding estimates of the error in an approximation. We lead up to its proof by summarizing the theory of differentiation. Note that we use the idea of Theorem A.1 to give a definition of limit of a function that makes use of our results on sequences. The usual ε–δ definition appears in Exercise 1.

Definition A.3 Suppose x_0 is a point in an interval I and f is a function whose domain includes all the points of I, except perhaps x_0. We write that

$$\lim_{x \to x_0} f(x) = y_0 \quad \text{or, alternatively,} \quad f(x) \to y_0 \quad \text{as } x \to x_0$$

if $f(t_n) \to y_0$ whenever $\{t_n\}$ is a sequence of points in the domain of f (but not equal to x_0) satisfying $t_n \to x_0$. We allow the values of ∞ and $-\infty$ for x_0 and y_0.

The proof of the next theorem is similar to that of Theorem A.3.

Theorem A.11 If $f(x) \to y_1$ and $g(x) \to y_2$ as $x \to x_0$, then
  (a) $f(x) + g(x) \to y_1 + y_2$
  (b) $f(x) - g(x) \to y_1 - y_2$
  (c) $f(x)g(x) \to y_1 y_2$
  (d) $f(x)/g(x) \to y_1/y_2$
except when the form of the limit is one of the meaningless expressions $\infty - \infty$, $0 \cdot \infty$, ∞/∞, or $y_1/0$.

Definition A.4 Suppose f is defined on the open interval (a, b) and x_0 is any point in (a, b). If

$$\lim_{x \to x_0} \frac{f(x) - f(x_0)}{x - x_0}$$

exists and is a real number, we say f is *differentiable* at x_0. Then the above limit is called the *derivative* of f at x_0 and is denoted $f'(x_0)$.

We omit the proofs of the next three elementary theorems of calculus.

Theorem A.12 If f is differentiable at x_0, f is continuous at x_0.

Theorem A.13 Suppose f and g are differentiable at x_0. Then
  (a) $(f + g)'(x_0) = f'(x_0) + g'(x_0)$
  (b) $(f - g)'(x_0) = f'(x_0) - g'(x_0)$
  (c) $(fg)'(x_0) = f(x_0)g'(x_0) + g(x_0)f'(x_0)$
  (d) $(f/g)'(x_0) = [g(x_0)f'(x_0) - f(x_0)g'(x_0)]/g^2(x_0)$, provided $g(x_0) \neq 0$.
  (e) $(cf)'(x_0) = cf'(x_0)$ for every real number c.

Theorem A.14 (Chain rule) If f is differentiable at x_0 and g is differentiable at $f(x_0)$, then

$$(g \circ f)'(x_0) = g'(f(x_0))f'(x_0)$$

Theorem A.15 Suppose f is defined on the open interval (a, b) and differentiable at the point x_0 in (a, b). If $f(x_0)$ is the maximum or minimum value of f on (a, b) then $f'(x_0) = 0$.

Proof We assume that $f(x_0)$ is the maximum value of f; the other case is proved similarly. But then $f(x_0) - f(x) \geq 0$ for all x in (a, b). Hence

$$\frac{f(x) - f(x_0)}{x - x_0} = \frac{f(x_0) - f(x)}{x_0 - x} \geq 0 \quad \text{when } a < x < x_0$$

and so, by Exercise 18 of Section 1.2,

$$f'(x_0) = \lim_{x \to x_0} \frac{f(x) - f(x_0)}{x - x_0} \geq 0$$

Also

$$\frac{f(x) - f(x_0)}{x - x_0} = \frac{f(x_0) - f(x)}{x_0 - x} \leq 0 \quad \text{when } x_0 < x < b$$

and so, by Exercise 17 of Section 1.2,

$$f'(x_0) = \lim_{x \to x_0} \frac{f(x) - f(x_0)}{x - x_0} \leq 0$$

Since $0 \leq f'(x_0) \leq 0$, it follows that $f'(x_0) = 0$.

The preceding theorem on critical points is just what we need to prove the mean value theorem. We deduce it as a corollary of the following more general result.

Theorem A.16 Suppose f and g are continuous at every point of the closed interval $[a, b]$ and differentiable at every point of the open interval (a, b). Then there exists a point x_0 in (a, b) such that

$$[f(b) - f(a)]g'(x_0) = [g(b) - g(a)]f'(x_0) \tag{A.2}$$

Proof We define a function h on $[a, b]$ by

$$h(x) = [f(b) - f(a)]g(x) - [g(b) - g(a)]f(x) \tag{A.3}$$

Then h is continuous on $[a, b]$ (Theorem A.3), and so h has a maximum value $h(c)$ and a minimum value $h(d)$ (Theorem A.9). If c is in the open interval (a, b), then $h'(c) = 0$ (Theorem A.15); so, when we differentiate (A.3) and substitute c for x, we get (A.2) with c as the value of x_0. Thus, our proof would be complete, as it would be if d were in (a, b). We proceed differently if c and d are the endpoints a and b. Since, by (A.3),

$$h(a) = f(b)g(a) - g(b)f(a) = h(b)$$

we conclude that the maximum and minimum values of h are equal. Therefore, h is a constant function and so $h'(x_0) = 0$ for every x_0 in (a, b).

Corollary A.17 (Mean value theorem) Suppose f is continuous at every point of $[a, b]$ and differentiable at every point of (a, b). Then there exists a point x_0 in (a, b)

such that

$$f(b) - f(a) = (b - a)f'(x_0)$$

Proof   The conclusion follows from Theorem A.16; just define the function g on $[a, b]$ by $g(x) = x$.

Theorem A.18 (L'Hospital's rule)   Suppose f and g are differentiable and $g'(x) \neq 0$ at every point x in (a, b), where $-\infty \leq a < b \leq \infty$. Suppose further that $f'(x)/g'(x) \to y$ as $x \to a$ and that either

$$f(x) \to 0 \quad \text{and} \quad g(x) \to 0 \quad \text{as } x \to a \qquad \text{(A.4)}$$

or

$$|f(x)| \to \infty \quad \text{and} \quad |g(x)| \to \infty \text{ as } x \to a \qquad \text{(A.5)}$$

Then $f(x)/g(x) \to y$ as $x \to a$. The theorem is also true with $x \to b$ in place of $x \to a$.

Proof   We will assume throughout that hypothesis (A.4) holds and that a is a real number. Our first step is to define $f(a)$ and $g(a)$ to be 0. Then, since $f(x) \to 0$ and $g(x) \to 0$ as $x \to a$, we have forced f and g to be continuous at a. Next, we choose any point x in (a, b). Since the hypotheses of Theorem A.16 hold on $[a, x]$, there exists a point z_x in (a, x) such that

$$[f(x) - f(a)]g'(z_x) = [g(x) - g(a)]f'(z_x)$$

or, since $f(a) = g(a) = 0$,

$$f(x)g'(z_x) = g(x)f'(z_x) \quad \text{or} \quad \frac{f(x)}{g(x)} = \frac{f'(z_x)}{g'(z_x)}$$

But $z_x \to a$ as $x \to a$ since $a < z_x < x$, and therefore

$$\frac{f(x)}{g(x)} = \frac{f'(z_x)}{g'(z_x)} \to y \quad \text{as } x \to a$$

The proof when $a = -\infty$ is only slightly different, but the proof under the hypothesis (A.5) is more intricate [for details, see Rudin (1976)].

EXERCISES

1. Suppose x_0 is a point in an interval I and f is a function whose domain includes all the points of I except perhaps x_0. Let y_0 be a real number. Prove that $\lim_{x \to x_0} f(x) = y_0$ in the sense of Definition A.3 iff for every positive real number ε there exists a positive real number δ such that $|f(x) - y_0| < \varepsilon$ whenever x is a point in the domain of f satisfying $0 < |x - x_0| < \delta$.

2. Prove Theorem A.11.   3. Prove Theorem A.12.

4. Prove Theorem A.13.   5. Prove Theorem A.14.

6. Suppose f is continuous at every point of an interval I and differentiable at every point of I except perhaps the end points.

(a) If $f'(x) \geq 0$ at all points of I other than the endpoints, prove that f is increasing on I. (That is, prove that $f(x_1) \leq f(x_2)$ whenever x_1 and x_2 are points of I satisfying $x_1 < x_2$.)

(b) If $f'(x) \leq 0$ at all points of I other than the endpoints, prove that f is decreasing on I. (That is, prove that $f(x_1) \geq f(x_2)$ whenever x_1 and x_2 are points of I satisfying $x_1 < x_2$.)

(c) If $f'(x) = 0$ at all points of I other than the endpoints, prove that f is a constant function on I.

7. If $f(x) = |x|$, show that f is not differentiable at $x = 0$.

8. If $f(x) = \sqrt{|x|}$, show that f is not differentiable at $x = 0$.

9. If $f(x) = x^2$ for $x \geq 0$ and $f(x) = -x^2$ for $x < 0$, show that f is differentiable at $x = 0$ and find $f'(0)$.

10. Let $f(x) = x \sin(1/x)$ if $x \neq 0$ and let $f(0) = 0$. Show that f is not differentiable at $x = 0$ but is continuous at $x = 0$.

11. Let $f(x) = x^2 \sin(1/x)$ if $x \neq 0$ and let $f(0) = 0$.

    (a) Show that f is differentiable at every point x, and find $f'(x)$ for all x.

    (b) Show that f' is not continuous at $x = 0$.

12. If f is differentiable at every point of an interval I and f' is a bounded function on I, prove that f satisfies the uniform continuity property on I.

13. Prove L'Hospital's rule (Theorem A.18) for the case when $a = -\infty$ and hypothesis (A.4) holds.

A.4 Integrals and Mean Value Theorems

There is no quick and easy definition of the integral of a function. The one given below is fairly standard. Roughly speaking, we define the integral to be the common limit of certain overestimates called "upper sums" and certain underestimates called "lower sums." These estimates and the integral have geometric interpretations when the given function is nonnegative. Then an upper sum is a sum of areas of rectangles whose union just barely encloses the region under the graph of the function (Figure A.4), and a lower sum is a sum of areas of rectangles whose union is just barely enclosed by the region under the graph of the function. Their common limit, the integral, is the area of the region under the graph of the function.

Definition A.5 We assume that f is a bounded function defined on the closed interval $[a, b]$. That is, we assume there exist numbers m and M such that $m \leq f(x) \leq M$ whenever $a \leq x \leq b$. We define a *partition* $\mathcal{P}$ of the interval $[a, b]$ to be a set of points

$$\mathcal{P} = \{x_0, x_1, \ldots, x_n\} \quad \text{where } a = x_0 \leq x_1 \leq \cdots \leq x_n = b$$

Given such a partition $\mathcal{P}$, we define $M_k(\mathcal{P}, f)$ and $m_k(\mathcal{P}, f)$ by

$$M_k(\mathcal{P}, f) = \text{lub}\{f(x)|x_{k-1} \leq x \leq x_k\}$$
$$m_k(\mathcal{P}, f) = \text{glb}\{f(x)|x_{k-1} \leq x \leq x_k\}$$

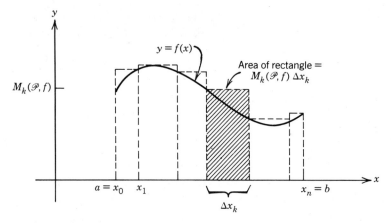

Figure A.4   Sum of areas of rectangles equals $U(\mathscr{P}, f)$.

for each value of k, $k = 1, 2, \ldots, n$. (These numbers exist since f is bounded; in fact, $m \le m_k(\mathscr{P}, f)$ and $M_k(\mathscr{P}, f) \le M$.) Using these numbers, we define the *upper sum* $U(\mathscr{P}, f)$ and *lower sum* $L(\mathscr{P}, f)$ by

$$U(\mathscr{P}, f) = \sum_{k=1}^{n} M_k(\mathscr{P}, f) \Delta x_k$$

$$L(\mathscr{P}, f) = \sum_{k=1}^{n} m_k(\mathscr{P}, f) \Delta x_k$$

where $\Delta x_k = x_k - x_{k-1}$. (Note that if f is non-negative, then $M_k(\mathscr{P}, f) \Delta x_k$ is the area of the rectangle with base Δx_k and height $M_k(\mathscr{P}, f)$; thus, $U(\mathscr{P}, f)$ is the sum of the areas of the rectangles in Figure A.4.) Finally, we define the following limits of the upper and lower sums:

$$U(f) = \mathrm{glb} \left\{ U(\mathscr{P}, f) \,|\, \mathscr{P} \text{ is a partition of } [a, b] \right\}$$

$$L(f) = \mathrm{lub} \left\{ L(\mathscr{P}, f) \,|\, \mathscr{P} \text{ is a partition of } [a, b] \right\}$$

(In Lemma A.20 we show that these limits exist.) If $L(f) = U(f)$, then f is said to be *integrable on* $[a, b]$; the common value of $L(f)$ and $U(f)$ is called the *integral of f on* $[a, b]$ and is denoted by $\int_a^b f(x)\, dx$.

The next definition suggests a slightly different way of defining the integral. In Theorem A.21 we show that, for continuous functions, the definitions are equivalent. (Actually, the equivalence is valid for all integrable functions; see Exercise 6.)

Definition A.6   We again assume that f is a bounded function on $[a, b]$, and we retain the notations of Definition A.5. Furthermore, we define the *norm* of a partition $\mathscr{P}$, denoted $\|\mathscr{P}\|$, to be the largest of the lengths $\Delta x_1, \ldots, \Delta x_n$. And, we define a

Riemann sum $R(\mathcal{P}, f)$ by

$$R(\mathcal{P}, f) = \sum_{k=1}^{n} f(z_k) \Delta x_k$$

where each z_k is chosen to be any point in the interval $[x_{k-1}, x_k]$, $k = 1, 2, \ldots, n$. (The value of the Riemann sum depends on the choice of $z_1, \ldots, z_n$ even though the symbol $R(\mathcal{P}, f)$ does not reflect that fact.) Finally, we say that a number I is the limit of these Riemann sums if for every positive number ε there exists a positive number δ such that

$$|R(\mathcal{P}, f) - I| < \varepsilon \quad \text{whenever} \quad \|\mathcal{P}\| < \delta$$

Our first major result is that if f is continuous on $[a, b]$ then f is integrable on $[a, b]$; the next two lemmas help us subdivide the proof into more manageable pieces. We retain the assumptions and notations of Definitions A.5 and A.6 in the lemmas.

Lemma A.19 For every partition $\mathcal{P}$ of $[a, b]$

$$L(\mathcal{P}, f) \leq R(\mathcal{P}, f) \leq U(\mathcal{P}, f) \tag{A.6}$$

Furthermore, for every pair of partitions $\mathcal{P}$ and $\mathcal{Q}$ on $[a, b]$ in which $\mathcal{P}$ is a subset of $\mathcal{Q}$

$$L(\mathcal{P}, f) \leq L(\mathcal{Q}, f) \quad \text{and} \quad U(\mathcal{Q}, f) \leq U(\mathcal{P}, f) \tag{A.7}$$

Proof The inequalities (A.6) follow from the fact that $m_k(\mathcal{P}, f) \leq f(z_k) \leq M_k(\mathcal{P}, f)$. The rest of this proof is devoted to the second of the two inequalities (A.7); the first can be proved similarly. Let $\mathcal{P} = \{x_0, x_1, \ldots, x_n\}$ and $\mathcal{Q} = \{t_0, t_1, \ldots, t_m\}$, and consider any interval of the form $[x_{k-1}, x_k]$. Since $\mathcal{P}$ is a subset of $\mathcal{Q}$, $x_{k-1} = t_{i-1}$ and $x_k = t_j$ for some values of i and j. Hence

$$[x_{k-1}, x_k] = [t_{i-1}, t_i] \cup \cdots \cup [t_{j-1}, t_j]$$

and so the length of the interval $[x_{k-1}, x_k]$ is the sum of the lengths of $[t_{i-1}, t_i], \ldots, [t_{j-1}, t_j]$. Furthermore, since the latter intervals are subsets of $[x_{k-1}, x_k]$, the least upper bound of f on such an interval cannot exceed the least upper bound of f on $[x_{k-1}, x_k]$. Therefore

$$M_k(\mathcal{P}, f) \Delta x_k = M_k(\mathcal{P}, f) \Delta t_i + \cdots + M_k(\mathcal{P}, f) \Delta t_j$$
$$\geq M_i(\mathcal{Q}, f) \Delta t_i + \cdots + M_j(\mathcal{Q}, f) \Delta t_j$$

Adding these inequalities together produces the inequality we set out to prove.

Lemma A.20 $U(f)$ and $L(f)$ exist, and $L(f) \leq U(f)$.

Proof Let $\mathcal{P}_1$ and $\mathcal{P}_2$ be any two partitions of $[a, b]$ and let $\mathcal{Q}$ denote their union. Then both $\mathcal{P}_1$ and $\mathcal{P}_2$ are subsets of $\mathcal{Q}$ and so, by inequalities (A.6) and (A.7),

$$L(\mathcal{P}_1, f) \leq L(\mathcal{Q}, f) \leq U(\mathcal{Q}, f) \leq U(\mathcal{P}_2, f)$$

Hence, if we hold $\mathcal{P}_1$ fixed and let $\mathcal{P}_2$ vary, we see that $L(\mathcal{P}_1, f)$ is a lower bound for all the upper sums $U(\mathcal{P}, f)$. Therefore, the greatest lower bound $U(f)$ exists and $L(\mathcal{P}_1, f) \leq U(f)$. Then, if we let $\mathcal{P}_1$ vary, we see that $U(f)$ is an upper bound for all the lower sums $L(\mathcal{P}, f)$. Therefore, the least upper bound $L(f)$ exists and $L(f) \leq U(f)$.

Theorem A.21   If f is continuous at every point of $[a, b]$, then f is integrable on $[a, b]$. Furthermore, the integral is the limit of the Riemann sums $R(\mathcal{P}, f)$ in the sense of Definition A.6.

Proof   Let ε be a given positive number. Since f has the uniform continuity property (Theorem A.10), there exists a positive number δ such that

$$|f(u_1) - f(u_2)| < \frac{\varepsilon}{b - a} \quad \text{whenever} \quad |u_1 - u_2| < \delta \tag{A.8}$$

Now let $\mathcal{P} = \{x_0, x_1, \ldots, x_n\}$ be any partition of $[a, b]$ such that $\|\mathcal{P}\| < \delta$. Then, for any choice of u_1 and u_2 from a single subinterval $[x_{k-1}, x_k]$,

$$f(u_1) - f(u_2) \leq |f(u_1) - f(u_2)| < \frac{\varepsilon}{b - a}$$

and so

$$f(u_1) < \frac{\varepsilon}{b - a} + f(u_2)$$

Hence, if we let u_1 vary throughout $[x_{k-1}, x_k]$, we see that

$$M_k(\mathcal{P}, f) \leq \frac{\varepsilon}{b - a} + f(u_2)$$

and so

$$M_k(\mathcal{P}, f) - \frac{\varepsilon}{b - a} \leq f(u_2)$$

And, if we next let u_2 vary, we see that

$$M_k(\mathcal{P}, f) - \frac{\varepsilon}{b - a} \leq m_k(\mathcal{P}, f)$$

and so

$$M_k(\mathcal{P}, f) - m_k(\mathcal{P}, f) \leq \frac{\varepsilon}{b - a}$$

Therefore

$$U(f) - L(f) \leq U(\mathcal{P}, f) - L(\mathcal{P}, f)$$

$$= \sum_{k=1}^{n} [M_k(\mathcal{P}, f) - m_k(\mathcal{P}, f)] \Delta x_k$$

$$\leq \frac{\varepsilon}{b - a} \sum_{k=1}^{n} \Delta x_k = \varepsilon$$

Since ε was an arbitrary positive number, it follows that $U(f) - L(f) \leq 0$. But since $0 \leq U(f) - L(f)$ by Lemma A.20, we conclude that $L(f) = U(f)$; so f is integrable on $[a, b]$.

To prove the last part of the theorem, we again choose δ so that it satisfies property (A.8). But we have just shown that if $\|\mathcal{P}\| < \delta$ then $U(\mathcal{P}, f) - L(\mathcal{P}, f) < \varepsilon$. Therefore, by inequality (A.6),

$$\int_a^b f(x)\, dx - R(\mathcal{P}, f) = U(f) - R(\mathcal{P}, f)$$

$$\leq U(\mathcal{P}, f) - L(\mathcal{P}, f) < \varepsilon$$

Similarly, $R(\mathcal{P}, f) - \int_a^b f(x)\, dx < \varepsilon$, and so our proof is complete.

Theorem A.21 has, among its many uses, the following application to finding limits of sequences. Suppose $\{\mathcal{P}_n\}$ is a sequence of partitions of $[a, b]$ such that $\|\mathcal{P}_n\| \to 0$. Then, if f is continuous on $[a, b]$, Theorem A.21 implies that

$$R(\mathcal{P}_n, f) \to \int_a^b f(x)\, dx \tag{A.9}$$

(Strictly speaking, for $\{R(\mathcal{P}_n, f)\}$ to be a sequence of numbers we must also choose a specific set of evaluation points z_k associated with each partition $\mathcal{P}_n$.) The following is a typical example of a sequence whose limit is given by formula (A.9).

EXAMPLE A.1 In beginning calculus it is shown that $\int_1^2 1/x\, dx = \log 2$. Let $\mathcal{P}_n$ be the partition of the interval $[1, 2]$ obtained by subdividing $[1, 2]$ into n congruent subintervals. So the points of $\mathcal{P}_n$ are $x_k = 1 + k/n$, $k = 0, 1, \ldots, n$. If we choose $z_k = x_k$, then

$$R(\mathcal{P}_n, f) = \sum_{k=1}^n \frac{1}{x_k}\, \Delta x_k = \sum_{k=1}^n \frac{1}{1 + k/n}\left(\frac{1}{n}\right) = \sum_{k=1}^n \frac{1}{n+k}$$

and therefore, by formula (A.9),

$$\sum_{k=1}^n \frac{1}{n+k} \to \log 2$$

The most important fact about integrals is given by the fundamental theorem of calculus (Theorem A.23). The hard part of the proof is contained in the next lemma.

Lemma A.22 Suppose f is continuous at every point of $[a, b]$. Let c be any point of $[a, b]$, and let M and m be upper and lower bounds, respectively, of f on the interval $[c, b]$. Then

$$m(b - c) \leq \int_a^b f(x)\, dx - \int_a^c f(x)\, dx \leq M(b - c) \tag{A.10}$$

Proof To help us distinguish between the two integrals in (A.10), we will use the symbol g to denote the function f with its domain restricted to the interval $[a, c]$. We use Theorem A.21 to approximate each of the two integrals by a Riemann sum. Specifically, for any positive number ε, there exist positive numbers δ_1 and δ_2 such that

$$\left| \int_a^b f(x)\, dx - R(\mathscr{P}, f) \right| < \varepsilon \quad \text{whenever } \|\mathscr{P}\| < \delta_1$$

$$\left| \int_a^c g(x)\, dx - R(\mathscr{Q}, g) \right| < \varepsilon \quad \text{whenever } \|\mathscr{Q}\| < \delta_2$$

where $\mathscr{P}$ denotes a partition of $[a, b]$ and $\mathscr{Q}$ a partition of $[a, c]$. We now choose one partition $\mathscr{P}$ and one partition $\mathscr{Q}$ so that $\|\mathscr{P}\| < \delta_1$ and $\|\mathscr{Q}\| < \delta_2$ and also so that $\mathscr{Q}$ is the "leading part" of $\mathscr{P}$. By this, we mean that if $\mathscr{P} = \{x_0, x_1, \ldots, x_n\}$, we choose $\mathscr{Q} = \{x_0, x_1, \ldots, x_m\}$ (where $m \leq n$ since $x_m = c \leq b = x_n$). Also, we choose the evaluation points z_k so that g is evaluated at exactly the same points of $[x_0, x_1], \ldots, [x_{m-1}, x_m]$ that f is. Therefore

$$\int_a^b f(x)\, dx - \int_a^c f(x)\, dx = \int_a^b f(x)\, dx - \int_a^c g(x)\, dx$$

$$< (R(\mathscr{P}, f) + \varepsilon) - (R(\mathscr{Q}, g) - \varepsilon)$$

$$= \sum_{k=1}^n f(z_k)\, \Delta x_k - \sum_{k=1}^m f(z_k)\, \Delta x_k + 2\varepsilon$$

$$= \sum_{k=m+1}^n f(z_k)\, \Delta x_k + 2\varepsilon$$

$$\leq \sum_{k=m+1}^n M\, \Delta x_k + 2\varepsilon$$

$$= M(b - c) + 2\varepsilon$$

Since this result holds for every positive number ε, we have proved the rightmost inequality in (A.10). The other inequality in (A.10) is proved similarly.

Theorem A.23 (Fundamental theorem of calculus)   If f is continuous at every point of $[a, b]$, then

(a) $\dfrac{d}{dx} \int_a^x f(t)\, dt = f(x)$ for every x in $[a, b]$.

(b) If F is an antiderivative of f (that is, $F'(x) = f(x)$ for all x in $[a, b]$), then $\int_a^b f(x)\, dx = F(b) - F(a)$.

Proof

(a) We define a function g by $g(x) = \int_a^x f(t)\, dt$. For any x_0 in $[a, b]$, we compute $g'(x_0)$ by finding the limit of the usual difference quotient:

$$\frac{g(x) - g(x_0)}{x - x_0} = \frac{1}{x - x_0} \left[\int_a^x f(t)\, dt - \int_a^{x_0} f(t)\, dt \right].$$

If we assume that $x > x_0$, we can apply Lemma A.22 to the expression in square brackets. First, we let ε be any positive real number and choose δ such that

$$|f(t) - f(x_0)| < \varepsilon \quad \text{whenever } |t - x_0| < \delta$$

and so

$$f(x_0) - \varepsilon < f(t) < f(x_0) + \varepsilon \quad \text{whenever } |t - x_0| < \delta$$

Therefore, assuming that $|x - x_0| < \delta$ and applying Lemma A.22, we get

$$(f(x_0) - \varepsilon)\frac{(x - x_0)}{(x - x_0)} < \frac{g(x) - g(x_0)}{x - x_0} < (f(x_0) + \varepsilon)\frac{(x - x_0)}{(x - x_0)}$$

or

$$-\varepsilon < \frac{g(x) - g(x_0)}{x - x_0} - f(x_0) < \varepsilon$$

or

$$\left|\frac{g(x) - g(x_0)}{x - x_0} - f(x_0)\right| < \varepsilon$$

A similar argument produces the same result when $x < x_0$; so we have proved that $g'(x_0) = f(x_0)$.

(b) See Exercise 12.

The next two theorems are true for integrable functions, not just continuous functions; but those proofs are much longer.

Theorem A.24 Suppose f and g are continuous at every point of $[a, b]$. Then

(a) $\int_a^b [f(x) + g(x)]\, dx = \int_a^b f(x)\, dx + \int_a^b g(x)\, dx$

(b) $\int_a^b [f(x) - g(x)]\, dx = \int_a^b f(x)\, dx - \int_a^b g(x)\, dx$

(c) $\int_a^b cf(x)\, dx = c\int_a^b f(x)\, dx$ for every number c

(d) $\int_a^b f(x)\, dx \leq \int_a^b g(x)\, dx$ if $f(x) \leq g(x)$ for all x in $[a, b]$

(e) $\left|\int_a^b f(x)\, dx\right| \leq \int_a^b |f(x)|\, dx$

Proof

(a) By Theorem A.23a, f and g have antiderivatives F and G, respectively. But, by Theorem A.13a, $F + G$ is then an antiderivative of $f + g$. So, by Theorem A.23b,

$$\int_a^b [f(x) + g(x)]\, dx = (F(b) + G(b)) - (F(a) + G(a))$$

$$= (F(b) - F(a)) + (G(b) - G(a))$$

$$= \int_a^b f(x)\, dx + \int_a^b g(x)\, dx$$

(b) This theorem is proved similarly.

(c) This theorem is proved similarly.

(d) $g(x) - f(x) \geq 0$, and so $L(\mathscr{P}, g - f) \geq 0$ for every partition $\mathscr{P}$ of $[a, b]$. Therefore, by part (b),

$$\int_a^b g(x)\, dx - \int_a^b f(x)\, dx = \int_a^b [g(x) - f(x)]\, dx$$

$$= L(g - f) \geq L(\mathscr{P}, g - f) \geq 0$$

which is what we needed to show.

(e) Since $f(x) \leq |f(x)|$ and $-f(x) \leq |f(x)|$, parts (c) and (d) imply that

$$\left| \int_a^b f(x)\, dx \right| = \pm \int_a^b f(x)\, dx$$

$$= \int_a^b \pm f(x)\, dx$$

$$\leq \int_a^b |f(x)|\, dx$$

The next theorem can be proved much as we did Theorem A.24a.

Theorem A.25   Suppose f is continuous at every point of $[a, b]$. If $a < c < b$, then

$$\int_a^b f(x)\, dx = \int_a^c f(x)\, dx + \int_c^b f(x)\, dx$$

Our main goal is to prove a pair of mean value theorems analogous to the pair in Section A.3. Lemma A.26 is not only useful in their proofs but has its own applications.

Lemma A.26   Suppose g is continuous and $g(x) \geq 0$ at every point x in $[a, b]$. If $\int_a^b g(x)\, dx = 0$, then $g(x) = 0$ for all x in $[a, b]$.

Proof   Suppose instead that $g(x_0) > 0$ for some x_0 in $[a, b]$. If we let $\varepsilon = g(x_0)/2$, there exists a positive number δ such that

$$|g(x) - g(x_0)| < \frac{g(x_0)}{2} \quad \text{whenever } |x - x_0| < \delta \text{ and } a \leq x \leq b$$

So if $[u, v]$ is a closed interval contained in the intervals $[a, b]$ and $(x_0 - \delta, x_0 + \delta)$, it follows that if $u \leq x \leq v$ then

$$g(x) = g(x_0) + (g(x) - g(x_0))$$

$$\geq g(x_0) - |g(x) - g(x_0)|$$

$$> g(x_0) - \frac{g(x_0)}{2} = \frac{g(x_0)}{2}$$

We now choose such an interval $[u, v]$ with $u < v$. Therefore, by Theorems A.25 and

A.24d,

$$0 = \int_a^b g(x)\, dx = \int_a^u g(x)\, dx + \int_u^v g(x)\, dx + \int_v^b g(x)\, dx$$

$$\geq \int_a^u 0\, dx + \int_u^v \frac{g(x_0)}{2}\, dx + \int_v^b 0\, dx$$

$$= 0 + (v - u)\frac{g(x_0)}{2} + 0 > 0$$

This contradiction proves that $g(x) = 0$ for all x in $[a, b]$.

Theorem A.27   Suppose f and g are continuous and $g(x) \geq 0$ at every point x in $[a, b]$. Then there exists a point x_0 in $[a, b]$ such that

$$\int_a^b f(x)g(x)\, dx = f(x_0)\int_a^b g(x)\, dx \qquad (A.11)$$

Proof   If $\int_a^b g(x)\, dx = 0$, then $g(x) = 0$ for all x in $[a, b]$ (Lemma A.26). Since (A.11) is then true for every x_0, we may assume that $\int_a^b g(x)\, dx > 0$. Next, we let M and m be the respective maximum and minimum values of f. Hence

$$mg(x) \leq f(x)g(x) \leq Mg(x)$$

for all x in $[a, b]$, since $g(x) \geq 0$. So, by Theorems A.24c and A.24d,

$$m\int_a^b g(x)\, dx \leq \int_a^b f(x)g(x)\, dx \leq M\int_a^b g(x)\, dx$$

or

$$m \leq \frac{\int_a^b f(x)g(x)\, dx}{\int_a^b g(x)\, dx} \leq M \qquad (A.12)$$

But $m = f(u)$ and $M = f(v)$ for some u and v in $[a, b]$. Therefore, by Corollary A.6 applied to f on the interval $[u, v]$ (or $[v, u]$ if $v > u$), there exists a point x_0 in $[u, v]$ such that

$$\frac{\int_a^b f(x)g(x)\, dx}{\int_a^b g(x)\, dx} = f(x_0)$$

since we showed in equation (A.12) that the quotient of the two integrals is between $f(u)$ and $f(v)$.

Corollary A.28   Suppose f is continuous at every point of $[a, b]$. Then there exists a point x_0 in $[a, b]$ such that

$$\int_a^b f(x)\, dx = f(x_0)(b - a)$$

Proof   The conclusion follows from Theorem A.27; just define the function g on $[a, b]$ by $g(x) = 1$.

EXERCISES

1. Prove, directly from Definition A.5, that any function that is constant on the interval $[a, b]$ is integrable on $[a, b]$. Find the value of the integral.

2. Let $f(x) = 0$ if $0 \le x \le 1$ and 1 if $1 < x \le 2$. Prove, directly from Definition A.5, that f is integrable on $[0, 2]$.

3. Let $f(x) = 1$ if $x = 1, \frac{1}{2}, \frac{1}{3}, \ldots,$ and 0 otherwise, where $0 \le x \le 1$. Prove, directly from Definition A.5, that f is integrable on $[0, 1]$.

4. Let $f(x) = 1$ if x is rational and 0 if x is irrational, where $0 \le x \le 1$. Prove that f is not integrable on $[0, 1]$. *Hint:* Between any two real numbers, there is both a rational number and an irrational number.

5. Give an example of a function f that is not integrable on $[0, 1]$ but whose absolute value is integrable on $[0, 1]$.

6. If f is integrable on $[a, b]$ in the sense of Definition A.5, prove that $\int_a^b f(x)\, dx$ is the limit of the Riemann sums $R(\mathscr{P}, f)$, where the limit is in the sense of Definition A.6. (Theorem A.21 is not applicable, since we are not assuming that f is continuous.)

7. If f and g are integrable on $[a, b]$, prove that $f + g$ is integrable on $[a, b]$ and
$$\int_a^b [\, f(x) + g(x)\,]\, dx = \int_a^b f(x)\, dx + \int_a^b g(x)\, dx$$
(Theorem A.24 is not applicable, since we are not assuming that f and g are continuous.)

8. Suppose $f(x) = F'(x)$ for all x in $[a, b]$, where F is any differentiable function with a bounded derivative. Prove that f is integrable on $[a, b]$ and that $\int_a^b f(x)\, dx = F(b) - F(a)$. *Hint:* Theorem A.23 is not applicable since we are not assuming that f is continuous; instead, use the mean value theorem (Corollary A.17).

9. (Integrable function with one discontinuity)   Suppose f is defined and bounded on $[a, b]$ and is continuous at every point of $(a, b]$. Prove that f is integrable on $[a, b]$.

10. Suppose that f and g are defined and bounded on $[a, b]$ and are continuous at every point of $(a, b]$. If $g(x) = f(x)$ for all x in $(a, b]$ but $g(a) \ne f(a)$, prove that $\int_a^b g(x)\, dx = \int_a^b f(x)\, dx$.

11. If f is increasing on $[a, b]$, prove that f is integrable on $[a, b]$.

12. Prove Theorem A.23b.

13. Prove Theorems A.24b and A.24c.

14. Prove Theorem A.25.

15. Give an example of a function g such that g is integrable on $[a, b]$, $\int_a^b g(x)\, dx = 0$, $g(x) \ge 0$ for all x in $[a, b]$, and $g(x) > 0$ for at least one value of x in $[a, b]$. (Compare Lemma A.26.)

In Exercises 16–19, find the limit of the given sequence by showing that the terms are Riemann sums that converge to an integral.

16. $\displaystyle\sum_{k=1}^{n} \frac{n}{(n+k)^2}$ *Hint:* Let $f(x) = 1/x^2$

17. $\displaystyle\sum_{k=1}^{n} \frac{1}{n} \sin\left(\frac{k\pi}{n}\right)$ *Hint:* Let $f(x) = \sin x$

18. $\displaystyle\sum_{k=1}^{n} \frac{k^4}{n^5}$ *Hint:* Let $f(x) = x^4$

19. $\displaystyle\sum_{k=1}^{n} \frac{n}{n^2 + k^2}$ *Hint:* Let $f(x) = 1/(1 + x^2)$

APPENDIX B

The binomial theorem

For every pair of integers k and n for which $0 \le k \le n$, the *binomial coefficient* $\binom{n}{k}$ is defined by

$$\binom{n}{k} = \frac{n!}{k!(n-k)!}$$

where $0! = 1$ and $n! = (n-1)!\,n$ for $n > 0$. The coefficient $\binom{n}{k}$ is sometimes read as "n choose k" since it equals the number of k-element subsets of an n-element set or, more informally, the number of ways one can choose k objects out of a collection of n objects. We prove this equality as follows. Suppose we select k of the n given objects, one at a time. Then the first of our k selected objects could be any of the n given objects; the next of the k selected objects could be any of the remaining $n-1$ given objects; and so on. Thus, we have counted $n(n-1)(n-2)\cdots(n-k+1) = n!/(n-k)!$ different ways to select k objects *in a specific order* out of a collection of n objects. Since we are only interested in counting how many different k-element subsets there are and do not want to distinguish between different orderings of the same k-element subset, we have counted most subsets too many times. Specifically, since there are $k!$ permutations of a k-element set, we have counted each such set $k!$ times. Therefore, the number of k-element subsets of an n-element set is actually

$$\frac{n!}{(n-k)!}\bigg/ k! = \binom{n}{k}$$

The Binomial Theorem For all real numbers x and y and for all nonnegative integers n,

$$(x+y)^n = \sum_{k=0}^{n} \binom{n}{k} x^k y^{n-k}$$

Proof $(x+y)^n = (x+y)(x+y)\cdots(x+y)$, where the factor $(x+y)$ occurs n times. To prove the theorem, we simply multiply out and combine like terms. If the factor x occurs exactly k times in some term, then y must occur exactly $n-k$ times in that term since each term is a product that includes an x or y from each factor $(x+y)$. Thus, all the terms have the form $x^k y^{n-k}$, where $0 \le k \le n$. Furthermore, for each value of k, there are exactly $\binom{n}{k}$ terms of this form since there are exactly $\binom{n}{k}$ ways to choose k occurrences of x from the n factors $(x+y)$.

Complex numbers

Definition C.1 A *complex number* is an ordered pair (a, b) of real numbers.

We can therefore think of a complex number as a point in the plane (Figure C.1), which, as we will see, is useful to do. It is also useful to write a complex number (a, b) in the traditional form $a + bi$, and our first theorem will help us justify doing that. Theorem C.1 describes the algebraic properties of sums and products of complex numbers; so we must first define these two operations. The definitions are chosen so that they mimic the following plausible way of computing the sum and product of two complex numbers written in the traditional form:

$$(a + bi) + (c + di) = (a + c) + (b + d)i$$

$$(a + bi)(c + di) = (ac - bd) + (bc + ad)i$$

The minus sign in the second of these two rules is plausible since i is often thought of as a "square root of -1" and so $i^2 = -1$. To avoid this ill-defined phrase, we make the unambiguous definitions

$$(a, b) + (c, d) = (a + c, b + d) \tag{C.1}$$

$$(a, b)(c, d) = (ac - bd, bc + ad) \tag{C.2}$$

Theorem C.1 For all complex numbers z, w, and v

(a) $z + w = w + z$; $\quad zw = wz$
(b) $(z + w) + v = z + (w + v)$; $\quad (zw)v = z(wv)$
(c) $z(w + v) = zw + zv$
(d) $z + 0 = z$; $\quad z1 = z$ where 0 denotes $(0, 0)$ and 1 denotes $(1, 0)$
(e) For each z, there exists a unique w such that $z + w = 0$
(f) For each $z \neq 0$, there exists a unique w such that $zw = 1$

Note: The number w in part (e) *is denoted by* $-z$, *and the number w in* (f) *is denoted by* z^{-1}. *Also* $v - z$ *denotes* $v + (-z)$ *and* v/z *denotes* vz^{-1}.

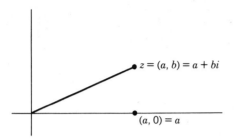

$z = (a, b) = a + bi$

$(a, 0) = a$

Figure C.1   Definition of a complex number

Some typical proofs   We write $z = (a, b)$, $w = (c, d)$, and $v = (e, f)$.

(a)

$$zw = (a, b)(c, d) = (ac - bd, bc + ad)$$
$$= (ca - db, da + cb)$$
$$= (c, d)(a, b) = wz$$

(c)

$$z(w + v) = (a, b)((c, d) + (e, f))$$
$$= (a, b)(c + e, d + f)$$
$$= (a(c + e) - b(d + f), b(c + e) + a(d + f))$$
$$= (ac - bd + ae - bf, bc + ad + be + af)$$
$$= (ac - bd, bc + ad) + (ae - bf, be + af)$$
$$= (a, b)(c, d) + (a, b)(e, f) = zw + zv$$

(f) If $z = (a, b) \neq (0, 0)$, we can let $w = \left(\dfrac{a}{(a^2 + b^2)}, \dfrac{-b}{a^2 + b^2} \right)$. Then

$$zw = \left(\frac{a^2}{a^2 + b^2} + \frac{b^2}{a^2 + b^2}, \frac{ba}{a^2 + b^2} - \frac{ab}{a^2 + b^2} \right) = (1, 0)$$

Furthermore, if $zv = (1, 0)$ then, by parts (d), (b), and (a),

$$w = w(1, 0) = wzv = zwv = (1, 0)v = v$$

which proves that the multiplicative inverse of z is unique.

If we identify the real numbers a and b with the respective complex numbers $(a, 0)$ and $(b, 0)$ and if we denote the complex number $(0, 1)$ by the symbol i, then by definitions (C.1) and (C.2)

$$a + bi = (a, 0) + (b, 0)(0, 1)$$
$$= (a, 0) + (0, b) = (a, b)$$

This enables us to write complex numbers in the traditional form $a + bi$, which we do from now on. Justifications for identifying the real number a with the complex number $(a, 0)$ can be found in Figure C.1 and Theorem C.1. In Figure C.1, we see that the complex numbers $(a, 0)$ are exactly the points on the horizontal axis; we identify this axis with the real line and we refer to it as the *real axis* in the complex plane. (The vertical axis is called the *imaginary axis* and the points $(0, b) = bi$ are called *imaginary numbers*.) Theorem C.1 gives the added assurance that the complex numbers $(a, 0)$ have the same algebraic properties that the real numbers a do. Furthermore, since $i^2 = (0, 1)(0, 1) = (-1, 0) = -1$, we are justified in thinking of i as a square root of -1.

Next, we define the *conjugate*, $\overline{a + bi}$, and the *absolute value*, $|a + bi|$, of a complex number $a + bi$ by

$$\overline{a + bi} = a - bi \tag{C.3}$$

$$|a + bi| = \sqrt{a^2 + b^2} \tag{C.4}$$

These two operations have the properties described in the next theorem.

Theorem C.2   For all complex numbers z and w

(a) $\overline{z + w} = \bar{z} + \bar{w}$; $\overline{zw} = \bar{z}\bar{w}$
(b) $\bar{\bar{z}} = z$
(c) $|z + w| \le |z| + |w|$; $|zw| = |z| |w|$
(d) $z\bar{z} = |z|^2$

Some typical proofs   We let $z = a + bi$ and $w = c + di$.

(a)
$$\overline{zw} = \overline{(a + bi)(c + di)} = \overline{(ac - bd) + (bc + ad)i}$$
$$= (ac - bd) - (bc + ad)i$$
$$= (a - bi)(c - di) = \bar{z}\bar{w}$$

(c)
$$|zw|^2 = |(a + bi)(c + di)|^2$$
$$= |(ac - bd) + (bc + ad)i|^2$$
$$= (ac - bd)^2 + (bc + ad)^2$$
$$= a^2c^2 - 2abcd + b^2d^2 + b^2c^2 + 2abcd + a^2d^2$$
$$= (a^2 + b^2)(c^2 + d^2) = (|a + bi| |c + di|)^2 = (|z| |w|)^2$$

Therefore, after taking the square root of each side, we get that $|zw| = |z| |w|$.

In Example 3.9, we defined e^z for complex numbers z and proved that

$$e^{it} = \cos t + i \sin t \tag{C.5}$$

for all real numbers t. This complex exponential function has the following properties.

Theorem C.3   For all real numbers t and u and all integers n,

(a) $e^{it}e^{iu} = e^{i(t+u)}$
(b) $(e^{it})^n = e^{int}$
(c) $\overline{e^{it}} = e^{-it}$
(d) $|e^{it}| = 1$
(e) $e^{in\pi} = (-1)^{|n|}$

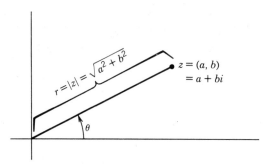

Figure C.2   Polar coordinates of a complex number

Some typical proofs

(a) $e^{it}e^{iu} = (\cos t + i \sin t)(\cos u + i \sin u)$

$$= (\cos t \cos u - \sin t \sin u) + i(\sin t \cos u + \cos t \sin u)$$

$$= \cos(t + u) + i \sin(t + u)$$

$$= e^{i(t+u)}$$

(b) For $n \geq 0$, the conclusion can be deduced from part (a) by induction on n. Also by part (a)

$$e^{it}e^{-it} = e^0 = 1 \quad \text{and so} \quad (e^{it})^{-1} = e^{-it}$$

which proves (b) for $n = -1$. Finally, the validity of (b) for $n < 0$ follows from its validity for $n > 0$ and $n = -1$.

The polar coordinates r and θ (where $r \geq 0$) of the point (a, b) can be used to write any complex number in the following useful form. Since $a = r \cos \theta$ and $b = r \sin \theta$ (see Figure C.2)

$$a + bi = r \cos \theta + ir \sin \theta$$

$$= r(\cos \theta + i \sin \theta)$$

$$= re^{i\theta}$$

The expression $re^{i\theta}$ is called the *polar form* of $a + bi$; it is especially useful for expressing products and powers of complex numbers. The following properties follow from the corresponding properties in Theorem C.3.

Theorem C.4   For all real numbers r, s, t, u (where $r \geq 0$ and $s \geq 0$) and all integers n

(a) $(re^{it})(se^{iu}) = (rs)e^{i(t+u)}$
(b) $(re^{it})^n = r^n e^{int}$
(c) $\overline{re^{it}} = re^{-it}$
(d) $|re^{it}| = r$

EXAMPLE C.1

$$(3 + 4i) + (2 - i) = 5 + 3i$$

$$(3 + 4i)(2 - i) = (6 + 4) + (8 - 3)i = 10 + 5i$$

$$\overline{3 + 4i} = 3 - 4i$$

$$|3 + 4i| = \sqrt{9 + 16} = 5$$

$$(3 + 4i)^{-1} = z^{-1} = \frac{\bar{z}}{z\bar{z}} = \frac{\bar{z}}{|z|^2} = \frac{3 - 4i}{25} = \frac{3}{25} - \frac{4i}{25}$$

$$3 + 4i = re^{i\theta} = \text{(approximately)} \ 5e^{i(0.9273)}$$

$$\text{since } r = |3 + 4i| = 5 \text{ and } \theta = \arctan\left(\tfrac{4}{3}\right)$$

APPENDIX D

Differential equations

D.1 First-Order Equations

A differential equation is an equation that describes a relationship between an unknown function and one or more of its derivatives. A "first-order" differential equation is one in which a first derivative, but no higher derivative, of the unknown function occurs. In practice, first-order differential equations can almost always be written in the form

$$\frac{dy}{dx} = F(x, y) \tag{D.1}$$

Sections D.1 and D.2 are devoted to equations of this form; in Sections D.3 and D.4 we consider second-order differential equations, which are equations in which a second derivative, but no higher derivative, occurs.

By a *solution* of the differential equation (D.1) on some interval I, we mean a function s that is differentiable at every point of I and that satisfies

$$s'(x) = F(x, s(x)) \quad \text{for all } x \text{ in } I$$

By the *general* solution of the differential equation, we mean the set of all its solutions. For example, the function $y = 2\sqrt{x^2 - 9}$ is a solution of the first-order differential equation $y' = 4x/y$ on the intervals $(-\infty, -3)$ and $(3, \infty)$, since $y' = 2x/\sqrt{x^2 - 9} = 4x/y$. And, as we will see in Example D.1, the general solution of $y' = 4x/y$ is $y = \pm 2\sqrt{x^2 + c}$, where c can be any constant. In other words, the set of all solutions is infinite, and different solutions are distinguished by the value of the parameter c. As we will see, it is typical and predictable that a formula for the general solution of a first-order differential equation should contain a single arbitrary parameter.

A physical example can give us insight into the role of this parameter. Suppose we wish to study the velocity of a falling object that is being acted on by the force of gravity (which we will assume is constant near the surface of the earth) and the resisting force of the air (which we will assume is proportional to the velocity of the object). Then, since Newton's second law says that the sum of these forces equals the mass times the acceleration of the object and since acceleration is the derivative of

327

velocity, we have

$$m\frac{dv}{dt} = -kv + mg \tag{D.2}$$

where m is the mass of the object, g is the constant acceleration due to gravity near the surface of the earth, and k is the positive constant of proportionality for the resisting force of the air. This differential equation can be put in the form (D.1) simply by dividing both sides by m. It can be solved using the techniques of Section D.2; its solutions are

$$v = c\exp\left(\frac{-kt}{m}\right) + \frac{mg}{k} \tag{D.3}$$

This formula shows that the velocity of the falling object depends on the values of the constants m, g, and k and on how much time, t, has elapsed. But the velocity at time t must also depend on the initial velocity of the falling object; the parameter c, therefore, enables us to adjust the solution to take account of any initial value of the dependent variable. For example, if $v = v_0$ when $t = 0$, substituting these values into (D.3) yields $c = v_0 - mg/k$; so (D.3) becomes

$$v = \left(v_0 - \frac{mg}{k}\right)\exp\left(\frac{-kt}{m}\right) + \frac{mg}{k}$$

Typically, when a first-order differential equation $y' = F(x, y)$ arises in a real-world problem, an initial condition $y(x_0) = y_0$ arises with it. So, in practice, it is an "initial-value problem"

$$\frac{dy}{dx} = F(x, y), \qquad y(x_0) = y_0 \tag{D.4}$$

that we want to solve and that we expect to have a unique solution. It is an important fact that an initial-value problem usually does have a unique solution, but it is not an easy fact to demonstrate. We state a precise theorem below and leave the proof of it, or rather the proof of a modified version, to Section 3.10. In the theorem, we assume that the function F in (D.4) is continuous in both its variables together and satisfies a Lipschitz condition in its second variable. These terms are defined as follows. Suppose $(x, \dot{y})$ is in the domain of F; then F is *continuous at* (x, y) if

for every positive real number ε, there exists a positive real number δ such that

$$|F(u, v) - F(x, y)| < \varepsilon$$

whenever (u, v) is a point in the domain of F satisfying

$$|u - x| < \delta \quad \text{and} \quad |v - y| < \delta$$

The function F satisfies a *Lipschitz condition* on the set D if

there exists a constant L (called a *Lipschitz constant* for F) such that

$$|F(x, y_1) - F(x, y_2)| \le L|y_1 - y_2|$$

for all points (x, y_1) and (x, y_2) in the set D.

For example, the function F defined by $F(x, y) = xy$ satisfies a Lipschitz condition on any set D in the plane which is bounded on the left and right by vertical lines, since

$$|F(x, y_1) - F(x, y_2)| = |xy_1 - xy_2|$$
$$= |x||y_1 - y_2|$$
$$\leq L|y_1 - y_2|$$

where L is any number such that $|x| \leq L$. (Recall that we assumed that the set D is bounded in the x direction.)

Theorem D.1 (Existence–uniqueness)   Suppose D is an open rectangle of the form

$$D = \{(x, y)|a < x < b \quad \text{and} \quad c < y < d\}$$

and F is a function of two variables that is defined on D, is continuous at every point of D, and satisfies a Lipschitz condition on D. Let (x_0, y_0) be a point in D. Then the initial-value problem

$$\frac{dy}{dx} = F(x, y), \qquad y(x_0) = y_0$$

has a solution on some open interval that includes the point x_0 and is contained in the interval (a, b). Furthermore, if s_1 and s_2 are solutions of this initial-value problem on some open interval I that includes the point x_0 and is contained in the interval (a, b), then $s_1(x) = s_2(x)$ for all x in I.

Theorem D.1 is just one instance of the many existence and uniqueness theorems found in advanced differential equations textbooks; it is not one of the most powerful. For example, the domain of F need not be a rectangle. Also, the function F need not satisfy a Lipschitz condition, although, as Exercise 12 shows, some restriction on F beyond continuity is necessary for the solution of an initial-value problem to be unique. Another avoidable limitation of Theorem D.1 is that it only guarantees the existence of a solution on *some* interval, possibly a quite small one. In fact, it can be shown that there exists a solution whose graph extends to the boundary of the domain of F [see Birkhoff and Rota (1978)].

EXERCISES

1. Verify that the functions $y = \pm 2\sqrt{x^2 + c}$ are solutions of the differential equation $y' = 4x/y$.

2. Verify that the functions (D.3) are solutions of the differential equation (D.2).

3. Show that all the solutions of the differential equation $y' = 2y^{3/2}$ are concave up. (Do not solve the equation; just differentiate it.)

4. Solve the initial-value problem $y' = 2x$, $y(3) = 4$.

5. Solve the initial-value problem $y' = x \log x$, $y(1) = 0$.

6. **(a)** Verify that the functions $y = x^2/(1 + cx^2)$ are solutions of the differential equation $y' = 2y^2/x^3$.

(b) Find the unique solution that satisfies the initial condition $y(1) = -1$.

(c) On what intervals is this unique solution valid?

7. Suppose D is a set of points in the plane such that, for every pair of points (x, y_1) and (x, y_2) in D, the line segment joining these points is also in D. If $\frac{\partial f}{\partial y}(x, y)$ exists for all (x, y) in D and if $\frac{\partial f}{\partial y}$ is bounded on D (that is, there exists some M such that $\left|\frac{\partial f}{\partial y}(x, y)\right| \leq M$ on D), show that f satisfies a Lipschitz condition on D. *Hint:* Use the mean value theorem.

In Exercises 8–11, find out whether the function satisfies a Lipschitz condition on its entire domain. If it does not, find as large a subset as you can on which it does. (Exercise 7 will help.)

8. $f(x, y) = 1 + x^2$

9. $f(x, y) = 1 + y^2$

10. $f(x, y) = y/(1 + x^2)$

11. $f(x, y) = x/(1 + y)$

12. **(a)** Show that $y = x^3$ and the zero function are two distinct solutions of the initial-value problem $y' = 3y^{2/3}$, $y(0) = 0$.

(b) What hypothesis of Theorem D.1 is false for the function F defined by $F(x, y) = 3y^{2/3}$?

D.2 Methods for Solving First-Order Equations

For many first-order differential equations, there are no known methods for finding exact solutions. However, there are a few useful methods that apply to certain special types of first-order equations that occur frequently in real-world applications. We consider two such special types of equations in this section: separable and linear.

A first-order differential equation $y' = F(x, y)$ is *separable* if the function F can be expressed as a product of a function of x and a function of y:

$$\frac{dy}{dx} = g(x)h(y) \tag{D.5}$$

The method of solution of such an equation is referred to as "separation of variables," since the first step of the method is to separate the variables x and y:

$$\frac{dy}{h(y)} = g(x)\, dx$$

Then we insert integral signs and antidifferentiate:

$$\int \frac{dy}{h(y)} = \int g(x)\, dx$$

or, equivalently,

$$H(y) = G(x) + c \tag{D.6}$$

where H is an antiderivative of $1/h$ and G is an antiderivative of g. The final step of the method is to solve for y as a function of x, if possible.

Although the above method of solution contains some steps in which the symbols dx and dy are manipulated more freely than the rules of calculus permit, the conclusion can be justified rigorously (Exercise 1). The method does not, however, always produce all the solutions of (D.5); a solution for which $h(y) = 0$ at some key point might not be found by separation of variables since the method involves dividing by $h(y)$.

EXAMPLE D.1   Find all the solutions of $y' = 4x/y$.

Separating the variables x and y produces

$$y\,dy = 4x\,dx$$

and so

$$\int y\,dy = \int 4x\,dx$$

or

$$\frac{y^2}{2} = 2x^2 + c$$

or

$$y = \pm\sqrt{4x^2 + 2c} = \pm 2\sqrt{x^2 + d}$$

where $d = c/2$; hence, d can have any possible value.

EXAMPLE D.2   Find all the solutions of $y' = -2y/x$.

Separating the variables x and y produces

$$\frac{dy}{y} = \frac{-2}{x}\,dx$$

and so

$$\int\frac{dy}{y} = -\int\frac{2}{x}\,dx$$

or

$$\log|y| = -2\log|x| + c$$

or

$$|y| = e^{-2\log|x|+c} = e^{-2\log|x|}e^c = e^c|x|^{-2}$$

and so

$$y = \pm e^c x^{-2} = dx^{-2}$$

where $d = \pm e^c$; hence, d can have every possible value other than 0. But, if we do let $d = 0$, y is then the zero function, which is easily seen to be a solution of the given differential equation. Thus, all the functions of the form $y = dx^{-2}$ are solutions of the given differential equation, and these are all the solutions.

Now we consider a second special type of first-order differential equation $y' = F(x, y)$. We call the equation *linear* if F can be expressed as a first-degree polynomial in y whose coefficients are functions of x:

$$\frac{dy}{dx} + p(x)y = q(x) \tag{D.7}$$

First we solve the even more special case of a *homogeneous* first-order, linear differential equation, which is when the function q in (D.7) is the zero function. Then the equation becomes $y' = -p(x)y$, which is separable; thus

$$\int \frac{dy}{y} = -\int p(x)\,dx$$

or, if P denotes an antiderivative of p,

$$\log|y| = -P(x) + c$$

or

$$y = de^{-P(x)} \tag{D.8}$$

But the solution (D.8) of the homogeneous equation can also give us insight into how we might solve the nonhomogeneous equation (D.7). Equation (D.8) can be interpreted as saying that $e^{P(x)}y$ is a constant function and thus has derivative equal to zero:

$$\frac{d}{dx}\left(e^{P(x)}y\right) = 0$$

or, by the product rule for derivatives,

$$e^{P(x)}\frac{dy}{dx} + e^{P(x)}p(x)y = 0$$

or

$$e^{P(x)}\left(\frac{dy}{dx} + p(x)y\right) = 0$$

The left side of this last equation, however, is exactly the left side of (D.7) multiplied by $e^{P(x)}$. So if we multiply both sides of (D.7) by $e^{P(x)}$, we can reverse the above three steps and thus write (D.7) as

$$\frac{dy}{dx}e^{P(x)} + p(x)e^{P(x)}y = e^{P(x)} \cdot q(x) \qquad \frac{d}{dx}\left(e^{P(x)}y\right) = e^{P(x)}q(x) \tag{D.9}$$

where P still denotes an antiderivative of p. Therefore, to solve for y we find all the antiderivatives of $e^{P(x)}q(x)$ and then divide by the coefficient $e^{P(x)}$ of y. That is, if $Q(x)$ denotes an antiderivative of $e^{P(x)}q(x)$, then

$$e^{P(x)}y = Q(x) + c$$

or

$$y = Q(x)e^{-P(x)} + ce^{-P(x)} \tag{D.10}$$

This is a form of solution that we will see again, in which the general solution of the nonhomogeneous equation is expressed as the sum of a particular solution $y = Q(x)e^{-P(x)}$ and the general solution $y = ce^{-P(x)}$ of the corresponding homogeneous equation.

EXAMPLE D.3   Find all the solutions of

$$\frac{dy}{dx} + \frac{2y}{x} = \frac{\sin x}{x}$$

Since the coefficient of y is $2/x$, we let $P(x) = 2\log|x|$. Thus $e^{P(x)} = e^{2\log|x|} = |x|^2 = x^2$. We multiply both sides of the given differential equation by this function, use the product rule for derivatives to rewrite the left side, and integrate:

$$x^2 \frac{dy}{dx} + 2xy = x \sin x$$

or

$$\frac{d}{dx}\left(x^2 y\right) = x \sin x$$

or, integrating by parts,

$$x^2 y = \int x \sin x \, dx$$

$$= -x \cos x + \int \cos x \, dx = -x \cos x + \sin x + c$$

Therefore

$$y = \frac{-\cos x}{x} + \frac{\sin x}{x^2} + \frac{c}{x^2}$$

Separable and linear equations are not the only special types of first-order differential equations for which there is an exact method of solution. Some textbooks, in fact, contain a large number of special methods [see, for example, Ford (1955)]. However, there are numerical methods, such as those in Sections 2.7 and 2.8, that apply to much larger classes of differential equations. And, although these numerical methods produce only approximate solutions, such solutions are often quite satisfactory for real-world problems.

EXERCISES

1. Let H be an antiderivative of $1/h$ and G be an antiderivative of g; assume that h is never zero.

   (a) If $y = s(x)$ is a differentiable function satisfying $H(y) = G(x) + c$ for all x in some interval I, differentiate this equation to show that $y = s(x)$ is also a solution of $y' = g(x)h(y)$ on I.

   (b) If $y = s(x)$ is a solution of $y' = g(x)h(y)$ on an interval I, show that there is a constant c such that $y = s(x)$ also satisfies the equation $H(y) = G(x) + c$ for all x in I. *Hint:* Show that the derivative of $H(y) - G(x)$ equals zero if $y = s(x)$.

In Exercises 2–15, (a) find all the solutions of the given differential equation, and (b) solve the given initial-value problem.

2. $y' = 3y^2 x^{-2}$, $\quad y(1) = -1$
3. $y^2 y' = 4x + 3$, $\quad y(0) = 2$
4. $y' + 3y = 2e^{-3x}$, $\quad y(0) = 4$
5. $y' - 3y = 2e^x$, $\quad y(0) = -3$
6. $y' = y^2 - 1$, $\quad y(0) = 0$
7. $(1 - x)y' = 1 - 2y$, $\quad y(-1) = -\frac{3}{2}$
8. $y' = (1 + y^2)\cos x$, $\quad y(\pi/2) = 0$

9. $y' + 2xy = 3x$, $y(0) = -1$

10. $y' = 2xy + x^3$, $y(0) = 1$

11. $y' = y - 4x^4$, $y(-1) = 3$

12. $y' + y/x = \cos(2x)$, $y(\pi) = 0$

13. $y' - (\tan x)y = 3x$, $y(0) = 2$

14. $y' = 1/(y + 2x)$, $y(-1) = 0$ *Hint:* Treat x as the dependent variable; solve for x as a function of y.

15. $y' = (x + 2y)/(2x + y)$, $y(2) = 1$ *Hint:* Find a new differential equation whose dependent variable is $u(x) = y(x)/x$.

16. (Bernoulli's equation)

    (a) Show that the substitution $u(x) = y(x)^{1-n}$ transforms

    $$\frac{dy}{dx} + p(x)y = q(x)y^n, \quad \text{where } n \neq 1$$

    into a linear equation in u.

    (b) Solve the initial-value problem $y' + y/x = y^3x^2$, $y(1) = -1$.

17. Suppose s_1 is a solution of the nonhomogeneous differential equation $y' + p(x)y = q(x)$ and s_2 is a solution of the corresponding homogeneous equation $y' + p(x)y = 0$. Verify, by direct substitution, that $s_1 + s_2$ is a solution of the nonhomogeneous equation.

18. Find every function whose derivative is proportional to its square.

19. Solve the initial-value problem $mv' = -kv + mg$, $v(0) = v_0$.

20. Radioactive carbon-14 has a half-life of 5570 years. If 1 g is present at time $t = 0$, find out how many grams are left after 1000 years. *Hint:* The rate y' at which the weight y of a radioactive substance decays is proportional to y; use the fact that the weight y decreases by one-half after 5570 years to find this constant of proportionality.

21. A radioactive isotope of thorium has a half-life of 590 hr. Find out how long it takes for 90% of the thorium to decay. (See the hint for the preceding exercise.)

22. Suppose that a colony of bacteria is growing at a rate proportional to the number present. If the number triples in 1 day, find

    (a) an expression for the number present after t hrs and

    (b) the time it takes for the number to double.

23. Suppose that a colony of bacteria has a limited food supply that prevents the number of bacteria from exceeding a certain maximum number M. Suppose that this fact then implies that the rate at which the number of bacteria is growing is proportional to the difference between M and the number present. If the size of the colony is $M/2$ when $t = 0$ and $3M/4$ when $t = 10$ hrs, find

    (a) an expression for the number present after t hr, and

    (b) the time it takes for the size to reach 90% of the maximum M.

24. A pot of water is brought to boiling ($100°C$) and then removed from the stove and set down in the kitchen where the air is $20°C$. Assume that the rate at which the water cools is proportional to the difference between the temperatures of the air and the water (Newton's law of cooling). If the water temperature falls to $80°C$ in 5 min, find

    (a) an expression for the water temperature after t min, and

    (b) the time it takes for the water to reach $50°C$.

25. The spread of certain epidemics can be described by observing that the rate of change of the number of infected persons is proportional to the product of the number of infected persons and the number not yet infected. Assume that the total population is a fixed number M and that the number of infected persons is $M/2$ when $t = 0$.

(a) Set up and solve this differential equation for the number of infected persons.

(b) Sketch the graph of the solution for $M = 5$ (million) and constant of proportionality 0.1.

(c) Find the number of infected persons when the rate of change is greatest.

26. Suppose two chemical substances A and B react to produce the substance C. Assume that one part of A and one part of B react to product two parts of C and that the rate at which C is produced is proportional to the product of the amounts of A and B currently present. If the reaction starts with 10 g of A, 10 g of B, and none of C, and if there are 15 g of C after 3 min, find

(a) an expression for the amount of C after t minutes, and

(b) the time it takes for 18 g of C to be produced.

27. Suppose two chemical substances A and B react to produce the substance C under the same assumptions as in the preceding exercise. If the reaction starts with 5 g of A, 10 g of B, and none of C, and if there are 6 g of C after 2 min, find

(a) an expression for the amount of C after t minutes, and

(b) the time it takes for 5 g of C to be produced.

28. A tank contains 100 liters of water with 10 kg of salt mixed in. Pure water pours in at the rate of 4 liters/min and the new mixture flows out of a tap in the tank at 6 liters/min. Find

(a) an expression for the amount of salt left in the tank after t min (where $t \leq 50$) and

(b) the time it takes for half the salt to run out.
Hint: For part (a) assume that the salt and water mix instantaneously; then the rate of decrease of the amount of salt equals the concentration of the salt times the rate at which the mixture flows out.

29. A tank contains 100 liters of water with 5 kg of salt mixed in. A salt and water mixture with 4 kg of salt per 100 liters of water pours in at the rate of 6 liters/min and the new mixture flows out of a tap in the tank at 6 liters/min. Find

(a) an expression for the amount of salt left in the tank after t minutes, and

(b) the time it takes for the amount of salt to decrease to 4.5 kg.
(Modify the hint in the preceding exercise.)

30. A block slides along a table top and is brought to a stop by a frictional force proportional to the square root of its velocity. If the block has the initial velocity of 12 ft/sec and travels 8 ft before it stops, find out how long it takes to stop.

D.3 Second-Order Equations

The second-order differential equations that occur in practice can almost always be written in the form

$$y'' = F(x, y, y') \tag{D.11}$$

By a *solution* of this equation on some interval I, we mean a function s that is twice differentiable at every point of I and that satisfies

$$s''(x) = F(x, s(x), s'(x)) \quad \text{for all } x \text{ in } I$$

For example, the function $y = 1/x$ is a solution of the second-order differential equation

$$y'' = \frac{-3y'}{x} - \frac{y}{x^2}$$

on the intervals $(0, \infty)$ and $(-\infty, 0)$, since $y' = -x^{-2}$, $y'' = 2x^{-3}$, and so

$$\frac{-3y'}{x} - \frac{y}{x^2} = \frac{3}{x^3} - \frac{1}{x^3} = \frac{2}{x^3} = y''$$

The set of all solutions (also called the *general* solution) of this differential equation is (see Theorem 4.2)

$$y = \frac{c_1 + c_2(\log |x|)}{x}$$

Note the presence of two independent parameters, c_1 and c_2, in the general solution; this suggests that in the analog of Theorem D.1 for differential equations of the form (D.11), we will have two initial conditions. Such a theorem does hold; but since we are interested almost exclusively in a special type of second-order equation known as "linear," the existence–uniqueness theorem we state applies only to them.

A second-order differential equation is *linear* if the function F in (D.11) is a first-degree polynomial in y and y' whose coefficients are functions of x:

$$y'' + p(x)y' + q(x)y = r(x) \tag{D.12}$$

When r is the zero function, equation (D.12) is said to be *homogeneous*:

$$y'' + p(x)y' + q(x)y = 0 \tag{D.13}$$

The first facts we wish to establish about such a linear equation concern the form of its general solution, which is similar to the form (D.10) of the general solution of a first-order linear equation.

Theorem D.2

    (a) Suppose f and g are solutions of the homogeneous linear equation (D.13) on some interval I. Then every function of the form $y = c_1 f(x) + c_2 g(x)$, where c_1 and c_2 are arbitrary constants, is also a solution of (D.13).

    (b) Suppose g is a particular solution of the nonhomogeneous linear equation (D.12) on I and f is the general solution of the corresponding homogeneous equation (D.13) on I. Then $f + g$ is the general solution of the nonhomogeneous equation on I.

Proof

    (a) We are given that

$$f''(x) + p(x)f'(x) + q(x)f(x) = 0$$

and

$$g''(x) + p(x)g'(x) + q(x)g(x) = 0$$

for all x in I. So if we multiply the first equation by c_1, the second by c_2, and then add, we get (using the differentiation rules $(cf)' = cf'$ and $(f + g)' = f' + g'$)

$$(c_1 f(x) + c_2 g(x))'' + p(x)(c_1 f(x) + c_2 g(x))'$$
$$+ q(x)(c_1 f(x) + c_2 g(x)) = 0$$

for all x in I. Thus, $c_1 f + c_2 g$ is also a solution of the homogeneous equation (D.13).

(b) First we show that every function of the form $f + g$ is a solution of the nonhomogeneous equation (D.12) if

$$g''(x) + p(x)g'(x) + q(x)g(x) = r(x)$$

and

$$f''(x) + p(x)f'(x) + q(x)f(x) = 0$$

for all x in I. But we can add these two equations and write the sum in the form

$$(f(x) + g(x))'' + p(x)(f(x) + g(x))' + q(x)(f(x) + g(x)) = r(x)$$

which is the desired conclusion. Finally, if h is any solution of the nonhomogeneous equation (D.12) and g denotes the given particular solution of (D.12), we must show that h can be written in form $f + g$ for some solution f of the corresponding homogeneous equation (D.13). To do this, it suffices to show that $h - g$ satisfies (D.13). But

$$h''(x) + p(x)h'(x) + q(x)h(x) = r(x)$$

and

$$g''(x) + p(x)g'(x) + q(x)g(x) = r(x)$$

and we can write the difference of these equations in the form

$$(h(x) - g(x))'' + p(x)(h(x) - g(x))' + q(x)(h(x) - g(x)) = 0$$

Thus, $h - g$ is a solution of the homogeneous equation, which completes the proof.

There is a phrase that is commonly used to describe the functions $c_1 f + c_2 g$ in Theorem D.2a; they are called the *linear combinations* of f and g.

Theorem D.3 (Existence–uniqueness) Suppose p, q, and r are continuous functions on an interval I. Let x_0 be a point in I. Then the initial-value problem

$$y'' + p(x)y' + q(x)y = r(x), \qquad y(x_0) = y_0, \qquad y'(x_0) = y_1$$

has a unique solution on I.

We prove below that the solution of this initial-value problem is unique, but we omit the proof that the solution exists. Instead, in all the types of problems of interest to us, we show how to construct the solution. The following lemma is used in the uniqueness proof. It gives an analog of the fact that the solutions of the separable differential equation $y' = ky$ are the functions $y = ce^{kx} = ce^{ka}e^{k(x-a)} = y(a)e^{k(x-a)}$.

Lemma D.4 Suppose h is a differentiable function on the interval $[a, b]$. If there exists a constant k such that $h'(x) \le kh(x)$ for all x in $[a, b]$, then

$$h(x) \le h(a)e^{k(x-a)} \quad \text{for all } x \text{ in } [a, b]$$

Proof

$$\frac{d}{dx}\left[h(x)e^{-kx}\right] = h'(x)e^{-kx} - kh(x)e^{-kx}$$

$$= e^{-kx}\left[h'(x) - kh(x)\right]$$

which is never positive since $h'(x) \le kh(x)$. Therefore, $h(x)e^{-kx}$ is a decreasing function on $[a, b]$, and so, for all x in $[a, b]$,

$$h(x)e^{-kx} \le h(a)e^{-ka}$$

or, after multiplying by e^{kx},

$$h(x) \le h(a)e^{kx}e^{-ka} = h(a)e^{k(x-a)}$$

Proof of Theorem D.3 (Uniqueness only) Suppose f and g are both solutions of (D.12) on the given interval I and suppose both satisfy the initial conditions $y(x_0) = y_0$, $y'(x_0) = y_1$. Let $s = f - g$. Then $s(x_0) = f(x_0) - g(x_0) = 0$ and $s'(x_0) = f'(x_0) - g'(x_0) = 0$. Also, by Theorem D.2b, s is a solution of the corresponding homogeneous equation:

$$s''(x) + p(x)s'(x) + q(x)s(x) = 0$$

Next, we let $h = s^2 + (s')^2$ and differentiate h:

$$h'(x) = 2s(x)s'(x) + 2s'(x)s''(x)$$

$$= 2s'(x)\left[s(x) + s''(x)\right]$$

$$= 2s'(x)\left[s(x) - q(x)s(x) - p(x)s'(x)\right]$$

$$= 2s'(x)s(x)\left[1 - q(x)\right] - 2p(x)\left[s'(x)\right]^2$$

For the moment, we assume that the interval I has the form $[x_0, b]$ for some $b > x_0$. Then, since $1 - q$ and $-2p$ are continuous functions on this closed interval, there exist constants M and N such that $|1 - q(x)| \le M$ and $|-2p(x)| \le N$ for all x in I (Theorem A.9). We also need the fact that $2|ab| \le a^2 + b^2$ for any real numbers a and b (Exercise 2). Then

$$h'(x) \le |h'(x)| \le 2M|s'(x)s(x)| + N\left[s'(x)\right]^2$$

$$\le M\left(\left[s(x)\right]^2 + \left[s'(x)\right]^2\right) + N\left(\left[s(x)\right]^2 + \left[s'(x)\right]^2\right)$$

$$= (M + N)h(x)$$

Therefore, by Lemma D.4,

$$h(x) \le h(x_0)e^{(M+N)(x-x_0)} \quad \text{for all } x \text{ in } I$$

But $h(x_0) = [s(x_0)]^2 + [s'(x_0)]^2 = 0$. Thus, since $h(x) \geq 0$, $h(x) = 0$ for all x in I; hence, $s(x) = 0$ for all x in I, since $h = s^2 + (s')^2$. But $s = f - g$, and so $f = g$ on I. Although we assumed that I has the special form $[x_0, b]$, an analogous proof holds for intervals I of the form $[b, x_0]$; and, if I is an open or half-open interval, our above conclusion shows that $f = g$ on every closed interval contained in I and therefore on all of I. This completes the proof of uniqueness.

Finally, we can combine this result about the uniqueness of solutions of second-order linear equations with Theorem D.2a to describe the general solution of homogeneous linear equations in more detail.

Theorem D.5 Suppose f and g are solutions of the homogeneous linear differential equation (D.13) on some interval I. Suppose further that

$$f(x_0)g'(x_0) \neq f'(x_0)g(x_0) \tag{D.14}$$

for some point x_0 in I. Then the general solution of (D.13) on I is the set of all linear combinations of f and g:

$$y = c_1 f(x) + c_2 g(x) \tag{D.15}$$

Proof By Theorem D.2a, every linear combination of the solutions f and g is a solution of the homogeneous equation (D.13) on I. We still must show, however, that every solution h of (D.13) is a linear combination of f and g. That is, we must find constants c_1 and c_2 such that $h(x) = c_1 f(x) + c_2 g(x)$ for all x in I. But the system of equations

$$\begin{aligned} h(x_0) &= c_1 f(x_0) + c_2 g(x_0) \\ h'(x_0) &= c_1 f'(x_0) + c_2 g'(x_0) \end{aligned} \tag{D.16}$$

can be solved for the unknowns c_1 and c_2, since the determinant of the coefficient matrix is $f(x_0)g'(x_0) - f'(x_0)g(x_0)$, which we assumed is not zero. Furthermore, equations (D.16) say that the functions h and $c_1 f + c_2 g$ satisfy the same two initial conditions. Therefore, the uniqueness part of Theorem D.3 says that $h = c_1 f + c_2 g$.

The condition (D.14) guarantees that the functions f and g are *linearly independent*; that is, it guarantees that f is not of the form cg and g is not of the form cf.

EXAMPLE D.3 Verify that $y = 2x$ is a solution of the second-order, linear, nonhomogeneous differential equation $y'' + 4y = 8x$ and that $y = \cos(2x)$ and $y = \sin(2x)$ are solutions of the corresponding homogeneous equation. Use these facts to find the general solution of the nonhomogeneous equation and the unique solution which satisfies the initial conditions $y(0) = -3$, $y'(0) = 8$.

If $y = \cos(2x)$, then $y' = -2\sin(2x)$, $y'' = -4\cos(2x)$, and so $y'' + 4y = 0$; similarly, $y = \sin(2x)$ is also a solution of the homogeneous equation. Furthermore, if $f(x) = \cos(2x)$, $g(x) = \sin(2x)$, and $x_0 = 0$, then

$$f(x_0)g'(x_0) - f'(x_0)g(x_0) = \cos(0)2\cos(0) + 2\sin(0)\sin(0) = 2 \neq 0$$

Therefore, by Theorem D.5, the general solution of the homogeneous equation is

$y = c_1 \cos(2x) + c_2 \sin(2x)$. Next, if $y = 2x$, then $y'' + 4y = 0 + 4(2x) = 8x$, and so $y = 2x$ is a particular solution of the nonhomogeneous equation. Therefore, by Theorem D.2b, the general solution of the nonhomogeneous equation is

$$y = c_1 \cos(2x) + c_2 \sin(2x) + 2x$$

Furthermore, the initial condition $y(0) = -3$ implies that $-3 = c_1 + 0 + 0 = c_1$, and the initial condition $y'(0) = 8$ implies that $8 = 0 + 2c_2 + 2$ and so $c_2 = 3$. Thus

$$y = -3\cos(2x) + 3\sin(2x) + 2x$$

is the unique solution of the initial-value problem.

EXERCISES

1. Verify that $y = (\log|x|)/x$ is a solution of

$$y'' = \frac{-3y'}{x} - \frac{y}{x^2}$$

2. Show that $2|ab| \le a^2 + b^2$ for all real numbers a and b. *Hint:* Expand the expression $(|a| - |b|)^2$.

In Exercises 3–6, a second-order, linear differential equation is given, along with two initial conditions, and three functions. (a) Show that the first function satisfies the differential equation and that the last two satisfy the corresponding homogeneous equation and the linear independence criterion (D.14). (b) Deduce the general solution of the differential equation. (c) Find the unique solution of the initial-value problem.

3. $y'' - 3y' + 2y = 10\cos x$, $y(0) = 3$, $y'(0) = 0$, $\cos x - 3\sin x$, e^x, e^{2x}

4. $y'' - 6y' + 9y = 9x + 12$, $y(0) = 2$, $y'(0) = 5$, $x + 2$, e^{3x}, xe^{3x}

5. $x^2 y'' + 2xy' - 2y = 8x^2$, $y(1) = 0$, $y'(1) = -1$, $2x^2$, x, x^{-2}

6. $x^2 y'' - 3xy' + 4y = 3x - 4$, $y(1) = -1$, $y'(1) = 2$, $3x - 1$, x^2, $x^2 \log x$

In Exercises 7–10, find the general solution of the given differential equation. *Hint:* In none of these differential equations does y appear explicitly, and so each equation can be thought of as a first-order equation in the function y'; the techniques of Section D.2 should be applicable.

7. $y'' + 3y' = 4e^{-3x}$ 8. $y'' = 2(y')^2$

9. $y'' = x\sqrt{1 - 2y'}$ 10. $y'' = 2y'/x + x^2 \cos x$

11. (a) Verify that $y = x$ is a solution of

$$(1 - x^2)y'' - 2xy' + 2y = 0, \qquad -1 < x < 1$$

which is Legendre's equation of order 1.

(b) Find the general solution of this equation. *Hint:* If f is one solution of a second-order, linear, homogeneous differential equation, there exists another of the form $y = f(x)u(x)$ for some function u.

12. (a) Verify that $y = e^x$ is a solution of

$$xy'' - (x + 2)y' + 2y = 0, \qquad x \ne 0$$

(b) Find the general solution of this equation; see the hint in the preceding exercise.

13. A ball is thrown upward at 64 ft/sec from the top of a building 80 ft high. Find the time it takes for the ball to reach the ground. *Hint:* Assume that the acceleration due to gravity is approximately 32 ft/sec^2, and recall that acceleration is the second derivative of position.

D.4 Methods for Solving Second-Order Equations

We will solve two types of second-order differential equations that arise often in applications: equations in which the independent variable does not occur explicitly and homogeneous linear equations with constant coefficients. If a differential equation has the form $y'' = F(y, y')$ (which says that the independent variable x does not occur), we can reduce it to a first-order equation as follows. First we introduce the new dependent variable $u = y'$, and then we change the independent variable associated with u to y instead of x. It may not always be possible to change the independent variable to y; but if the differential equation has a solution $y = f(x)$ which has an inverse function $x = g(y)$, then

$$u = y'(x) = y'(g(y))$$

and so u becomes a function of y. Then we can use the chain rule to rewrite y'':

$$y'' = \frac{dy'}{dx} = \frac{du}{dx} = \frac{du}{dy}\frac{dy}{dx} = u\frac{du}{dy} \qquad (D.17)$$

The differential equation $y'' = F(y, y')$ is now of first order in u as a function of y.

EXAMPLE D.4 Find the general solution of $y'' + 4y = 0$.
  We let $u = y'$. Then, by (D.17),

$$u\frac{du}{dy} + 4y = 0$$

and so

$$\int u\, du = -4\int y\, dy$$

or

$$\tfrac{1}{2}u^2 = -2y^2 + c$$

or

$$\frac{dy}{dx} = u = \pm\sqrt{d - 4y^2} \qquad \text{where } d = 2c$$

and so

$$\int \frac{dy}{\sqrt{d - 4y^2}} = \pm \int dx$$

or

$$\frac{1}{\sqrt{d}} \int \frac{dy}{\sqrt{1 - 4y^2/d}} = \pm x + c \qquad \text{where this } c \text{ is new}$$

or

$$\frac{1}{2} \arcsin\left(\frac{2y}{\sqrt{d}}\right) = c \pm x$$

or

$$\frac{2y}{\sqrt{d}} = \sin(2c \pm 2x)$$

or

$$y = \frac{\sqrt{d}}{2} \sin(2c) \cos(2x) \pm \frac{\sqrt{d}}{2} \cos(2c) \sin(2x)$$

$$= c_1 \cos(2x) + c_2 \sin(2x)$$

EXAMPLE D.5   Suppose that a rocket is shot up from the surface of the earth, that all its fuel soon burns out, and that it coasts thereafter. Express its position as a function of t, starting at the instant the fuel burns out. Assume that the rocket's velocity at burnout is exactly the so-called "escape velocity" $\sqrt{2aR}$, where a is the magnitude of the acceleration of the rocket at burnout and R is the distance from the rocket to the center of the earth at burnout.

By limiting our consideration of the rocket's behavior to the time after burnout, we have made the analysis much simpler. For then the only force acting on the rocket is gravity. But, since the rocket is likely to coast quite far from earth, we cannot assume that the force of gravity is constant. Instead, we must use Newton's law of gravitational attraction, which says that the force of gravity is inversely proportional to the square of the distance y between the rocket and the center of the earth:

$$\text{Force} = \frac{-k}{y^2}, \qquad k \text{ a positive constant}$$

But, by Newton's second law of motion, the total force acting on the rocket equals the product of its mass m and its acceleration y''. Thus

$$my'' = \frac{-k}{y^2}$$

The second derivative y'' is with respect to time t, and we let $t = 0$ represent the time at burnout. Thus, since $y = R$ and $y'' = -a$ when $t = 0$, the above force equation implies that

$$-ma = \frac{-k}{R^2}$$

and so

$$k = maR^2$$

Therefore, the differential equation may be written

$$my'' = \frac{-maR^2}{y^2}$$

or

$$y'' = -aR^2 y^{-2}$$

We solve this differential equation by letting $u = y'$ and then replacing y'' by $u\dfrac{du}{dy}$ [see (D.17)]. Thus

$$u\frac{du}{dy} = -aR^2 y^{-2}$$

and so

$$\int u\,du = -\int aR^2 y^{-2}\,dy$$

or

$$\frac{u^2}{2} = aR^2 y^{-1} + c$$

or

$$u^2 = 2aR^2 y^{-1} + 2c$$

Also, $y = R$ and $u = y' = v_0$ (the velocity at burnout) when $t = 0$, and so

$$v_0^2 = 2aR + 2c$$

or

$$2c = v_0^2 - 2aR$$

Therefore, when we substitute this value of $2c$ back into the equation for u^2, we get

$$(y')^2 = u^2 = 2aR^2 y^{-1} + v_0^2 - 2aR \qquad \text{(D.18)}$$

But we assumed that $v_0 = \sqrt{2aR}$ (the escape velocity), which makes the differential equation (D.18) separable. Thus

$$y' = \sqrt{2a}\,R y^{-1/2}$$

which implies

$$\int y^{1/2}\,dy = \int \sqrt{2a}\,R\,dt$$

or

$$\tfrac{2}{3} y^{3/2} = \sqrt{2a}\,Rt + c$$

Finally, $y = R$ when $t = 0$, and so

$$\tfrac{2}{3} y^{3/2} = \sqrt{2a}\,Rt + \tfrac{2}{3} R^{3/2}$$

or

$$y = \left(\tfrac{3}{2}\sqrt{2a}\,Rt + R^{3/2}\right)^{2/3}$$

In particular, this result shows that the rocket coasts arbitrarily far from earth since its position y is an unbounded function of time t.

The other type of second-order differential equation that we will solve is the homogeneous linear equation

$$y'' + py' + qy = 0 \tag{D.19}$$

in which the coefficients p and q are constants. Actually, these equations belong to the class of equations we just finished solving, because our assumption that p and q are constants implies that the independent variable x does not appear explicitly in this equation. In particular, the differential equation $y'' + 4y = 0$, which we solved in Example D.4, is of this type. However, there is a less cumbersome method for solving equations of the form (D.19). We already know a good bit about the solutions from Theorem D.5; namely, once we find two solutions that satisfy the linear independence criterion (D.14), the general solution can then be expressed as the set of all their linear combinations. For example, the general solution of $y'' + 4y = 0$ is the set of all linear combinations of $\cos(2x)$ and $\sin(2x)$ (see Example D.4). But the solutions of (D.19) are not always trigonometric functions. For example, suppose we use the method for separable differential equations to solve (D.19) when $q = 0$. The result is (Exercise 1)

$$y = c_1 e^{-px} + c_2$$

which is the set of all linear combinations of e^{-px} and 1.

Now let us put together these pieces of knowledge to find a systematic method for solving all homogeneous, linear, second-order differential equations with constant coefficients. It seems promising to start by looking for solutions of the form $y = e^{wx}$. The solution e^{-px} that we found above has the form; so does the solution 1, since $1 = e^{0x}$. Even the trigonometric functions are related to these exponential functions, since (see Example 3.9 and Appendix C)

$$e^{ix} = \cos x + i \sin x \tag{D.20}$$

We therefore substitute $y = e^{wx}$ into (D.19). Since $y' = we^{wx}$ and $y'' = w^2 e^{wx}$, we get

$$w^2 e^{wx} + pwe^{wx} + qe^{wx} = 0$$

or, after dividing by e^{wx},

$$w^2 + pw + q = 0 \tag{D.21}$$

This last equation is called the *characteristic equation* associated with the differential equation (D.19). Since it is a quadratic equation, it has either two distinct real roots or a single real root (called a "double" root since it appears in both linear factors of (D.21)) or two nonreal roots. If w denotes a root, then, since our derivation of (D.21) is reversible, $y = e^{wx}$ is a solution of the differential equation (D.19). In particular, if (D.21) has two distinct real roots w_1 and w_2, then $e^{w_1 x}$ and $e^{w_2 x}$ are both solutions of (D.19) and they satisfy the linear independence criterion (D.14) (Exercise 2a). Thus

$$y = c_1 e^{w_1 x} + c_2 e^{w_2 x} \tag{D.22}$$

is the general solution of (D.19).

If, however, the characteristic equation (D.21) has a double root w, we have found only one solution $y = e^{wx}$ and must seek another. But Exercise 11 of Section D.3 contains a hint. It suggests that if f is one solution of a linear, homogeneous differential equation, there exists another of the form $y = u(x)f(x)$. We try this. Along the way we use the fact that, since w is a double root of (D.21),

$$p = -2w \quad \text{and} \quad q = w^2 \qquad (D.23)$$

Thus, if $y = ue^{wx}$, then $y' = u'e^{wx} + wue^{wx}$ and $y'' = u''e^{wx} + 2wu'e^{wx} + w^2ue^{wx}$, and so the differential equation (D.19) becomes

$$\left[(u'' + 2wu' + w^2u) + p(u' + wu) + qu\right]e^{wx} = 0$$

or, after dividing by e^{wx} and combining like terms,

$$u'' + (2w + p)u' + (w^2 + pw + q)u = 0$$

or, after using (D.23),

$$u'' = 0$$

and so

$$u = c_1 + c_2 x$$

which implies

$$y = ue^{wx} = c_1e^{wx} + c_2xe^{wx} \qquad (D.24)$$

Finally, we solve the differential equation (D.19) when the characteristic equation (D.21) has two conjugate complex roots $w_1 = a + ib$ and $w_2 = a - ib$. We can find real-valued solutions of (D.19) from the complex-valued solutions $e^{w_1 x}$ and $e^{w_2 x}$ by computing certain linear combinations of them:

$$\frac{1}{2}e^{w_1 x} + \frac{1}{2}e^{w_2 x} = \frac{1}{2}e^{ax+ibx} + \frac{1}{2}e^{ax-ibx}$$
$$= \frac{1}{2}e^{ax}\left[\cos(bx) + i\sin(bx)\right] + \frac{1}{2}e^{ax}\left[\cos(bx) - i\sin(bx)\right]$$
$$= e^{ax}\cos(bx)$$

and

$$\frac{-ie^{w_1 x}}{2} + \frac{ie^{w_2 x}}{2} = \frac{-ie^{ax}}{2}\left[\cos(bx) + i\sin(bx)\right] + \frac{ie^{ax}}{2}\left[\cos(bx) - i\sin(bx)\right]$$
$$= e^{ax}\sin(bx)$$

These two real-valued functions are solutions of the linear, homogeneous differential equation (D.19), since they are linear combinations of solutions. Therefore, the general solution is

$$y = c_1e^{ax}\cos(bx) + c_2e^{ax}\sin(bx) \qquad (D.25)$$

We summarize the above results in the following theorem.

Theorem D.6

(a) If the characteristic equation (D.21) has two distinct real roots w_1 and w_2, then the general solution of the linear, homogeneous differential equation (D.19) is

$$y = c_1e^{w_1 x} + c_2e^{w_2 x}$$

(b) If (D.21) has a double root w, then the general solution of (D.19) is

$$y = c_1e^{wx} + c_2xe^{wx}$$

(c) If (D.21) has complex roots $w = a \pm ib$, then the general solution of (D.19) is

$$y = c_1 e^{ax} \cos(bx) + c_2 e^{ax} \sin(bx)$$

EXAMPLE D.6 Find the general solution of $y'' - 6y' + 9 = 0$.

The characteristic equation is $w^2 - 6w + 9 = 0$, which has the double root $w = 3$. So, by Theorem D.6b,

$$y = c_1 e^{3x} + c_2 x e^{3x}$$

EXAMPLE D.7 Find the general solution of $y'' - 6y' + 13 = 0$.

The characteristic equation is $w^2 - 6w + 13 = 0$, which has the complex roots

$$w = \frac{6 \pm \sqrt{36 - 52}}{2} = 3 \pm 2i$$

So, by Theorem D.6c,

$$y = c_1 e^{3x} \cos(2x) + c_2 e^{3x} \sin(2x)$$

It is also useful to be able to solve some linear, constant-coefficient, nonhomogeneous linear equations:

$$y'' + py' + qy = r(x)$$

We have already seen that the general solution of this equation may be found by adding one particular solution to the general solution of the corresponding homogeneous equation (Theorem D.2b). We will not, however, study methods for finding a particular solution of the nonhomogeneous equation; instead, each example contains a hint on how to find one.

EXAMPLE D.8 Find the general solution of $y'' - 8y' + 15y = 6e^{2x}$.

The characteristic equation of the corresponding homogeneous equation is $w^2 - 8w + 15 = 0$ which has the roots $w = 3$ and 5. So, by Theorem D.6a, the general solution of the homogeneous equation is

$$y = c_1 e^{3x} + c_2 e^{5x}$$

We now guess that the nonhomogeneous equation has a solution of the form $y = d e^{2x}$. When we substitute this guess into the differential equation, we get

$$4 d e^{2x} - 16 d e^{2x} + 15 d e^{2x} = 6e^{2x}$$

or, after dividing by e^{2x} and combining like terms, $3d = 6$, which implies that $d = 2$. Therefore, the general solution of the nonhomogeneous equation is

$$y = c_1 e^{3x} + c_2 e^{5x} + 2e^{2x}$$

Linear equations arise often in the sciences. Our final example is a typical application in physics.

EXAMPLE D.9 A spring is pulled one-half foot downward by a 5-lb weight. If the spring is then pulled 3 in. below this equilibrium point and released, find its equation of motion.

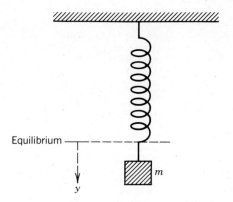

Figure D.1 Spring and weight

We assume that Hooke's law holds for this spring; namely, that the restoring force acting on the spring is proportional to the signed difference between the current length of the spring and its natural length. Thus, if k denotes the constant of proportionality for this spring, $5 = k(\frac{1}{2})$ and so $k = 10$. Now we let y denote the position of the end of the spring, where $y = 0$ denotes its equilibrium position with the 5-lb weight attached (see Figure D.1). So we may express the total force acting on the spring, which is the restoring force plus the force of gravity, as

$$-10\left(y + \tfrac{1}{2}\right) + 5 = -10y$$

But this force, according to Newton's second law of motion, equals the product of mass and acceleration. Thus, since the mass corresponding to the weight of 5 lb is $\frac{5}{32}$, we have

$$\tfrac{5}{32}y'' = -10y$$

or

$$y'' + 64y = 0$$

The characteristic equation of this differential equation is $w^2 + 64 = 0$, which has the roots $w = \pm 8i$, and so

$$y = c_1 \cos(8x) + c_2 \sin(8x)$$

The initial conditions $y(0) = \frac{1}{4}$ and $y'(0) = 0$ then imply that $c_1 = \frac{1}{4}$, $c_2 = 0$ and so

$$y = \frac{\cos(8x)}{4}$$

EXERCISES

1. Use the methods of Section D.2 to solve $y'' + py' = 0$.

2. Verify the linear independence criterion (D.14) with $x_0 = 0$ for each of the following pairs of functions:

   (a) $f(x) = e^{w_1 x}$, $g(x) = e^{w_2 x}$ where $w_1 \neq w_2$
   (b) $f(x) = e^{wx}$, $g(x) = xe^{wx}$

(c) $f(x) = e^{ax} \cos(bx)$, $g(x) = e^{ax} \sin(bx)$ where $b \neq 0$

In Exercises 3–16, find the general solution of the given differential equation. For each nonhomogeneous linear equation, a hint is given that describes the form of a particular solution h.

3. $y'' = -2(y')^2/y$

4. $y^3 y'' = 4$

5. $y'' - 7y' + 10y = 0$

6. $y'' - 10y' + 25y = 0$

7. $y'' - 16y = 0$

8. $y'' - 2y' + y = 0$

9. $y'' + 9y = 0$

10. $y'' + 2y' + 2 = 0$

11. $2yy'' = 1 + (y')^2$

12. $yy''(1 + y) = (y')^2$

13. $y'' - 7y' + 10y = 8$, $h(x) = c$

14. $y'' - 10y' + 25y = 16e^{3x}$, $h(x) = ce^{3x}$

15. $y'' - 16y = 24e^{4x}$, $h(x) = cxe^{4x}$

16. $y'' + 2y' + 2y = 3x + 5$, $h(x) = c_1 x + c_2$

17.   In Example D.5, assume that the initial velocity v_0 is greater than $\sqrt{2aR}$. Show that the rocket again coasts arbitrarily far from earth, and find the limit of its velocity as t gets arbitrarily large. *Hint:* Use (D.18) but do not solve it.

18.   A 4-lb object is attached to a spring with spring constant $k = 2$ lb/ft. The object moves up and down in a viscous oil bath which creates a friction force proportional to the velocity of the object; assume that this constant of proportionality is $c = 1$ lb-sec/ft.

(a) Find the equation of motion of the weight.

(b) Find the specific equation of motion if the spring is initially pulled down 1 ft and given no initial velocity. Also graph the equation of motion and describe the motion in words.

(c) Find the specific equation of motion if the spring is again pulled down 1 ft but is given an upward initial velocity of 5 ft/sec. Also graph the equation of motion and describe the motion in words.
Hint: The force equation is

$$my'' = -ky - cy'$$

where the mass m is the weight of the object divided by the acceleration due to gravity, since weight is a measure of force.

The principle of
mathematical induction

The Principle of Mathematical Induction Let $S_1, S_2, \ldots, S_n, \ldots$ be a sequence of statements and suppose

1. S_1 is true;
2. whenever S_k is true for some integer $k \geq 1$, then S_{k+1} is also true.

Then every one of the statements $S_1, S_2, \ldots, S_n, \ldots$ is true.

We accept this principle as an axiom for the positive integers. But the following argument seems to suggest that we could actually prove it. S_1 is true by hypothesis 1. Therefore, by hypothesis 2 with $k = 1$, S_2 is true. Again, by hypothesis 2 but with $k = 2$, we see that S_3 is true; and so on. Of course, this argument cannot be made into a proof of the principle, since it would require infinitely many steps to deduce that every S_n is true. The point of these observations is that they suggest that the principle of mathematical induction ought to be a powerful proof technique when we do have an infinite sequence of statements to prove.

EXAMPLE E.1 The Fibonacci sequence $\{F_n\}$ is defined recursively by

$$F_{n+2} = F_{n+1} + F_n \quad \text{for } n \geq 0, \, F_0 = 0, \, F_1 = 1$$

Prove the identity

$$F_n^2 = F_{n+1} F_{n-1} + (-1)^{n-1} \quad \text{for all } n \geq 1 \tag{E.1}$$

For each positive integer n, we let S_n denote the statement which asserts that the identity (E.1) is valid. When $n = 1$, the right side of the identity becomes $F_2 F_0 + 1 = 1$, which does equal F_1^2. Thus, S_1 is true. Next we make the so-called "induction hypothesis": suppose S_k is true for some positive integer k; that is, suppose

$$F_k^2 = F_{k+1} F_{k-1} + (-1)^{k-1}$$

We show that S_{k+1} is true:

$$F_{k+2}F_k + (-1)^k = (F_{k+1} + F_k)F_k + (-1)^k$$
$$= F_{k+1}F_k + F_k^2 + (-1)^k$$
$$= F_{k+1}F_k + F_{k+1}F_{k-1} + (-1)^{k-1} + (-1)^k$$
$$= F_{k+1}(F_k + F_{k-1}) = (F_{k+1})^2$$

Thus, the truth of S_k implies the truth of S_{k+1}. Therefore, by the principle of mathematical induction, every one of the statements S_n is true.

References

Birkhoff, Garrett, and Gian-Carlo Rota. *Ordinary Differential Equations*, 3rd ed. New York: Wiley, 1978.

Boas, Mary L. *Mathematical Methods in the Physical Sciences*. New York: Wiley, 1966.

Boas, Ralph P., Jr. "Partial sums of infinite series, and how they grow." *Amer. Math. Monthly* **84**, 237–258 (1977).

Clark, Allan. *Elements of Abstract Algebra*. Belmont, Calif.: Wadsworth, 1971.

Eberhart, James G., and Thomas R. Sweet. "The numerical solution of equations in chemistry." *J. Chem. Ed.* **37**, 422–428 (1960).

Ford, Lester R. *Differential Equations*, 2nd ed. New York: McGraw-Hill, 1955.

Forsythe, George E. "Pitfalls in computation, or why a math book isn't enough." *Amer. Math. Monthly* **77**, 931–956 (1970).

Goldberg, Samuel. *Introduction to Difference Equations*. New York: Wiley, 1958.

Grattan-Guinness, Ivor. *The Development of the Foundations of Mathematical Analysis from Euler to Riemann*. Cambridge, Mass.: MIT Press, 1970.

Hatcher, R. S. "Some little-known recipes for π." *Math. Teacher* **66**, 470–474 (1973).

Johnsonbaugh, Richard. "Summing an alternating series." *Amer. Math. Monthly* **86**, 637–648 (1979).

Kline, Morris. *Mathematical Thought from Ancient to Modern Times*. New York: Oxford University Press, 1972.

Knopp, Konrad. *Infinite Sequences and Series*. New York: Dover, 1956.

Marsden, Jerrold E. *Elementary Classical Analysis*. San Francisco: Freeman, 1974.

Nayfeh, Ali-Hasan. *Perturbation Methods*. New York: Wiley, 1973.

Ralston, Anthony, and Philip Rabinowitz. *A First Course in Numerical Analysis*, 2nd ed. New York: McGraw-Hill, 1978.

Rheinboldt, Werner C. "Algorithms for finding zeros of functions" (UMAP unit 264). Newton, Mass.: COMAP, 1978, 1980.

Rosenlicht, Maxwell. "Integration in finite terms." *Amer. Math. Monthly* **79**, 963–972 (1972).

Rudin, Walter. *Principles of Mathematical Analysis*, 3rd ed. New York: McGraw-Hill, 1976.

Smith, David A. *Interface: Calculus and the Computer*, 2nd ed. Philadelphia: Saunders, 1984.

Smith, Kennan T. *Primer of Modern Analysis*. Tarrytown-on-Hudson, N.Y.: Bogden & Quigley, 1971.

Stewart, Gilbert W. *Introduction to Matrix Computations*. New York: Academic Press, 1973.

Wilf, Herbert S. *Mathematics for the Physical Sciences*. New York: Dover, 1978.

Wilkinson, James H., and C. H. Reinsch (Eds.). *Handbook for Automatic Computation II, Linear Algebra*. Berlin: Springer, 1971.

Wrench, J. W., Jr. "The evolution of extended decimal approximations to π." *Math. Teacher* **53**, 644–650 (1960).

Solutions to selected exercises

SECTION 1.1

1. -4 3. 0 5. $1/2$ 7. $3/2$

9. No limit (sequence oscillates between 1 and -1).

11. 1 13. 1 15. e

17. 0 19. $\sqrt{3}$ 21. $\sqrt[3]{2}$

23. No limit (sequence tends toward infinity).

25. 7 27. Choose $N > 2/\varepsilon$

29. Choose $N > 25/\varepsilon^2$ 31. Choose $N > \log \varepsilon / \log a$

33. Choose $N > \sqrt{(2 + B)/3}$ (where $-B$ is the given bound)

SECTION 1.2

11. $\sqrt{3}$ 13. $\sqrt[3]{2}$

SECTION 1.3

1. $2/3^n$ 3. $n!$ 5. $10 \cdot 5^n - 35 \, n \, 5^{n-1}$

7. $3 + 2n$ 9. $5 \cdot 2^{n/2} \cos(3\pi n/4) + 3 \cdot 2^{n/2} \sin(3\pi n/4)$

11. $2 \cdot 5^n - 7 n \, 5^{n-1} + 4 \cdot 3^n$ 13. $5 - 10 \, n + 3 \, n^2$

15. $\sin(n\theta)$ 17. $2/3$

19. $x_n = 3(\sqrt{3})^n/2 + (-\sqrt{3})^n/2$
    $y_n = 3(\sqrt{3} - 1)(\sqrt{3})^n/2 - (\sqrt{3} + 1)(-\sqrt{3})^n/2$

25. **(a)** If $|w| < 1$ for both roots w of the characteristic equation, then $x_n \to 0$ for every solution $\{x_n\}$. If one root equals 1 and the other is strictly between -1 and 1, then every solution $\{x_n\}$ converges but the limit can have any value. Otherwise, not every solution $\{x_n\}$ converges.

(b) If the roots of the characteristic equation are nonreal, then none of the ratios $\{x_{n+1}/x_n\}$ converges. If the two roots are the negatives of one another, then not every ratio $\{x_{n+1}/x_n\}$ converges. Otherwise, every ratio $\{x_{n+1}/x_n\}$ converges and the limit is always the larger root (in absolute value) of the characteristic equation.

27. **(a)** $(1 + r)^n P$ **(b)** $(P + M/r)(1 + r)^n - M/r$

29. $\dfrac{1}{\sqrt{5}}\left(\dfrac{1 + \sqrt{5}}{2}\right)^{n+1} - \dfrac{1}{\sqrt{5}}\left(\dfrac{1 - \sqrt{5}}{2}\right)^{n+1}$ pairs

33. **(a)** $r_{n+1} - r_n = k_1 r_n - k_3 r_n f_n$

     $f_{n+1} - f_n = -k_2 f_n + k_4 k_3 r_n f_n$

SECTION 1.4

1. **(a)** $\displaystyle\sum_{k=0}^{[n/2]} x^{2k}/(2k)!$

3. **(a)** $\displaystyle\sum_{k=0}^{p/2} (-1)^k x^{2k}/(2k)!$ where p is the largest even integer not exceeding n.

5. **(a)** $\displaystyle\sum_{k=0}^{n} (-1)^k (k + 1)(x - 1)^k$

9. 0.08   11. 0.05

13. $1 + x + x^2/2 + x^3/6 + x^4/24 + x^5/120$

15. $x^2 - x^6/6 + x^{10}/5! - x^{14}/7!$ 17. $\displaystyle\int_0^{0.5}(x - x^2/2)\,dx = 0.104$

SECTION 1.5

1. **(a)** x **(b)** 0.25

3. **(a)** $\dfrac{(x - \frac{3}{2})(x - 2)}{(-\frac{1}{2})(-1)} + \dfrac{\frac{2}{3}(x - 1)(x - 2)}{(\frac{1}{2})(-\frac{1}{2})} + \dfrac{\frac{1}{2}(x - 1)(x - \frac{3}{2})}{(1)(\frac{1}{2})}$ **(b)** 0.05

5. **(a)** $\dfrac{\log(\frac{4}{3})(x - 1)(x - \frac{5}{3})(x - 2)}{(\frac{1}{3})(-\frac{1}{3})(-\frac{2}{3})} + \dfrac{\log(\frac{5}{3})(x - 1)(x - \frac{4}{3})(x - 2)}{(\frac{2}{3})(\frac{1}{3})(-\frac{1}{3})}$

     $+ \dfrac{\log 2\,(x - 1)(x - \frac{4}{3})(x - \frac{5}{3})}{(1)(\frac{2}{3})(\frac{1}{3})}$ **(b)** 0.0031

7. $\dfrac{e\,(x + \frac{1}{3})(x - \frac{1}{3})(x - 1)}{(-\frac{2}{3})(-\frac{4}{3})(-2)} + \dfrac{e^{1/3}\,(x + 1)(x - \frac{1}{3})(x - 1)}{(\frac{2}{3})(-\frac{2}{3})(-\frac{4}{3})}$

     $+ \dfrac{e^{-1/3}\,(x + 1)(x + \frac{1}{3})(x - 1)}{(\frac{4}{3})(\frac{2}{3})(-\frac{2}{3})} + \dfrac{e^{-1}\,(x + 1)(x + \frac{1}{3})(x - \frac{1}{3})}{(2)(\frac{4}{3})(\frac{2}{3})}$

9. $\displaystyle\int_1^2 \left[\dfrac{(x - \frac{3}{2})(x - 2)}{(-\frac{1}{2})(-1)} + \dfrac{\frac{2}{3}(x - 1)(x - 2)}{(\frac{1}{2})(-\frac{1}{2})} + \dfrac{\frac{1}{2}(x - 1)(x - \frac{3}{2})}{(1)(\frac{1}{2})}\right] dx = 0.69$

13. $y_0 + z_0 x + (-3 y_0 + 3 y_1 - 2 z_0 - z_1) x^2 + (2 y_0 - 2 y_1 + z_0 + z_1) x^3$

SECTION 1.6

3. **(a)** $f(x) = 0$ if $x > 0$; $f(0) = 1$

(b)

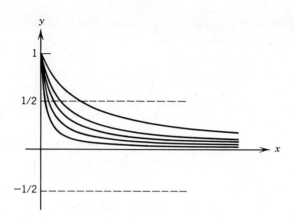

5. **(a)** $f(x) = 0;$ **(c)** $[-1, 1]$ and all its subintervals.
7. **(a)** $f(x) = 0$ if $x < \frac{\pi}{2}; f\left(\frac{\pi}{2}\right) = 1$
    (c) Every $[0, c]$ $\left(\text{where } 0 \le c < \frac{\pi}{2}\right)$ and all their subintervals.
9. **(a)** $f(x) = 0;$ **(c)** $[0, 1]$ and all its subintervals.
11. **(a)** $f(x) = 0$ if $0 \le x < 1; f(1) = \frac{1}{2}; f(x) = 1$ if $x > 1$
    (c) Every $[0, c]$ (where $0 \le c < 1$), every $[c, \infty)$ (where $c > 1$), and all their subintervals.
13. **(a)** $f(x) = 0;$
    (c) Every $[c, \infty)$ (where $c > 0$) and all their subintervals.
15. **(a)** $f(x) = 0;$
    (c) Every $[c, 1]$ (where $c > 0$) and all their subintervals.

SECTION 2.1

1.	$[-2, -1]$	3.	$[1, 2]$		
5.	$(0, 1]$	7.	$[1, 2]$		
9.	2.656	11.	0.594	13.	1.0
15.	-1.7692919	17.	1.4987020		
19.	0.9117689	21.	1.0187216		
23.	1.417	25.	3.145		

SECTION 2.2

1.	1.415	3.	0.567
5.	2.6505179	7.	0.5671433
9.	Algorithm fails		
11.	1.414	13.	0.567
15.	2.6505179	17.	0.5671433

19. Algorithm fails

23. 0.77 percent   **25.** 1.63 and 13.55 days

27. Radius is 0.65 ft. and height is 0.75 ft.

SECTION 2.3

1. (a) 53.49;   **(b)** 0;   **(c)** Does not compute.

3. .000049, for example

5. Multiply and divide by the conjugate of the numerator, $b \pm \sqrt{b^2 - 4ac}$, but only for the root in which $-b$ and $\pm \sqrt{b^2 - 4ac}$ have opposite signs.

7. Multiply and divide by the conjugate of the numerator, $\sqrt[3]{a^2} + \sqrt[3]{a}\,\sqrt[3]{b} + \sqrt[3]{b^2}$

9. 1.499 and 1.500, respectively. The latter is more accurate, since fewer significant digits are lost when the smaller numbers are added first.

11. (a) $\mu_3 = 1$, $\sigma_3^2 = 0$; $\mu_3 = 1$, $\sigma_3^2 = 0$

   (b) $\mu_3 = 1$, $\sigma_3^2 = 0$; $\mu_3 = 1$, $\sigma_3^2 = -.0003$

   (c) $\mu_3 = 1$, $\sigma_3^2 = .0001$; $\mu_3 = .9998$, $\sigma_3^2 = 0$

   The third set of formulas is most reliable.

SECTION 2.4

1. -1.7692924 **3.** 1.4987011

5. 0.9117686   **7.** 1.0187208

SECTION 2.5

1. 0.71, using $n = 2$ **3.** 3.07, using $n = 5$

5. 0.92, using $n = 2$ in both terms

7. 0.694, using $n = 2$ **9.** 3.059, using $n = 4$

11. 0.921, using $n = 2$ in both terms

13. 0.69316   **15.** 3.14156

17. 0.92526   **19.** 1.85192

21. 0.78540   **23.** 0.74683

25. 0.37855   **33.** 0.88621, using $m = 3$

35. $M_n = h[\, f(a + h/2) + f(a + 3h/2) + \cdots + f(a + (2n - 1)h/2)]$
   where $h = (b - a)/n$.

37. 1.85194   **39.** 5.81 calories/mole.

SECTION 2.6

1. 0.6932   **3.** 0.7468

5. 1.8519

7. 0.74682   **9.** 1.85194

SECTION 2.7

1.

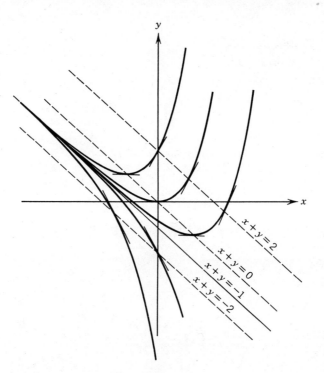

3.

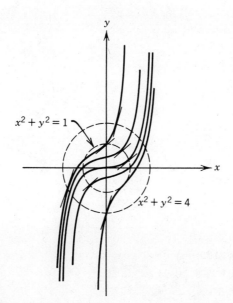

5   (a) $y = 0.000, 0.003, 0.015, 0.042, 0.090$ at
        $x = 0.1, 0.2, 0.3, 0.4, 0.5$.
     (b) $y = x^3$

7.   (a) $y = 0.40, 0.82, 1.28, 1.85, 2.59$ at
        $x = 0.1, 0.2, 0.3, 0.4, 0.5$
     (b) $y = 2\tan(2x)$

9.   (a) $y = 1.10, 1.19, 1.28, 1.36, 1.44$ at
        $x = 0.1, 0.2, 0.3, 0.4, 0.5$
     (b) $y = \sqrt{2x + 1}$

11. $y = 0.0080, 0.0640, 0.2160, 0.5120, 0.9999$ at
     $x = 0.2, 0.4, 0.6, 0.8, 1.0$.

13. $y = 0.4054, 0.8456, 1.3683, 2.0593, 3.1148$ at
     $x = 0.1, 0.2, 0.3, 0.4, 0.5$.

15. $y = 1.2649, 1.4833, 1.6733, 1.8439, 2.0001$ at
     $x = 0.3, 0.6, 0.9, 1.2, 1.5$

17. $y = 1.9435, 1.8794, 1.8051, 1.7157, 1.6000, 1.4079$ at
     $x = 1.04, 1.08, 1.12, 1.16, 1.20, 1.24$.

19. $e_k = (hM/2L)\left[(1 + hL)^k - 1\right]$

SECTION 2.8

1. $z = y_0 + hf(x_0, y_0)$
     $y_1 = y_0 + h\left[f(x_0, y_0) + f(x_1, z)\right]/2$
     $y_k = y_{k-2} + 2hf(x_{k-1}, y_{k-1})$ for $k \geq 2$

3. $y = 0.005, 0.021, 0.049, 0.091, 0.147$ at
     $x = 0.1, 0.2, 0.3, 0.4, 0.5$.

5. $y = 1.937, 1.882, 1.832, 1.788, 1.747$ at
     $x = 1.1, 1.2, 1.3, 1.4, 1.5$.

7. $y = 1.843, 1.598$ at $x = 1.1, 1.2$.

9. $y = 0.0052, 0.0214$ at $x = 0.1, 0.2$.

11. $y = 0.0214, 0.0918, 0.2221, 0.4255, 0.7182$ at
     $x = 0.2, 0.4, 0.6, 0.8, 1.0$.

13. $y = 0.2392, 0.5461, 0.8919, 1.2406, 1.5661$ at
     $x = 1.2, 1.4, 1.6, 1.8, 2.0$.

15. $y = 1.1328, 2.3198, 0.7935, 2.6895, 1.0864, 2.6022$ at
     $x = 0.5, 1.0, 1.5, 2.0, 2.5, 3.0$.

17. $y = 1.4397, 2.6440, 14.3048$ at $x = 0.3, 0.6, 0.9$.

19. $y = 0.4054, 0.8456, 1.3683, 2.0593, 3.1148$ at
     $x = 0.1, 0.2, 0.3, 0.4, 0.5$.

21. $y = 1.2649, 1.4832, 1.6733, 1.8439, 2.0000$ at
$x = 0.3, 0.6, 0.9, 1.2, 1.5.$

23. $y = 1.9282, 1.8437, 1.7399, 1.6000, 1.2500$ at
$x = 1.05, 1.10, 1.15, 1.20, 1.25.$

(The y value 1.2500 is the exact value at 1.25; it is not likely to be obtained by a student program.)

SECTION 3.1

1. $\dfrac{3}{7}$ **3.** $-\dfrac{65}{18}$ **5.** $\dfrac{1}{2}$

7. $-\infty$ (divergent)   **9.** $1/(1 - \log x)$ if $e^{-1} < x < e$

11. $\dfrac{1}{3}$ **13.** 4357/999   **15.** 60 miles

23. Diverges   **25.** Converges to approximately 2.6

SECTION 3.2

1. $\displaystyle\sum_{k=0}^{\infty} (-1)^k x^{2k}/(2k)!$ **3.** $\displaystyle\sum_{k=1}^{\infty} (-1)^{k-1}(x-1)^k/k$

5. $\displaystyle\sum_{k=0}^{\infty} (-1)^k\left(x - \frac{\pi}{2}\right)^{2k}/(2k)!$ **7.** $-\displaystyle\sum_{k=1}^{\infty} x^k/k$

9. $\displaystyle\sum_{k=0}^{\infty} x^{2k}/k!$ **11.** $1 + \displaystyle\sum_{k=1}^{\infty} (-1/2)(-3/2) \cdots (1/2 - k)(-1)^k x^{2k}/k!$

13. $\displaystyle\sum_{k=0}^{\infty} (-1)^k x^{2k+1}/k!(2k+1)$ **15.** $x e^x$

17. $(e^x - 1)/x$ **19.** $(e^x + e^{-x})/2 = \cosh x$

21. $\dfrac{d}{dx}(x \cos x) = \cos x - x \sin x$ **25.** 2   **27.** $(e^2 - 1)/2$

SECTION 3.3

1. Bounded, monotonically decreasing, convergent to 1.

3. Bounded, not monotonic, divergent.

SECTION 3.4

1. Convergent   **3.** Divergent

5. Divergent   **7.** Convergent

9. Convergent   **11.** Divergent

13. Convergent

15. Let $b_k = -a_k$, where Σa_k is any divergent series

SECTION 3.5

1. Absolutely convergent
3. Divergent
5. Absolutely convergent
7. Conditionally convergent
9. Conditionally convergent
11. Absolutely convergent
13. $-1 \le x \le 1$
15. $-10 \le x < 10$
17. $-\infty < x < \infty$
19. $-4 < x < 2$
21. $-1 \le x < 1$
23. $x > 0$
25. $-1 < x \le 1$
29. Divergent

SECTION 3.6

1. (b) $\displaystyle\sum_{k=1}^{71} 1/k^3 = 1.20196$

3. (b) $\displaystyle\sum_{k=2}^{1178} 1/k(\log k)^5 = 3.42972$

5. (b) $\displaystyle\sum_{k=1}^{25} 2^k/(1 + 3^k) = 1.77512$

7. (b) $\displaystyle\sum_{k=0}^{10} (1 + 2^k)/k! = 10.10728$

9. (b) $\displaystyle\sum_{k=2}^{22026} (-1)^k/(\log k)^4 = 3.82446$

13. $\displaystyle\sum_{k=0}^{6} (-1)^k/k!(2k + 1) = 0.74684$

15. $\displaystyle\sum_{k=0}^{4} (-1)^k x^{2k}/(2k)!$

SECTION 3.7

1. (b) $\displaystyle\sum_{k=0}^{99} 4(-1)^k/(2k + 1) + 2/201 = 3.14154$

3. (b) $\displaystyle\sum_{k=1}^{22} (2k + \sqrt{k})/k^3 + \int_{22.5}^{\infty} (2x + \sqrt{x})/x^3 \, dx = 4.63138$

5. (b) $\displaystyle 1 + \sum_{k=1}^{9} 1/k^2(k + 1) + \int_{9.5}^{\infty} 1/x^2(x + 1) \, dx = 1.64495$

7. (b) $\displaystyle\frac{1}{4} + \sum_{k=1}^{22} (3k + 2)/k^3(k + 1)(k + 2) = 1.20198$

SECTION 3.8

3. $\left| f_k(x) \right| \le 1/(1 + k^2)$
5. $\left| f_k(x) \right| \le 1/k^2$
7. $\left| f_k(x) \right| \le t^k/2^k$
9. $\left| f_k(x) \right| \le 1/t^k$
11. (c) 0.5771

SECTION 3.9

1. **(a)** $\frac{1}{2}$; **(b)** $[-1/2, 1/2]$
3. **(a)** ∞; **(b)** $(-\infty, \infty)$
5. $(x - 1)/\left[1 - (x - 1)^3\right]$
7. $\log \sqrt{(1 + x)/(1 - x)}$
9. $\dfrac{d}{dx}(x \sin x) = \sin x + x \cos x$
11. e^x 13. $\sin x$
15. $(e^x - e^{-x})/2 = \sinh x$

SECTION 3.10

3. $g_0(x) = 1$; $g_1(x) = 1 + x^2$; $g_2(x) = 1 + x^2 + x^4/2$
5. $g_0(x) = 1$; $g_1(x) = 1 + x + x^3/3$;
   $g_2(x) = 1 + x + x^2 + 2x^3/3 + x^4/6 + 2x^5/15 + x^7/63$

SECTION 4.1

1. **(a)** $y_1 = \displaystyle\sum_{k=0}^{\infty} (-1)^k x^{2k}/2^k k!$; $y_2 = \displaystyle\sum_{k=0}^{\infty} (-1)^k x^{2k+1}/1 \cdot 3 \cdot 5 \cdots (2k + 1)$
   (b) $R_1 = R_2 = \infty$

3. **(a)** $y_1 = 1$; $y_2 = x + \displaystyle\sum_{k=2}^{\infty} (-3)(-1)(1) \cdots (2k - 7) x^k/k!$
   (b) $R_1 = \infty$; $R_2 = \frac{1}{2}$

5. **(a)** $y_1 = 1 + \displaystyle\sum_{k=1}^{\infty} (-1)^k(-3)(-1) \cdot 1 \cdot 3 \cdots (2k - 5) x^{2k}/2^k 2^k k!$;
   $y_2 = x + x^3/3$; **(b)** $R_1 = \sqrt{2}$; $R_2 = \infty$

7. **(a)** $y_1 = 1 + \displaystyle\sum_{k=1}^{\infty} (-p)(4 - p) \cdots (4k - 4 - p) x^{2k}/(2k)!$;
   $y_2 = x + \displaystyle\sum_{k=1}^{\infty} (2 - p)(6 - p) \cdots (4k - 2 - p) x^{2k+1}/(2k + 1)!$
   (b) $R_1 = R_2 = \infty$

9. **(a)** $y_1 = 1 + \displaystyle\sum_{k=1}^{\infty} (-p)(2 - p) \cdots (2k - 2 - p)p(2 + p) \cdots (2k - 2 + p) x^{2k}/(2k)!$;
   $y_2 = x + \displaystyle\sum_{k=1}^{\infty} (1 - p)(3 - p) \cdots (2k - 1 - p)(1 + p)(3 + p) \cdots$
   $(2k - 1 + p) x^{2k+1}/(2k + 1)!$
   (b) $R_1 = R_2 = 1$

11. If n is even, $y = 1$
    $+ \displaystyle\sum_{k=1}^{n/2} (-n)(2 - n) \cdots (2k - 2 - n)(1 + n)(3 + n) \cdots (2k - 1 + n) x^{2k}/(2k)!$;

If n is odd, $y = x + \displaystyle\sum_{k=1}^{(n-1)/2} (1-n)(3-n) \cdots$

$(2k-1-n)(2+n)(4+n) \cdots (2k+n) \, x^{2k+1}/(2k+1)!$

13. $y = 1 + x^3/6 - x^5/120 + \cdots$

15. $y = x^2/2 + x^3/6 + x^4/24 + \cdots$

17. 1

19. At the recommended values of x, $y = 1.0152,\ 1.0631,\ 1.1519,\ 1.2989,\ 1.5395,\ 1.9530,$ $2.7456,\ 4.6295,\ 12.0744.$

21. At the recommended values of x, $y = 0.9800,\ 0.9211,\ 0.8253,\ 0.6960,\ 0.5380,\ 0.3558,$ $0.1548,\ -0.0603,\ -0.2848,\ -0.5142$

SECTION 4.2

1. $y = c_1 x^2 + c_2 x^{-1}$

3. $y = c_1 x + c_2 x \log x$

5. $y = c_1 x^3 \cos(2 \log x) + c_2 x^3 \sin(2 \log x)$

9. $x = 1$ is regular singular; all others are ordinary.

11. $x = -\frac{1}{2}$ is regular singular; $x = 2$ is irregular singular; all others are ordinary.

13. **(b)** $w = 1, \frac{3}{2};$

    (c), (d) $y_1 = x + \displaystyle\sum_{k=1}^{\infty} x^{k+1}/k! \cdot 1 \cdot 3 \cdot 5 \cdots (2k-1);$

    $y_2 = x^{3/2} + \displaystyle\sum_{k=1}^{\infty} x^{k+3/2}/k! \cdot 1 \cdot 3 \cdot 5 \cdots (2k+1)$

15. **(b)** $w = 0;$ **(c)** $y_1 = \displaystyle\sum_{k=0}^{\infty} (-1)^k x^{2k}/2^{2k} (k!)^2$

17. **(b)** $w = 0, \frac{1}{2};$ **(c), (d)** $y_1 = 1 + \displaystyle\sum_{k=1}^{\infty} 2^k k! x^k/1 \cdot 3 \cdot 5 \cdots (2k-1);$

    $y_2 = x^{1/2} + \displaystyle\sum_{k=1}^{\infty} 1 \cdot 3 \cdot 5 \cdots (2k+1) x^{k+1/2}/2^k k!$

19. $y = \displaystyle\sum_{k=0}^{n} (-1)^k n! x^k/(k!)^2 (n-k)!$

21. $y = (c_1 \cos x + c_2 \sin x)/\sqrt{x}$

SECTION 4.3

5. $a = 0,\ b = nr/s,\ c = s/n,\ p = n.$

    $u(t) = c_1 J_n(x) + c_2 K_n(x),$ where $x = nrt^{s/n}/s.$

11. **(a)** $x^2 u'' + x u' + 4 \pi^2 w^2 x u/g = 0;$

    $u = c_1 J_0(4 \pi w \sqrt{x/g}) + c_2 K_0(4 \pi w \sqrt{x/g}).$

    (c) 0.5413 cycles per second.

SECTION 4.4

1. $y_1 = \sum_{k=0}^{\infty} x^k/(k!)^2$; $y_2 = y_1 \log x - \sum_{k=1}^{\infty} 2 H_k x^k/(k!)^2$

3. $y_1 = \sum_{k=0}^{\infty} x^{k+1}/k!\,(k+1)!$;

   $y_2 = y_1 \log x + 1 - \sum_{k=1}^{\infty} \left(2 H_{k-1} + \dfrac{1}{k} \right) x^k/k!\,(k-1)!$

5. See Example 4.11.

7. $y_1 = \sum_{k=0}^{\infty} (-1)^k x^{2k+2}/k!\,(k+2)!\,2^{2k-1} = 8 J_2$;

   $y_2 = -y_1 (\log x)/16 + x^{-2} + 1/4$

   $\qquad + \sum_{k=0}^{\infty} (-1)^{k+1} x^{2k+2}(1 - H_k - H_{k+2})/k!\,(k+2)!\,2^{2k+4}$

9. x

11. $-1 - \left(\dfrac{x}{2} \right) \log(1 - x) + \left(\dfrac{x}{2} \right) \log(1 + x)$

SECTION 4.5

11. $x^{-1}/2 + \sum_{k=1}^{\infty} (-1)^k 1 \cdot 3 \cdot 5 \cdots (2k-1) x^{-2k-1}/2^{k+1}$

SECTION 5.1

1. Period $\dfrac{\pi}{2}$ and frequency $\dfrac{2}{\pi}$.

7. (a) Fourth example in Table 5.1.

   (b) $y = 4[\sin x + \sin(3x)/3]/\pi$

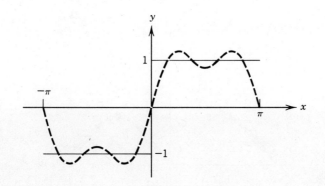

9. **(a)** Fifth example in Table 5.1.

   (b) $y = \dfrac{\pi}{2} - 4(\cos x)/\pi$

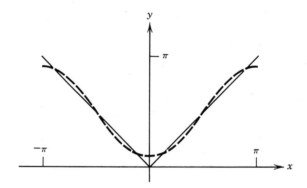

11. **(a)** Sixth example in Table 5.1.

    (b) $y = \dfrac{\pi}{4} - 2(\cos x)/\pi + \sin x$

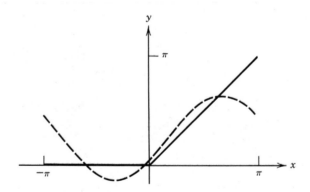

13. **(a)** Seventh example in Table 5.1.

    (b) $y = \dfrac{2}{\pi} - 4\cos(2x)/3\pi$

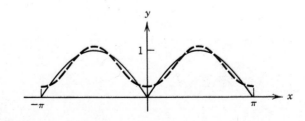

SECTION 5.2

5.

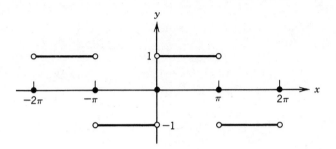

7.

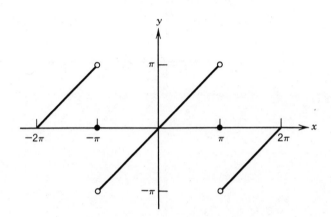

9.

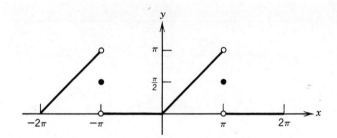

11.

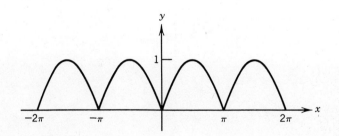

15. $\pi^2/12$ 17. $(\pi - 2)/4$

19. $\pi/4$ 21. $(1 - \pi \operatorname{csch} \pi)/2$

SECTION 5.3

1. **(a)** $\frac{1}{4} - \sum\limits_{k=0}^{\infty} 2 \cos\left((2k+1)\pi x\right)/(2k+1)^2 \pi^2 - \sum\limits_{k=1}^{\infty} (-1)^k \sin(\pi k x)/\pi k$

   (b)

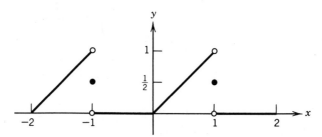

3. **(a)** $\dfrac{2}{\pi} - \sum\limits_{k=1}^{\infty} 4 \cos(2kx)/(4k^2 - 1)\pi;$ **(b)** $\sin x$

   (c)

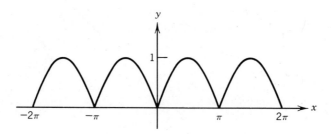

   (d)

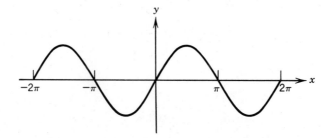

5. **(a)** $\frac{1}{2} + \sum\limits_{k=0}^{\infty} 2(-1)^k \cos\left((2k+1)x\right)/(2k+1)\pi;$

   (b) $- \sum\limits_{k=1}^{\infty} 2\left(\cos(\pi k/2) - 1\right) \sin(kx)/\pi k;$

(c)

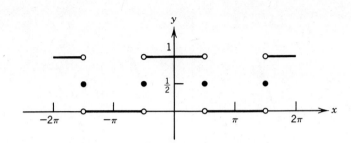

(d)

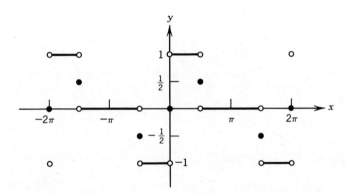

7. (a) $\frac{1}{3} + \sum_{k=1}^{\infty} 4(-1)^k \cos(\pi kx)/\pi^2 k^2$;

   (b) $\sum_{k=1}^{\infty} 2(-1)^{k+1} \sin(\pi kx)/\pi k - \sum_{k=0}^{\infty} 8 \sin((2k+1)\pi x)/\pi^3 (2k+1)^3$

   (c)

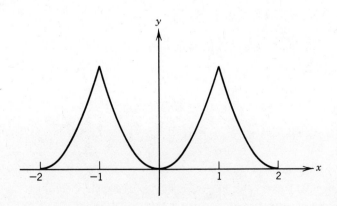

(d)

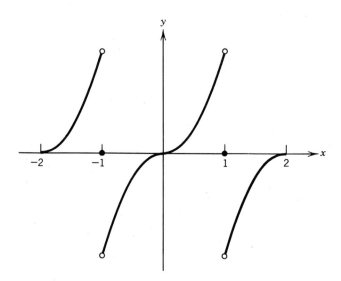

SECTION 5.4

3. $300, 600, 900, 1200$ cycles per second.

5. (a) $b_{2k} = 0$, $b_{2k+1} = 4(-1)^k/\pi^2 (2k+1)^2$ for $k \geq 0$.

   (b) $y = 0$ when $t = 1$, and $y = f(x - 2)$ when $t = 2$.

7. (a) $y = a_0/2 + \sum_{k=1}^{\infty} a_k \cos(\pi k x/L) \cos(\pi k c t/L)$

   where $a_k = (2/L)\int_0^L f(x) \cos(\pi k x/L) \, dx$, and $y = [\, f(x + ct) + f(x - ct)\,]/2$

   where f is even and has period $2L$.

9. (a) $u = a_0/2 + \sum_{k=1}^{\infty} a_k \cos(\pi k x/L) \exp(-\pi^2 k^2 ct/L^2)$

   where $a_k = (2/L)\int_0^L f(x) \cos(\pi k x/L) \, dx$.

11. (a) $u = \sum_{k=1}^{\infty} b_k \sin(\pi k x/L) \sinh(\pi k y/L)$, where

   $b_k = (2/L \sinh(\pi k M/L))\int_0^L f(x) \sin(\pi k x/L) \, dx$

13. Add them.

SECTION 5.5

7. Any nonzero vector (x_1, x_2, x_3, x_4) such that $x_1 + 2x_2 - 3x_4 = 0$ and $2x_1 + 3x_3 - x_4 = 0$.

9. $f(x) = x$, for example

27. $\mathbf{u} \cdot \mathbf{v} = 6 + 3i$; $\|\mathbf{u}\| = 2\sqrt{3}$; $\|\mathbf{u} - \mathbf{v}\| = \sqrt{11}$; projection of $\mathbf{v}$ on $\mathbf{u} = (1/4 + i/2, 3/4 + i/4, 3/2 - 3i/4)$.

SECTION 5.6

1. **(b)** $c_1 = \frac{1}{2}$, $c_2 = \frac{1}{2}$, $c_3 = 1/\sqrt{10}$.
3. **(b)** $c_n = 1/\sqrt{L}$
15. No   17. No

SECTION 5.7

5. $\pi^2/8$ 7. $\pi^4/96$

SECTION 5.8

1. $(1, -1/2, 0)$
3. **(a)** $(1/3, 2/3, 2/3)$, $\left(0, 1/\sqrt{2}, -1/\sqrt{2}\right)$

   (b) $A = \begin{bmatrix} \dfrac{1}{3} & 0 \\[2mm] \dfrac{2}{3} & \dfrac{1}{\sqrt{2}} \\[2mm] \dfrac{2}{3} & -\dfrac{1}{\sqrt{2}} \end{bmatrix} \begin{bmatrix} 3 & 3 \\ 0 & \sqrt{2} \end{bmatrix}$

   (c) $\left(\frac{5}{9}, 0\right)$
7. $1/\sqrt{2}, \sqrt{\frac{3}{2}}\, x, \sqrt{\frac{5}{2}}\, (\frac{3}{2})(x^2 - \frac{1}{3})$.
13. $y = 4.76 - 0.51\, x$
15. $y = 0.32\, x^{1.48}$
17. $z = 2.21\, x + 0.45\, y + 60.44$

SECTION A.2

3. $I_n = \left(0, \dfrac{1}{n}\right]$
7. $f(x) = \sin(1/x)$

SECTION A.4

5. $f(x) = 1$ if x is rational; $f(x) = -1$ if x is irrational.
15. $a = 0$, $b = 1$, $f(x) = 0$ if $0 < x \le 1$, $f(0) = 1$.
17. $2/\pi$ 19. $\pi/4$

SECTION D.1

5. $x^2 (\log x)/2 - x^2/4 + \frac{1}{4}$
9. Lipschitz on every $\{(x, y)| c \le y \le d\}$, where $c < d$.
11. Lipschitz on every $\{(x, y)| c \le x \le d$ and $|y + 1| \ge e\}$ where $c < d$ and $e > 0$.

SECTION D.2

3. **(a)** $y^3 = 6x^2 + 9x + c$;   **(b)** $c = 8$

5. **(a)** $y = -e^x + ce^{3x}$;   **(b)** $c = -2$

7. **(a)** $y = c(1-x)^2 + \frac{1}{2}$;   **(b)** $c = -\frac{1}{2}$

9. **(a)** $y = c\exp(-x^2) + \frac{3}{2}$;   **(b)** $c = -\frac{5}{2}$

11. **(a)** $y = 4x^4 + 16x^3 + 48x^2 + 96x + 96 + ce^x$;   **(b)** $c = -33/e$

13. **(a)** $y = 3x\tan x + 3 + c\sec x$;   **(b)** $c = -1$

15. **(a)** $x + y = c(x-y)^3$;   **(b)** $c = 3$

19. $(v_0 - gm/k)e^{-kt/m} + gm/k$

21. 1960 hours

23. **(a)** $M(1 - 2^{-1-t/10})$;   **(b)** 23.22 hours

25. **(a)** $M/(1 + e^{-Mkt})$;   **(c)** $M/2$

27. **(a)** $20\left(\left(\frac{7}{4}\right)^{t/2} - 1\right)\Big/\left(2\left(\frac{7}{4}\right)^{t/2} - 1\right)$;   **(b)** 1.45 minutes

29. **(a)** $4 + \exp(-0.06\,t)$;   **(b)** 11.55 minutes

SECTION D.3

3. **(b)** $y = \cos x - 3\sin x + c_1 e^x + c_2 e^{2x}$;
   (c) $y = \cos x - 3\sin x + e^x + e^{2x}$

5. **(b)** $y = 2x^2 + c_1 x + c_2 x^{-2}$;
   (c) $y = 2x^2 - 3x + x^{-2}$

7. $y = -4e^{-3x}(3x + 1)/9 + c_1 e^{-3x} + c_2$

9. $y = (20x - x^5)/40 + c_1(x^3 - 18c_1 x) + c_2$

11. **(b)** $y = c_1 x + c_2\left(1 + x\log\sqrt{(1-x)/(1+x)}\,\right)$

13. 5 seconds

SECTION D.4

1. $y = c_1 e^{-px} + c_2$

3. $y^3 = c_1 x + c_2$

5. $y = c_1 e^{2x} + c_2 e^{5x}$

7. $y = c_1 e^{4x} + c_2 e^{-4x}$

9. $y = c_1 \cos(3x) + c_2 \sin(3x)$

11. $y = cx^2/4 + dx + (1 + d^2)/c$

13. $y = 0.8 + c_1 e^{2x} + c_2 e^{5x}$

15. $y = 3xe^{4x} + c_1 e^{4x} + c_2 e^{-4x}$

17. $\lim\limits_{t\to\infty} v = \sqrt{v_0^2 - 2aR}$

Index

Absolute convergence, 126
Absolute value:
  of complex number, 14, 323
  of real number, 10
Accelerated alternating series method, 138, 172
Accelerated integral method, 138–141
Acceleration method of Kummer, 143
Accumulated round-off error, 66
Additivity of integral, 316
Airy's equation, 169, 178
Algebra of:
  complex numbers, 321
  continuous functions, 300
  derivatives, 306
  integrals, 315
  limits of functions, 306
  limits of sequences, 12
  sums of series, 108, 275
Algorithm, 52
  accelerated alternating series method, 172
  bisection method, 53
  classical Runge-Kutta method, 101
  Euler method, 92
  hybrid method, 70, 72
  improved Euler method, 98
  Newton's method, 58, 59
  ratio test method, 173
  Romberg method, 86
  secant method, 56
  Simpson's rule, 77
  trapezoid rule, 77
Alternating harmonic series, 128, 130, 134, 138
Alternating series test, 127, 134, 138, 172

Analytic function, 166–167. *See also* Power series
Angle between vectors, 270
Antiderivative, 314
Approximation of:
  Bessel functions, 82
  continuous functions, 283, 305
  definite integrals, 29, 73–87
  elliptic integrals, 81
  error function, 81
  exponential function, 66, 135
  logarithm function, 135
  period of pendulum, 211
  π, 9, 134–135
  reciprocals, 62
  roots of equations, 49–61, 70–73
  solutions of differential equations, 88–102, 170–173, 213–216
  square roots, 37, 62
  sums of series, 131–141, 170–173
  see also Algorithm; Error bound; Error estimate
Approximation:
  asymptotic, 207–209
  least-squares, 292–294
Arctangent function, 113, 129, 146, 152
Asymptotic sequence, 206
Asymptotic series, 206
Asymptotic solution, 213–216
Average, 69, 116, 295
Axiom of completeness, 117

Bessel function(s), 177, 187–190
  approximation of, 82
  asymptotic formula for, 188, 205

Bessel function(s) (*Continued*)
  graph of, 190
  integral form of, 147, 204
  of order n, 187–190, 205
  of order one, 201–202
  of order zero, 147, 196–197
  of second kind, 190, 197, 202
  zeros of, 188–190, 205
Bessel's equation, 177, 178, 184–190
  change of variables for, 185, 188
  of order n, 186, 204
  of order $n + 1/2$, 184
  of order one, 200–202
  of order one-half, 184
  of order p, 185
  of order zero, 195–197
Bessel's inequality, 236, 275, 279
Big oh notation, 204
Binomial coefficient, 319
Binomial distribution, 116
Binomial series, 113, 114, 153
Binomial theorem, 4, 319
Bisection method, 51–53
  convergence of, 52
Boas, M., 184
Bolzano-Weierstrass theorem, 302
Boundary conditions, 220
Bounded function, 309
Bounded sequence, 117

Cancellation error, 65
Catenary, 62
Cauchy convergence criterion, 119
Cauchy-Euler differential equation, 175
Cauchy-Euler method, *see* Euler method
Cauchy polygon approximation, 93
Cauchy-Schwarz inequality, 267, 268
Chain rule, 306
Change of variable:
  for Bessel's equation, 185, 188
  for Cauchy-Euler equation, 175
  log-log, 296
  in pendulum equation, 185, 209–215
  to reduce order of differential equation, 341
  semilog, 295
Characteristic equation:
  of Cauchy-Euler equation, 175
  of difference equation, 16
  of differential equation, 344
  see also Indicial equation
Classical differential equations, 178
Classical Runge-Kutta method, 99–102
  error estimate for, 95

Comparison test, 122, 132. *See also* M-test
Completeness axiom, 117
Complex exponential Fourier series, 234, 276
Complex exponential function(s), 115, 276, 323
Complex inner product, 268
Complex n-space, 268
Complex number, 321–325
  absolute value of, 14, 323
  algebra of, 321
  conjugate of, 323
  imaginary, 322
  illustration of, 322, 324
  polar form of, 324
Composite Simpson's rule, 77
Composite trapezoid rule, 76
Computer programs, 53. *See also* Algorithm;
    Programming note; Simulated computer
Condensation test, 125
Conditional convergence, 126
Conjugate of complex number, 323
Constant coefficient difference equation, 16
Constant coefficient differential equation, 344–346
Continuity of power series, 149
Continuous function(s), 299
  approximation of, 283, 305
  integral of, 312
  maximum and minimum of, 303
  piecewise, 227
  of two variables, 328
  uniform approximation of, 283
  and uniform convergence, 45, 145
  uniformly, 303
Convergence:
  absolute, 126
  of bisection method, 52
  Cauchy criterion for, 119
  conditional, 126
  of Euler method, 94
  of Fourier series, 238, 240, 241, 279
  inequality for, 11, 121
  interval of, 148
  of Newton's method, 59
  in norm, 273–274
  of Picard sequence, 121, 157
  pointwise, 40, 144
  quadratic, 61
  radius of, 148
  of Riemann sums, 312
  of secant method, 61
  of sequence, 7, 40, 273
  of series, 106, 144, 274
  of subsequences, 302

of trapezoid rule, 77–78
uniform, 42, 144. *See also* Uniform
    convergence
Convergence tests, 122–127. *See also* Tests for
    convergence
Cosines, law of, for vectors, 270
Cosine series, Fourier, 247, 248
CPC[*a, b*], 269
Critical point of a function, 307

d'Alembert's solution, 220
Derivative(s), 306
    anti-, 314
    chain rule for, 306
    at critical point, 307
    of Fourier series, 245
    fundamental theorem for, 314
    mean value theorem for, 26, 35, 307
    one-sided, 240
    of power series, 149
    product rule for, 306
    quotient rule for, 306
    and uniform convergence, 147, 258
Difference equation(s), 15–20
    characteristic equation of, 16
    constant coefficient, 16
    first-order, 16
    general solution of, 17
    homogeneous, 16
    linear, 16
    nonhomogeneous, 19
    second-order, 16
    system of, 21
Difference of two sequences, 12
Differentiable function, 306. *See also*
    Derivative
Differential equation(s), 155–159, 164–203,
    209–216, 327–347
    Airy's, 169, 178
    analytic solutions of, 167
    approximate solution of, 88–102, 170–173,
        213–216
    asymptotic solution of, 213–216
    Bessel's, 177, 178, 184–190. *See also* Bessel's
        equation
    Cauchy-Euler, 175
    characteristic equation of, 344
    constant coefficient, 344–346
    direction field of, 90
    existence-uniqueness theorem for, 157, 329,
        337
    first-order, 327–333
    Frobenius' method for, 177–178, 180–182
    Gauss', 178

general solution of, 327. *See also* General
    solution
Hermite's, 173, 178
homogeneous, 332, 336, 339, 345
independent variable missing from, 341
indicial equation of, 181
initial-value problem for, 88, 155, 157, 328,
    329, 337
irregular singular point of, 176
Laguerre's, 178, 184
Legendre's, 173, 178, 193, 340
Lindstedt's procedure for, 214–216
linear, 331, 336
Lipschitz condition for, 328
nonhomogeneous, 332, 336
ordinary point of, 176
perturbation solution for, 213
phase-space portrait of, 211–212
power series solutions of, 164–173
reduction of order of, 203, 340, 341
regular singular point of, 176
second-order, 164–203, 209–216, 335–347
separable, 330
series solutions of, 164–183, 194–203. *See
    also* Series solutions
singular point of, 176
solution of, 327, 335
for summing power series, 153
table of classical, 178
Tchebycheff's, 173, 178
see also Partial differential equation
Direction field, 90
Dirichlet kernel function, 238
Discontinuity, Fourier series at a, 241–244
Discontinuous function, integral of, 227, 318
Distance between vectors, 264, 266
Divergence, 7, 107
Divergent series, approximation by a, 208
Dot product, *see* Inner product
Double root, 16, 50, 344
Doubling rule for error estimates, 79, 95

Equation, *see* Difference equation; Differential
    equation; Linear equations
Error bound, 59
    accelerated alternating series method, 138
    accelerated integral method, 139
    from alternating series test, 134
    bisection method, 51
    classical Runge-Kutta method, 102
    from comparison test, 132
    Euler method, 94, 102
    improved Euler method, 102
    from integral test, 132

Error bound (*Continued*)
  Newton's method, 59
  from ratio test, 133
  Simpson's rule, 79
  trapezoid rule, 78
Error estimate, 59
  for approximating integrals, 79
  for approximating roots of equations, 59
  for approximating solutions of differential
    equations, 95
  from ratio test, 133
  trapezoid rule, 87
Errors, 63–68
  accumulated round-off, 66
  cancellation, 65
  data, 68
  how to minimize, 67–68
  propagation, 64
  representation, 64
  round-off, 65
  sensitivity, 64
Escape velocity, 342
Euclidean *n*-space, 262–263, 266, 268
Euler-Maclaurin formula, 140
Euler method, 91–95, 102
  convergence of, 94
Euler's constant, 126, 147
Even extension, 247
Even function, 81, 229
Existence:
  of integral, 312
  of least-squares solutions, 285–286
  of maximum and minimum values, 303
  of power series solutions, 167
  of series solutions, 167, 182
  of solution of initial-value problem, 157,
    329, 337
  of solutions of normal equations, 285
  of solutions at ordinary points, 167
  of solutions at regular singular points, 182
Existence-uniqueness theorem:
  for first-order equations, 157, 329
  for second-order equations, 337
  see also General solution
Exponential function:
  approximation of, 66, 135
  complex, 115, 323
  inequality for, 30
  Taylor series for, 113, 115

Factorial, 319. *See also* Stirling's formula
Fibonacci sequence, 5, 61, 205
Figure, *see* Graph; Illustration
First-order difference equations, 16

First-order differential equations, 327–333
Fitting equations to data, 292–294
Fourier series, 227–260, 278–282
  Bessel's inequality for, 236, 275, 279
  coefficients, 227
  complex exponential, 234, 276
  convergence of, 238, 240, 241, 279
  cosine series, 247, 248
  derivative of, 245
  at discontinuity, 241–244
  Gibbs phenomenon for, 243
  graph of partial sums of, 230, 243
  Parseval's equation for, 280
  of period 2*p*, 246, 248–249
  sine series, 248, 249
  table of, 232
  uniform convergence of, 279
  uniqueness of coefficients of, 279
Frequency, 219
  fundamental, 224, 255
  higher harmonic, 224, 255
Frobenius' method, 177–178, 180–182
Fundamental theorem of calculus, 314

Gauss' equation, 178
General solution, 17, 327
  of Cauchy-Euler equation, 175
  of first-order difference equation, 16
  of first-order differential equation, 327
  of nonhomogeneous difference equation,
    19
  of nonhomogeneous differential equation,
    336
  of second-order difference equation, 17
  of second-order differential equation, 167,
    175, 182, 336, 339, 345
  see also Existence
Geometric series, 108, 113
Gibbs phenomenon, 243
Gram-Schmidt process, 288
Graph of:
  Bessel functions, 190
  Lagrange polynomial, 34
  partial sums of Fourier series, 230, 243
  Taylor polynomials, 28
  see also Illustration
Gravitational attraction, law of, 342
Greatest lower bound, 117

Harmonics, 224, 255
Harmonic series, 110, 124
Heat equation, 250, 256–260
Hermite interpolation, 39
Hermite polynomials, 173

Hermite's equation, 173, 178
Homogeneous difference equation, 16
Homogeneous differential equation, 332, 336, 339, 345
Hooke's law, 347
Horner's method, 66
Householder transformations, 292, 295
Hybrid method, 70–72
Hyperbolic cosine, 31
Hyperbolic sine, 31

Illustration:
  of accelerated integral method, 139
  of bisection method, 52
  of classical Runge-Kutta method, 100
  of complex number, 322, 324
  of convergence of Fourier series, 230, 243
  of convergence of Lagrange polynomials, 34
  of convergence of Taylor polynomials, 28
  of direction field, 90
  of Euler method, 91
  of Gibbs phenomenon, 243
  of Gram-Schmidt process, 288
  of hybrid method, 70
  of improved Euler method, 99
  of integral test, 124
  of least-squares line, 291
  of least-squares solution, 285
  of limit of sequence, 2, 3
  of midpoint method, 103
  of Newton's method, 58, 60
  of nonuniform continuity, 304
  of orthogonal functions, 265
  of periodic extension, 241
  of phase space, 212
  of piecewise continuous function, 228
  of projections, 266
  of secant method, 55, 57
  of sum of series, 107
  of temperature distribution, 259
  of trapezoid rule, 76
  of travelling wave, 256
  of uniform convergence, 43
  of upper sum, 310
  see also Graph of
Imaginary numbers, 322
Improved Euler method, 97–99, 102
  error estimate for, 95
Independent variable missing from differential equation, 341
Indicial equation, 181
Induction, 349

Inequality(ies):
  accelerated alternating series method, 138
  accelerated integral method, 139
  alternating series test, 128, 134
  Bessel's, 236, 275, 279
  from binomial theorem, 4
  bisection method, 51
  Cauchy-Schwarz, 267, 268
  comparison test, 132
  for convergent sequences, 11, 121
  for convergent series, 121
  Euler method, 94
  for exponential function, 30
  for integrals, 315
  integral test, 123, 132
  for logarithm function, 30
  M-test, 145
  Newton's method, 59
  Pythagorean, 267
  ratio test, 133
  for remainder formula for Lagrange polynomials, 36
  Simpson's rule, 79
  trapezoid rule, 78
  triangle, 10, 132, 267–268, 315
  for uniform convergence of sequences, 43
  for uniform convergence of series, 145
Infimum, 117
Infinity as limit, 8, 306
Initial-value problem, 88, 155, 157, 328, 329, 337
  existence-uniqueness theorem for, 157, 329, 337
  integral form of, 155
  see also Differential equations
Inner product, 262
  complex, 268
  for $CPC[a, b]$, 269
  for Euclidean n-space, 263
  integral, 263, 269
  for $PC[a, b]$, 263
  of two Fourier series, 282
Integrable function, 310
Integral(s), 309–310
  additivity of, 316
  approximation of, 29, 73–87
  of discontinuous function, 227, 318
  of even function, 81, 229
  existence of, 312
  fundamental theorem for, 314
  inequality for, 315
  lower sum for, 310
  mean value theorem for, 317
  of odd function, 81, 229

Integral(s) (*Continued*)
  of piecewise continuous function, 227
  of power series, 150
  Riemann sum for, 311
  and uniform convergence, 46, 145
  upper sum for, 310
Integral form of initial-value problem, 155
Integral inner product, 263, 269
Integral test, 123, 132, 138
Intermediate value theorem, 50, 300–301
Internal numbers in computer, 64
Interpolation, 32, 39, 93. *See also* Lagrange
  polynomials
Interval of convergence, 148
Irregular singular point, 176
Iteration, *see* Picard iteration

Kummer's acceleration method, 143

Lagrange polynomial(s), 32–38
  remainder formula for, 34
Laguerre polynomials, 184
Laguerre's equation, 178, 184
Laplace's equation, 250
Law of cosines for vectors, 270
Least-squares line, 292
Least-squares solutions, 284
  existence of, 285–286
  illustration of, 285
  for linear fitting of data, 292
  and multiple-regression, 293
  and normal equation, 286
  for orthogonal columns, 288
  from a *QR*-factorization, 290
  uniqueness of, 285
Least upper bound, 117
Legendre polynomials, 173, 193, 277
Legendre's equation, 173, 178, 193, 277, 340
Leibnitz' alternating series test, 127, 134
Length of vector, 264
L'Hospital's rule, 308
Limit(s), 3, 7, 306
  algebra of, 12, 306
  of constant sequence, 7
  of function, 306
  illustration of, 2
  infinity as, 8, 306
  one-sided, 227
  of Riemann sums, 311–313
  of sequence, 3, 7
  of subsequence, 302
  uniqueness of, 10
  see also Convergence; Sequence
Limit comparison test, 122
Lindstedt's procedure, 214–216

Linear combination of functions, 337
Linear combination of sequences, 16
Linear fitting of data, 292–294
Linear difference equation, 16. *See also*
  Difference equation
Linear differential equation, 331, 336. *See also*
  Differential equation
Linear equations, system of, 283–294
  least-squares solution of, 284
  normal equations for, 285
Linear independence of functions, 339
Linear independence of sequences, 17
Lipschitz condition, 328
Lipschitz constant, 328
Little oh notation, 204
Logarithm function, 8
  approximation of, 135
  base of, 8
  inequality for, 30
  Taylor series for, 113
Log-log plot, 296
Lower sum, 310

Machin's formula, 135
Maclaurin series, 112. *See also* Taylor series
Mathematical induction, 349
Matrix:
  orthogonal, 295
  positive definite, 269
  QR-factorization of, 290
  reflection, 295
  rotation, 295
  symmetric, 269
Matrix form of normal equations, 287
Maximum of function, 303
Mean, 69, 116, 295
Mean value theorem, 26, 35, 307
  for integrals, 317
Method of approximation, *see* Algorithm;
  Approximation of
Midpoint method, 102
Midpoint rule, 82
Minimum of function, 303
Monotonicity of integral, 315
Monotonic sequence, 117
M-test, 145
Multiple-regression coefficients, 293
Multiple root, 50

Newton-Cotes approximation, 74–75
Newton's law of gravitational attraction, 342
Newton's method, 57–61
  convergence of, 59
Newton's second law, 327
Nonhomogeneous difference equation, 19

Nonhomogeneous differential equation, 332, 336
Norm:
    convergent in, 273–274
    of a partition, 310
    of vector, 264
Normal equations, 285, 287. *See also* Least-squares solutions

Odd extension, 247
Odd function, 81, 229
Oh notation, 204
One-sided derivatives, 240
One-sided limits, 227
Order of Bessel's equation, 185
Order of method of convergence, 61, 102
Ordinary point, 176
Orthogonal functions, illustration of, 265
Orthogonality:
    of complex exponential functions, 276
    by Gram-Schmidt process, 288
    of Legendre polynomials, 277
    of trigonometric functions, 225, 271
Orthogonal matrix, 295
Orthogonal sequence of vectors, 271
Orthogonal vectors, 264
Orthonormal sequence of vectors, 271

Parseval's equation, 280
Partial differential equation(s), 250–260
    boundary conditions for, 220
    heat equation, 250, 256–260
    Laplace's, 250
    separation of variables in, 251
    wave equation, 220–224, 250, 251–256
Partial sum of series, 106
Partition of interval, 309
$PC[a, b]$, 263, 266, 268, 274
Pendulum:
    period of, 163, 191–192, 210–212, 214–216
    problem, 161–163, 184–192, 209–216
Period, 219
    of Fourier series, 246, 248–249
    of pendulum, 163, 191–192, 210–212, 214–216
    of vibrating string, 224, 256
Periodic extension, 241, 242
Permutation, 319
Perturbation solution, 213
Phase-space, 211
π, approximation of, 9, 134–135
Picard iteration, 54, 121, 156
Picard method, 156
Picture, *see* Graph; Illustration
Piecewise continuous function, 227

Piecewise linear function, 48, 305
Pointwise convergence, 40, 144. *See also* Convergence
Poisson distribution, 116
Polar form of complex number, 324
Polynomials:
    Hermite, 173
    Horner's method of evaluating, 66
    Lagrange, 32–38
    Laguerre, 184
    Legendre, 173, 193, 277
    Taylor, 23–31
    Tchebycheff, 21, 173
    trigonometric, 224
Positive definite matrix, 269
Power series, 112, 147–154
    of complex variable, 154
    continuity of, 149
    derivative of, 149
    integral of, 150
    interval of convergence of, 148
    radius of convergence of, 148, 155
    as solutions of differential equations, 164–173
    and Taylor series, 152
    undetermined coefficients of, 164
    uniform convergence of, 149
    uniqueness of coefficients of, 152
Principle of mathematical induction, 349
Product rule for derivatives, 306
Product of two sequences, 12
Programming note:
    on arrays, 87
    on comparing successive approximate solutions of a differential equation, 96
    on loops, 56, 59
    on running sums, 77, 87
    on summing series, 111
Projection onto vector, 264
P-series, 110, 124
Punctured neighborhood, 204
Pythagorean inequality for vectors, 267
Pythagorean theorem for vectors, 267

QR-factorization, 290
Quadratic convergence, 61
Quadratic formula, 69
Quotient rule for derivatives, 306
Quotient of two sequences, 12

Radius of convergence, 148, 155
Range of function, 305
Ratio test, 127, 133, 172
Rearrangement of series, 130
Recursion, 5–6, 15–20, 24, 37, 54, 121, 156

Reduction of order, 203, 340, 341
Reflection matrix, 295
Regression coefficients, 293
Regular singular point, 176
Remainder formula:
   for Lagrange polynomials, 34
   for Taylor polynomials, 26
Repeating decimals, 109
Report on output of computer program, 53
Rheinboldt, W., 49
Riemann-Lebesgue lemma, 238
Riemann sum, 311
Romberg method, 85–86
Root(s), 49
   approximation of, 49–61, 70–73
   double, 16, 50, 344
   multiple, 50
   see also Zero
Root test, 131
Rotation matrix, 295
Round-off error, 65
Runge-Kutta methods, 97–102

Secant method, 55–57, 59, 61
   error estimate for, 59
Second law of Newton's, 327
Second-order difference equations, 16–19
Second-order differential equations, 164–203,
   209–216, 335–347
Second solution at regular singular point,
   194–203
Semilog plot, 295
Separable differential equation, 330
Separation of variables, 251, 330
   for first-order differential equations, 330
   for heat equation, 256–258
   for wave equation, 251–255
Sequence(s), 2, 40
   asymptotic, 206
   bounded, 117
   Cauchy criterion for, 119
   of complex numbers, 14
   of constants, 7
   convergent, 7, 40, 273
   divergent, 7
   Fibonacci, 5, 61, 205
   of functions, 40
   limit of, 3, 7. *See also* Limit
   monotonic, 117
   norm convergent, 273
   of numbers, 2
   orthogonal, 271
   orthonormal, 271

Picard, 54, 121, 156
   pointwise convergent, 40
   recursively defined, 5–6, 15–20, 24, 37, 54,
   121, 156
   of Riemann sums, 313
   sub-, 302
   term of, 2
   uniformly convergent, 42. *See also* Uniform
   convergence
Series, 106, 144
   absolutely convergent, 126
   alternating harmonic, 128, 130, 134, 138
   asymptotic, 206
   binomial, 113, 114, 153
   Cauchy criterion for, 119
   of complex numbers, 110
   conditionally convergent, 126
   convergent, 106, 144, 274
   divergent, 107
   Fourier, 227–260, 278–282. *See also* Fourier
   series
   of functions, 144
   geometric, 108, 113
   harmonic, 110, 124
   Maclaurin, 112
   norm convergent, 274
   of numbers, 106
   p-, 110, 124
   partial sum of, 106
   pointwise convergent, 144
   power, 112, 147–154. *See also* Power series
   rearrangement of, 130
   sum of, 106
   Taylor, 111–115. *See also* Taylor series
   telescoping, 107, 111
   tests for convergence, 122–127. *See also*
   Tests for convergence
   trigonometric, 224
   uniformly convergent, 144. *See also*
   Uniform convergence
Series solutions, 164–183, 194–203
   existence of, 167, 182
   Frobenius' method for, 177–178, 180–
   182
   indicial equation for, 181
   at ordinary points, 164–173
   at regular singular points, 174–183, 194–
   203
   undetermined coefficients of, 164, 195
Simpson's rule, 75–77, 79
   error bound for, 79
Simulated computer, 64
Sine series, Fourier, 248, 249
Singular point, 176

Solution, *see* Least-squares solutions; Root
Solution of differential equation, 327, 335.
    See also Differential equation; General
    solution; Series solution
Square roots, approximation of, 37, 62
Standard deviation, 69
Standing wave, 222, 255
Stirling's formula, 143
Subsequence, 302
Substitution by power series, 164
Sum of series, 106. *See also* Convergence;
    Series
Sum of two sequences, 12
Supremum, 117
Symmetric matrix, 269
System of linear equations, 283–294

Table:
    of classical differential equations, 178
    of Fourier series, 232
    of periods of a pendulum, 192, 211
    of Taylor series, 113
    of zeros of Bessel functions, 190
Taylor polynomials, 23–31
    remainder formula for, 26
Taylor series, 111–115
    and asymptotic series, 206
    and power series, 152
    table of, 113
Tchebycheff polynomials, 21, 173
Tchebycheff's equation, 173, 178
Telescoping series, 107, 111
Temperature distribution, illustration of,
    259
Term of a sequence, 2
Tests for convergence:
    alternating series, 127, 134
    comparison, 122, 132
    condensation, 125
    discussion of, 128–129
    integral, 123, 132
    limit comparison, 122
    M-, 145
    ratio, 127, 133, 172
    root, 131
Torque, 163
Transformation, *see* Change of variable
Trapezoid rule, 75–79
    convergence of, 77–78
Travelling wave, 256
Triangle inequality, 10, 132, 267, 268, 315
Trigonometric functions, 8
    orthogonality of, 225, 271

    radian measure and, 8
    Taylor series for, 113
Trigonometric identities, 225
Trigonometric polynomials, 224
Trigonometric series, 224. *See also* Fourier
    series

Undetermined coefficients of series, 164, 195
Uniform approximation of continuous
    functions, 283
Uniform continuity, 303
Uniform convergence, 42, 144
    and continuous functions, 45, 145
    and derivatives, 147, 258
    of Fourier series, 279
    illustration of, 43
    inequality for, 43, 145
    and integrals, 46, 145
    M-test for, 145
    of power series, 149
    of sequences, 42
    of series, 144
Uniqueness:
    of asymptotic series coefficients, 206
    of Fourier series coefficients, 279
    of least-squares solution, 285
    of limit of sequence, 10
    of power series coefficients, 152
    of solution of initial-value problem, 157,
      329, 337
    of solution of normal equations, 285
Upper sum, 310

Vectors:
    angle between, 270
    Cauchy-Schwarz inequality for, 267, 268
    complex inner product of, 268
    in complex n-space, 268
    convergence of, 273–274
    in $CPC[a, b]$, 269
    distance between, 264, 266
    in Euclidean n-space, 262–263, 266, 268
    inner product of, 262, 263, 268, 269
    law of cosines for, 270
    length of, 264
    norm of, 264
    norm convergent, 273–274
    orthogonal, 264
    orthogonal sequence of, 271
    orthonormal sequence of, 271
    in $PC[a, b]$, 263, 266, 268, 274
    projection onto, 264
    Pythagorean inequality for, 267

Vectors (*Continued*)
    Pythagorean theorem for, 267
    triangle inequality for, 267, 268
Vibrating string, 219–224, 251–256

Wallis' formula, 121, 245
Wave equation, 220–224, 250, 251–256

Waves, 225, 255–256
Weierstrass' *M*-test, 145

Zero(s):
    of Bessel functions, 188–190, 205
    of function, 49
    see also Root